AF530466

# Geoinformatics for Decentralized Planning and Governance

# Geoinformatics for Decentralized Planning and Governance

*Edited by*
M.S. Nathawat
A.C. Pandey

RAWAT PUBLICATIONS
Jaipur • New Delhi • Bangalore • Mumbai • Hyderabad • Guwahati

ISBN 81-316-0117-X
© Contributors, 2008

No part of this book may be reproduced or transmitted in any form or by any means, electronic or mechanical, including photocopying, recording or by any information storage and retrieval system, without permission in writing from the publishers.

*Published by*
Prem Rawat for *Rawat Publications*
Satyam Apts., Sector 3, Jawahar Nagar, Jaipur 302 004 (India)
Phone: 0141 265 1748 / 7006 Fax: 0141 265 1748
E-mail: info@rawatbooks.com
Website: www.rawatbooks.com

*Delhi Office*
4858/24, Ansari Road, Daryaganj, New Delhi 110 002
Phone: 011-23263290

*Also at Bangalore, Mumbai, Hyderabad and Guwahati*

Typeset by Rawat Computers, Jaipur
Printed at Nice Printing Press, New Delhi

# Contents

## III
## Geoinformatics for Rural and Urban Planning

# Foreword

The spatial information is fast becoming an important prerequisite in development, planning in government as well as in utility/facility provider companies. Natural resources management, disaster mitigation, environmental restoration and land-use planning are essential areas in which decision makers benefit from spatial information. The prerequisite for planning is the availability of database based on which the future potential growth zones are categorized. Informative data related to the society whether it is water, power, natural resources, transport or agriculture, minerals and demography, etc., are important for developing better strategies for planning. The Geoinformatics has emerged as an effective technology towards proper planning and better governance. The state of Jharkhand celebrated its 4th birth anniversary on 15, November, 2004 with focus on the developmental issues related to the state. It is therefore a matter of immense privilege that a national seminar on "Role of Geoinformatics in Decentralized Planning for Better Governance" organized by Department of Remote Sensing, BIT, Mesra during 7-8 October was focused on decentralized planning for better governance. Several eminent scientist from esteemed organizations like Department of S&T, New Delhi; NRSA, Hyderabad; RRSSC, Jodhpur; JSAC, Ranchi presented their work and discussed the futuristic trend in application of geoinformatics technology in rural and urban sectors. The thrust areas of deliberation were recent trends in geoinformatics, environmental management, natural resources and decentralized planning.

The various technical papers presented during the seminar were reviewed by the editorial board and compiled in this proceeding volume. I am sure that this volume will be useful to researchers, academicians, planners and professionals working in the field of geoinformatics, decentralized planning and resource management.

I am thankful to sponsoring agencies, namely, Department of Science and Technology, (DST), GoI, Department of Information Technology (DIT), GoJ, PCI Geomatics Private Ltd., Kolkata, ADCC Infocad Pvt. Ltd., Nagpur and ERDAS India Pvt. Ltd., Hyderabad for their support in organizing the seminar, as also organizers of the seminar and members of the editorial board for bringing out this informative seminar proceedings volume in a book form.

**S.K. Mukherjee**
Vice Chancellor
Birla Institute of Technology, Mesra

# Preface

Geoinformatics is a technology of extreme importance in the current scenario and still more important for a developing nation like India. The Information Technology (IT) is playing a major role in everyday activities and spatial information using Remote Sensing, Global Positioning System (GPS) and Geographic Information System (GIS) is no exception of it. While the last three decades have witnessed an impressive growth in the use of aerial and satellite imagery, the real explosive growth has been observed during last five years when the use of GIS and GPS has increased in this field.

Spatial information is vital in making precise decision at the local, regional and state level planning, implementation of action plans, infrastructure development, disaster management and business development. The awareness and utilization of the Geoinformatics technology and the power of spatial information systems specially oriented towards decision-making or resource management is growing rapidly in the country. A major challenge over the next decade will be to increase the use of spatially referenced data to support a wide variety of decisions at all levels of society. Using an effective, efficient, and widely accessible infrastructure, spatial data could be readily transported and easily integrated both thematically (across environmental, economic, and institutional databases) and hierarchically (from local to national and eventually to global levels).

In the view of rapid growth and application of Geoinformatics technology in planning and development and initiation of large number of development projects in the newly created Jharkhand State, the Birla institute of Technology, Mesra organized the National Seminar on "Role of Geoinformatics in Decentralized Planning for Better Governance". The seminar provided a suitable platform wherein eminent scientist, academicians, planners, professionals and bright young research fellows keenly participated and fruitful deliberations highlighting judicious use of this technology for rural-urban planning and resource management was done.

This volume consists of select papers dealing with recent trends in Geoinformatics, resource mapping, disaster management and rural-urban planning. This will be useful to planners, professionals and researchers in identifying potential and futuristic trends in application of Geoinformatics technology in decentralized planning and resource management.

We express our sincere thanks to Dr. H.C. Pande, Vice Chancellor Emeritus and Dr. S.K. Mukherjee Vice Chancellor, Birla Institute of Technology, Mesra for their guidance and support to organize the seminar. We are also thankful to the various sponsoring agency namely Department of Science and Technology, (DST), GoI, Department of Information Technology (DIT), GoJ, PCI Geomatics Private Ltd., Kolkata, ADCC Infocad Pvt. Ltd., Nagpur and ERDAS India Pvt. Ltd., Hyderabad for their financial support in organizing the seminar. We are thankful to M/s Rawat Publications, Jaipur for publishing the proceedings in such a good volume. We further acknowledge the help rendered by Brig. (Dr.) R. Sivakumar, Head NRDMS and NSDI, Ministry of S&T, GoI, Dr. J.R. Sharma, Head RRSSC, Jodhpur, Dr. N.C. Guatam, Director CLUMA, Hyderabad. Last but not the least I thank all my colleagues in BIT and students in the department for their support during the seminar.

**M.S. Nathawat**
Professor & Head
Department of Remote Sensing
Birla Institute of Technology, Mesra

# I

# Recent Trends in Geoinformatics

# 1

# Advanced Digital Image Processing for the Analysis of High Resolution Images

K.M.M. Rao

Remote Sensing is the state of the art for collecting information about the earth objects, without physically coming in contact using electromagnetic radiation and emission. Earth observation (EO) by remote sensing is the interpretation and understanding of measurements of electromagnetic energy that is reflected from or emitted by the Earth's surface. The two main sources of the data are air-borne camera systems and the space-borne systems. During the last more than four decades, the space-borne systems have advanced from hot air balloons to reusable space probes.

The major characteristics of imaging remote sensing instrument operating in the visible and infrared spectral region are described in terms of its spatial, spectral, radiometric and temporal resolutions. With the advancement of technology, all these four resolutions are continuously improving for producing higher resolution images. The high-resolution data is readily available of the shelf for commercial applications from various satellite operators.

Spatial resolution is defined as the Ground Sampling Distance. The spatial resolution historically evolved from few kilometres to sub-metre level over a period of few decades, which helps in identifying the smaller objects on the earth surface.

The term spectral resolution refers to the width of the spectral bands. The position in the spectrum, width and number of spectral bands will determine the degree to which individual targets (vegetation species, crop or rock types) can be discriminated.

Radiometric Sensitivity refers to the number of digital levels used to express data collected by the sensor. The number of grey levels is commonly expressed in terms of the number of binary digits (bits) needed to store the value to the maximum grey level. As the radiometric resolution of a sensor increases, the details in the image would be much clear.

Temporal Resolution refers to the frequency with which images of a given geographic location is acquired. To carry out the repetitive monitoring of the same geographical area, it needs to be covered frequently. This can be achieved either by acquiring the data from more number of satellites or by tilting the cameras when the satellite is in the nearby orbits.

The satellites that acquire the high spatial resolution data are IRS–PAN, LISS-IV, TES[3], IKONOS, Quick Bird and Spot-5. The commercial satellite operators provide the data with the spatial resolution of around one metre for any part of the globe.

Since 1970s multi-spectral image data has been used for countless studies in a wide variety of fields. Multi-spectral sensors measure multiple, wide, separated wavelength bands, and hyper spectral sensors measure multiple, narrow and contiguous wavelength bands. We consider two specific minerals, that have an absorption feature near 2,200 Nm, the exact position and shape of the feature in the spectral response is different for each mineral. If the differences are consistent and predictable, the minerals can be distinguished reliably on the basis of their spectra in the corresponding wavelength region. For example to distinguish snow from cloud, Short Wave Infra

Red band is used. Similarly Thermal band data is used to identify the forest fire and underground coal fire. By compromising the spatial resolution, one can accommodate more number of spectral bands. There are platforms that can accommodate more number of spectral bands. FTHSI, Fourier Transform Visible Hyper spectral Imager on Mighty Sat II has 256 hyper spectral bands. Today NRSA disseminates the products from MODIS, which has 36 spectral bands.

Using the texture information of the objects, additional classes can be derived. As the radiometric resolution increases, it improves the class seperability. AWiFS data, from IRS RESOURCESAT has 10-bit radiometric resolution with the spatial resolution of 56 metres. IKONOS and QUICK BIRD acquire the data in 11 bit radiometric resolution.

Various image processing techniques are used to put high-resolution data for large number of applications. Some of these techniques are:

1. Ortho rectification.
2. Image fusion.
3. Mosaic.
4. Natural colour composite.

## Ortho Rectification

Aerial and satellite images commonly contain spatial distortions due to terrain relief and off-vertical imaging geometry. Ortho rectification is a procedure that removes these distortions and creates an ortho image with features as they would in a planimetric map. Map derived thematic data layers register more accurately with an Ortho image and the spatial information extracted from an Ortho image is more accurate. Satellite images can be Ortho rectified with a mathematical model of the imaging geometry in the form of a set of Rational Polynomial Coefficients (RPC), and with a Digital Elevation Model (DEM).

Perspective view images provide the terrain relief information along with surface features. IRS Stereo pairs are acquired with different view angles on different dates. DEM is derived

**Figure 1**

*(a) Standard Corrected*

*(b) Ortho Corrected*

from stereo pairs. As shown in Figure 1, Ortho rectified images can exactly register with the map features of any terrain.

## Image Fusion

Due to the technological limitations, it is highly difficult to achieve the high spatial resolution and high spectral resolution from the same satellite sensors. The high spatial and high spectral data can help in enhancing the information content (1+1=3), increase the confidence, reduce ambiguity, improve reliability and improve interpretation capability. IRS-1C/1D and IRS-P6 have high resolution panchromatic and multi-spectral cameras. TES and IKONOS and Quick Bird have 1 m panchromatic cameras. The data sets from these sensors are merged (fused) to achieve high spatial and high spectral resolution data sets. Image fusion is a tool to combine multi-source imagery using advanced image processing techniques. Image fusion is performed at the lowest processing level, i.e., pixel. These images represent the information content in much more detail while preserving the spectral resolution with high spatial details. There are number of merging techniques such as Intensity Hue Saturation (IHS) method, Brovey Transformation method, Arithmetic Band Combination method, Wavelet method.

### *Wavelets Approach*

A tool originally developed in the field of signal processing is applied to fuse image data under the concept of Multi-Resolution Analysis (MRA). The wavelet transform creates a summation of elementary functions (wavelets), using these it is possible to create a synthetic image from the lower resolution image at the higher spatial resolution while preserving the spectral content. The forward transforms of wavelet and colour transformations are used for fusion of LISS-III and PAN data sets in the wavelet domain of YIQ space and inverse transforms of both are used to create the original data sets in the RGB space. The use of cubic spline wavelets for merging high spatial content of panchromatic (PAN) data and spectral contents of multi-spectral (LISS-3) data of IRS-1C/1D and IRS-P6 preserve the spatial content of the original PAN data and the spectral content of the LISS-3 better than the conventional approaches. It is also suggested that the spatial content of the merged data can further be enhanced by first correcting the PAN data for overall modulation transfer function (MTF) of the sensor.

### *IHS Transformation*

IHS transformation separates spatial (I) and spectral (H, S) information from an RGB image. A panchromatic higher resolution channel replaces the I and an inverse transformation is applied to generate the fused image. Variations of this method like weighted HIS is also being used in image processing applications.

The following method is used to convert RGB model into IHS model and reciprocating, which are found in several commercial software. The RGB to IHS transform can be performed using the following procedure:

$$\begin{pmatrix} I \\ v_1 \\ v_2 \end{pmatrix} = \begin{pmatrix} 1/\sqrt{3} & 1/\sqrt{3} & 1/\sqrt{3} \\ 1/\sqrt{6} & 1/\sqrt{6} & -2/\sqrt{6} \\ 1/\sqrt{2} & -1/\sqrt{2} & 0 \end{pmatrix} \begin{pmatrix} R \\ G \\ B \end{pmatrix}$$

and $H = \tan^{-1}(v_2/v_1)$ and $S = \sqrt{(v_1^2 + v_2^2)}$

H is not defined if $v_1 = 0$, i.e. if $R + G = 2B$.

Reciprocally, the IHS to RGB transform can be performed as follows:

$$\begin{pmatrix} R \\ G \\ B \end{pmatrix} = \begin{pmatrix} 1/\sqrt{3} & 1/\sqrt{3} & 1/\sqrt{3} \\ 1/\sqrt{6} & 1/\sqrt{6} & -2/\sqrt{6} \\ 1/\sqrt{2} & -1/\sqrt{2} & 0 \end{pmatrix} \begin{pmatrix} I \\ v_1 \\ v_2 \end{pmatrix}$$

$$v_1 = S\cos(H)$$
$$v_2 = S\sin(H)$$

## *Arithmetic Combinations*

The possibilities of combining the data using multiplication, ratios, summation or subtraction are manifold. Brovey transform is one of such methods with arithmetic combinations including ratio. Multiplication and additive approaches are also available.

The choice of weighing and scaling factors may improve the resulting image. The following equation gives an example of summation:

$$DN_f = A(w_1 * DN_a + w_2 * DN_b) + B$$

Here A and B are the scaling and additive factors respectively and $w_1$ and $w_2$ are weighting parameters. $DN_f, DN_a, DN_b$ refers to the digital numbers of the final fused image and the input images a and b respectively.

## *Brovey Transformation*

The Brovey transform uses a formula that normalizes multi-spectral bands used for an RGB display, and multiplies the result by any other higher resolution image to add the intensity or brightness component to the image. The formula is as follows

$$R = [R/(R + G + B)] * I$$

$$G = [G/(R + G + B)] * I$$

$$B = [B/(R + G + B)] * I$$

where R= Red, G= Green, B= Blue, I= Intensity of Panchromatic.

Most of the fusion techniques can perform better when the ratio of the spatial resolutions is within 1:6. The geo-coding plays an essential role because mis-registration causes artificial colours or features in multi-sensor data sets, which falsify the interpretation later on. In many image-processing applications including fusion it is necessary to compare multiple images of the same scene acquired by different sensors, or images taken by the same sensor but at different times. All these applications demand the co-registration of the images. Advantages of merging can be observed in Figure 2.

**Figure 2**

*(a) LISS Image (23.5 m)*

*(b) PAN Image (5 m)*

*(c) Merged Image (5 m)*

## Image Mosaic

The area of interest of the application study is covered in multiple scenes or may span to multiple passes. In this case it is necessary to combine the images to create one large file. Mosaicking is the joining of two such geographically adjacent scenes so that the combined scene does not show geometric inconsistency across the border. Usually there is an overlap in the two scenes, the analysis of whose contents helps in building the mosaicking model.

A pair of registered images covering adjacent areas with sufficient overlap is used as input for mosaicking. Mosaicking of images involves three stages, viz., Registration, Radiometric Normalization and Stitching.

### *Registration*

Pairs of tie points are used to build a geometric model whose parameters are found by least square estimation to transform one image geometry to the other. The models usually considered are polynomials of up to degree 2. One of the images taken as reference and the polynomial transformation model is applied to transform the other image.

### *Radiometric Normalization*

The grey values of the corresponding pixels in the scenes may have different digital numbers. This is due to the difference in the acquisition date or due to difference in illumination angle, if the scene is acquired with a tilted angle. If the seasons are different then there is significant change in feature contents between the images, then it is difficult to mosaic.

There are two approaches to radiometric normalization. One is to use ramp function, where the weightage of the left image is most at the left most point of overlap and decrease the weightage to zero as one moves to the right end of the overlap for each scan line.

$$O(x) = (d - x) * 1(x) + x * r(x)$$

where $d$ is the size of the overlap, $x$ is the distance of pixel from left end of overlap, $l(x)$ and $r(x)$ are the left and right image grey values at $x$, and $O(x)$ the output (mosaicked) grey value at $x$.

The other approach is to calculate statistics of the two images like mean, variance or histogram, and match them. In histogram matching the mapping between grey levels is determined to transform the histogram of the right image to the left. This mapping is applied to the right image.

### *Stitching*

This is the process of joining the two images at suitable pixel in each scan line so that the mosaic is seamless. This is not required for the first approach for radiometric normalization. In the second approach, however, it is required. At any given scan line, the stitch points is basically chosen where the difference in the normalized grey levels is the least. To make the procedure robust, the difference is calculated over a small window. Typical mosaic products are shown in Figure 3 and Figure 4. Figure 3

**Figure 3**
*Mosaicking Process*

*(a) Uncorrected*

*(b) Reference*

*(c) Mosaicked Image*

**Figure 4**

*(a) Left Side Image*

*(b) Right Side Image*

*(c) Mosaicked Image*

demonstrates the stitching of two images and the Figure 4 demonstrates the effective cut-line usage for high resolution images to eliminate the shadows in one of the images. Size $8 \times 1$ or $8 \times 3$ and sliding the window across.

In the case of medium resolution images, mosaicking is fairly simple task except for matching the radiometry. High-Resolution images have shadow effect. To overcome this effect a proper cut-line procedures may be adopted. But the limitation is within the building or the disparity in the shadows. There is no way to make all the shadows uniform.

## Natural Colour Composite

The spectral band combinations are used to get natural colour to the image, i.e., water in blue, vegetation in green, soil in brown, etc. This combination is also identified as a 'true colour' composite. In some of the satellites where the blue band data is not available, by synthesising the green and near infra-red, the green and blue bands are derived. However, this process is used only for displaying and presenting the images to administrators and planners and not for scientific analysis.

The IRS LISS – III multi-spectral sensor does not have a blue band. The first three bands of LISS – III correspond to the green (BAND 2), red (BAND 3), and NIR (BAND 4) bands respectively. By using the following combination of the spectral bands, one can simulate natural colour composite:

$$R = BAND3$$

$$G = (3.BAND2 + BAND4)/4$$

$$B = (3.BAND2 - BAND4)/4$$

R, G and B are the simulated respective spectral bands.

## Automatic Image Registration

In many image processing applications including fusion, mosaic, geocoding, etc., it is necessary to align multiple images of the same scene acquired by different sensors, or images taken by the same sensor but at different times. In such cases registration of one image with respect to another is essential step. The major challenge in registration lies with images of different resolutions.

Currently the image-to-image registration is carried out based on local methods that require manual selection of ground control points (GCPs) over the image and these points are used to establish the mathematical model. Mathematical modelling techniques are used to correct the geometric distortions like translation, scaling and rotation of the input image to that of the reference image, so that these images can be used in various applications like change detection, image fusion etc. The selection of these control points is subjective and can lead to inconsistencies, as it is interactive with the operator. Automatic registration algorithms to correct the geometric distortions of the input image with respect to the reference image, by increasing the accuracy level of the registration and reducing the RMS error to less than a pixel are currently under development. Some of the available auto-registration techniques are based on Fourier Transform, Wavelet, Morphological Pyramids and Genetic Algorithm approaches.

## Conclusion

During the last decade advanced image processing techniques have been developed to meet user demand for value addition to various applications of remote sensing. These techniques are required to combine small data sets to generate Area of Interest (AOI) precision corrected products, to get value addition of Multi-Spectral Sensor information onto high resolution PAN data. The applications are Precision agriculture, Urban mapping and Cadastral mapping, etc.

### *Acknowledgements*

I would like to thank Dr. R. Navalgund, Director, NRSA for encouragement and support for submission of this paper. I wish to acknowledge Ch. Venkateswara Rao for his assistance while writing this paper and collecting the information.

## Select References

Kumar, A.S., B. Kartikeyan and K.L. Majumdar, "Merging multispectral and MTF corrected IRS imagery using Wavelet Transformation", *Int. Journal of Remote Sensing,* Vol. 21, 581-594; 2000.

Ortho rectification using rational polynomials with TNT Mips, Randall B.Smith. www.microimages.com.

Rao, Ch.Venkateswara, K.M.M. Rao, A.S. Manjunath and R.V.N. Srinivas, "Optimization of Automatic Image Registration Algorithms and Characteristion", *ISPRS,* Vol. XXXV, Part B3, Istanbul 2004, pp. 698-703.

Rao, Ch.Venkateswara, G. Amar Jyothi and K.M.M. Rao, "Automatic Image Registration using Genetic Algorithm", National Seminar at Birla Institute of Technology, Ranchi, October 2004.

Rao, K.M.M. et al., "IRS-1C Data Products Generation and Dissemination", *Current Science,* Vol. 70, 551, 1996.

Rao, K.M.M. et al., "A Novel Technique to Resample High Resolution Remote Sensing Satellite Data", Int. Geoscience and Remote Sensing Symp-2002(IGRASS-02), June 24-28, Canada.

Reddy, B.S. and B.N. Chatterji, "An FFT-based technique for translation, rotation and scaleinvariant image registration", *IEEE Transactions on Image Processing,* 5(8): 1266-1271, 1996.

www.isro.org

www.nrsa.gov.in

# 2

# Enhancing Pulse Production in Dryland Areas: A Perspective Based on Geoinformatics Applications

S. Bandyopadhyay, S.K. Srivastava, V.S. Hegde and V. Jayaraman

Pulses, such as pigeonpea, lentil, chickpea and beans are the major source of protein for most of the 1 billion people in India. Till recent times there is marginal increase in production of pulses (around 3.0 million tonnes) from mid-fifties which is considered to unsynchronized growth in pulse production with the increase in population. The per capita availability of pulses has been declined from estimated 64 g/capita/day (1951-56) to less than 40 g/capita/day as against the FAO/WHO's recommendation of 80 g/capita/day. The average net sown area under pulse crops remain more or less stagnant (about 23 million hectare) with the marginal increase in productivity from 500 kg/ha to 600 kg/ha (Directorate of Economics and Statistics, 1999). The very fact is that the overall growth rate of production of pulses declines to 1.19 per cent per annum during the decade of nineties (1988-90 to 1998-99) as compared to 2.78 per cent per annum during the earlier decade, i.e., 1979-80 to 1989-90 (Economic Survey of India, 1999-2000). While cereal crops associated with the green revolution have exclusively contributed the rapid increase of food production over the last fifty years, predominantly from mid-sixties. Therefore, the pulse production as well as its consumption rate remains stationary or decreased because the land for pulse cultivation was partially used for cereals or oil seeds cultivation. The potential resources, indispensable for intensive and extensive agricultural production, by and large diverted towards production of cereal crops to feed the ever-growing people and thereby neglecting the pulse production. As a matter of fact, the irrigation and fertilizer application, pest management and other cultural operations under pulse agro ecosystem, so far could not gain the appropriate momentum. The impact of green revolution and technology mission for oilseeds and pulses are yet to penetrate to a level by breaking the stagnancy in pulse production (Figure 1).

## Pulse Production Scenario

In India, pulses are rarely grown as a sole crop and generally grown in rainfed conditions. Most often these form part of the intercropping, sequence cropping or traditional mixed seedling systems provided much needed stability to the farming community. Pigeonpea, being a long duration crop, is often intercropped with short duration cereals so that the land equivalent ratio is optimized. The pulses like chick pea, lathyrus, rajmah in *rabi* and pigeonpea, mungbean, urdbean, mothbean cowpea and horsegram in *kharif* are the main crops (Singh, et al., 2000) The irrigated and command areas are mostly cultivated with principal crops like wheat, rice sugarcane, cotton, tobacco, maize, etc. However, pegionpea and chickpea are occasionally grown with irrigation. In India, around 11 per cent pulse grown area is under irrigated ecosystem, which contribute about 21 per cent of the total pulse production.

**Figure 1**

*Pulse Productivity Over Last Fifty Years*

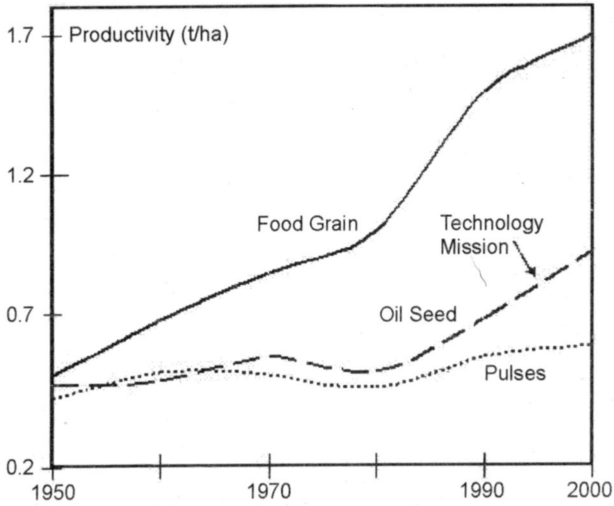


In spite of being largest producer of pulses in the world, India has to import nearly 0.7 million tonnes of pulses to meet the growing demand of present population. Projection for domestic demand by the end of first decade of this new millennium, i.e., by the year 2010, arrived at by adding the direct demand (human consumption and indirect demands (seed, feed, wastage and other uses), under the scenario of 5 per cent growth in GDP, works out to be 21.4 million tonnes of pulses. This means that an increase in yield by 66.4 per cent in pulses is required over the yield attained during 1994-95 (Yadav, et al., 2000). Therefore, to meet the growing demand of pulses judicious planning, optimum management practices and blending of modern technologies are very essential. This is where a need for convergence of technologies is called for. Agriculture of new millennium is transitioning towards – convergence of technologies and institutions. Core technologies, viz., plant breeding, genetics, bio-technologies, molecular biology, etc.; enabling technologies (soil and water conservation, plant pathology, entomology, chemistry, etc.); and facilitating technologies (geoinformatics, remote sensing, modelling, databases, networking etc.), need to be closely coupled to turn agriculture an information rich, extremely dynamic enterprise.

## Information Need: Role of Geoinformatics

Geoinformatics essentially involves capture, integrating, analysing, managing and depicting geospatial information. Geoinformatics is an interdisciplinary field which makes use of various independent subjects like surveying, photogrammetry, GIS, satellite remote sensing, etc. The satellite information technology, through application of remote sensing, has shown tremendous progress during the last two decades and have established the potential of using this technology for natural resource management and environmental monitoring. Timely and accurate information on agricultural production is of paramount importance for effective management of agricultural economy. In India, acreage and production forecasting of certain crops, crops yield modelling and crop stress detection is being done using remote sensing satellite data. Crop acreage estimation has been quite successful through both visual and digital analysis in many areas. However, a major constraint is cloudy sky in the kharif season when normal optical remote sensing is unable to acquire good data. With the emergence of microwave remote sensing data, this difficulty can be overcome, as such instrument can see through clouds.

Various factors such as intensive cultivation, monocropping, mechanization and indiscriminate use of fertilizer and pesticides frequently resulted in unfavourable conditions for crop growth (pest attack, nutrient toxicity, etc.) leading to the huge crop loss. Minimizing such losses through crop monitoring is one of the ways to increasing crop production, and remote sensing as a tool is found to be very useful in frequent monitoring and assessing the large areas under cultivation. Besides the remote sensing satellite, the constellations of INSAT – series satellites provide the vital information on periodical weather fluctuation affecting the agro-meteorological condition of the crops.

## Managing Pulse Agro-ecosystem

The pulse agro-ecosystem constitutes of interlinked biological and physical processes pertaining to crop, soil and atmospheric continuum. The understanding of such agro-ecosystems in terms of dynamic interaction and equilibrium between biotic and abiotic components, is crucial in the context of sustainable agriculture; it is important that the process of harmonization between such agro-ecosystems and natural environment is focused appropriately. The real strategy for developing sustainable pulse production therefore, lies in harmonization of the agro-ecosystem with natural ecology by integrating natural processes into management strategy. This strategy aims at (i) integrating the concepts and practices of natural ecosystem into the agricultural practices, such as integrated pest management, optimal rate and timing of nutrient cycle within system, increase energy and water use efficiency and (ii) reducing wasteful inputs, optimizing material inputs and increasing biotic and landscape diversity.

## Pulse Crop Inventory through Remote Sensing

In agriculture, the application of remote sensing varies widely ranging from crop condition assessment, crop discrimination, crop acreage and yield estimation to characterization of crop environment such as canopy temperature, status of soil conditions (e.g., moisture, fertility, salinity and alkalinity) and other soil degradation processes. Presently, there are five Indian Remote Sensing Satellites in the orbit to provide valuable information towards the natural resource inventory.

### *Remote Sensing in Pulse Crop Condition Assessment*

The crop phonological stages interact with the soil and atmospheric conditions throughout the crop growth cycle. Therefore, regular crop monitoring at a specified interval is most important to assess the production loss due to the environmental and biotic stress. The satellite based remote sensing has proved to be a useful tool to retrieve the information on aerial extent of crop stress (Rao, et al., 1997). Indeed, high-resolution spectral signature is most appropriate in such studies. However, detection of stress caused by multiple factors still remains a major limitation as the situation become complex. Although, with the help of local weather condition and ground truth information like soil nutrient deficiency/toxicity, infestation by pest and diseases, etc., would be the useful inputs towards the assessment of crop stress. More recently, use of hyper spectral remote sensing data to characterize crop signature under various stresses, are in the experimental stage.

### *Remote Sensing in Pulse Crop Identification and Classification*

The remote sensing satellite data have been proved to be quite effective in identifying and classifying the field crops (Navalgund, et al., 1996). In a major crop inventory study in Andhra Pradesh, 188 metre resolution IRS-WiFS data of January 25, 1996 covering the state were used to identify and delineate major crops grown in various parts of the state. It was possible to visually identify paddy and pulses that are grown in Godavari and Krishna Delta regions and cotton and other minor crops spread over in other districts on two-band False Colour Composite (FCC). Paddy and pulse crops in Godavari and Krishna delta and the standing kharif cotton in Guntur, Khaman, Warangal and Kurnool districts were classified. The FCC generated by assessing red colour to NIR band (B4) and green and blue colours in red band (B3) with district boundaries and crop distribution from classified image overlaid on it (Figure 2). The estimated acreage of paddy and pulses are 5.09 lakh ha and 2.69 lakh ha, respectively.

The temporal spectral behaviour of crops is related to their phenology and agronomic practices can be monitored for various applications like crop rotation, cropping pattern practices, crop discrimination, crop assessment, etc., using multidate WiFS data. A cropping pattern analysis was conducted in Burdwan District, West Bengal using IRS-WiFS data. The seasonal analysis shows

**Figure 2**

*IRS-1C WIFS Data of Part Godavari and Krishna Delta, Andhra Pradesh*

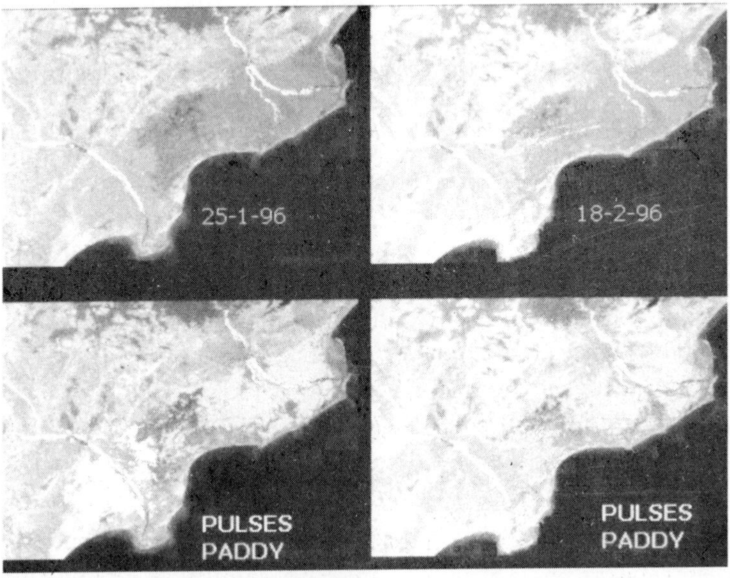


**Figure 3**

*Classified Map of Cropping Pattern in Summer Season, Burdwan District West Bengal*

that the rice is the main crop in the Kharif season and pulses occupy limited area in the district during summer season. The classified map of minor crop type represented in different colour effectively demarcates the area under respective crops (Figure 3).

### *Pulse Crop Acreage Estimation*

Identification and discrimination of land cover classes require quantitative magnitude of the spectral reflectance of the respective classes. The acreage estimation procedure broadly consist of identifying representative site of various crops/land cover classes on image based on ground truth collection, generation of spectral signature for different training sites and classifying the image using training statistics. Single date data corresponding to maximum vegetative stage are mostly used for such estimation. Sometimes two/three forecasts are also made as and when required. For district level acreage estimation the administrative boundary of the district is digitized, a mask is generated and superposed on the scene. All the picture elements (pixel) within the scene are extracted for further analysis. Such procedure has been successfully used for some major crops and can be applied for pulse crops also.

### *Discrimination of Pulse Crop*

Spectral signature in multiple cropping region is a product of the crops present in the area. However, the phenological growth stages of the crops differs with one another which forms the basis of crop discrimination in the spectral reflectance zone. With the availability of high resolution temporal and spectral data it is possible to identify and separate the spectral signature of the crops grown together.

### *Management of Pest and Disease in Pulses*

The stress caused by pest and disease affect the crop tissues and chlorophyll content of the leaves thereby influence spectral signature of the crop. Remote sensing can play important role in detecting the pest and disease in pulse crop particularly in high spectral resolution. The information on local weather condition and the ground condition assessment of the crop in combination with crop spectral signature improves the detection of pest and disease to a great extent.

## Enhancing Pulse Production

As the pulses mostly grown in marginal land with low inputs particularly, in terms of water, nutrient, plant protection measures, etc., the existing yield potential of the crops have not yet been realized. However, major strength of the pulse production system are rich in germ-plasm diversity and adaptability of pulses to variety of agro-climatic conditions and cropping system. Moreover, the pulses being leguminous crop help in fixing nitrogen in the rhizosphere and thereby, improve soil health.

In order to break the stagnation in pulse production calls for a paradigm shift of the cultivation strategy by making pulse agro-ecosystem as knowledge and technology intensive, turning it to information rich enterprise, ensuring total quality management in pulse cultivation as well as management and by integrating the principle of economics with ecology and equity – called eco-technology. The revival phase of pulse agro-ecosystem can thus only be built by close coupling of core, enabler and facilitator technologies (Figure 4). The very basis of new paradigms lies in fusion and blending of technologies, especially

**Figure 4**

*Technology Convergence in Agriculture*

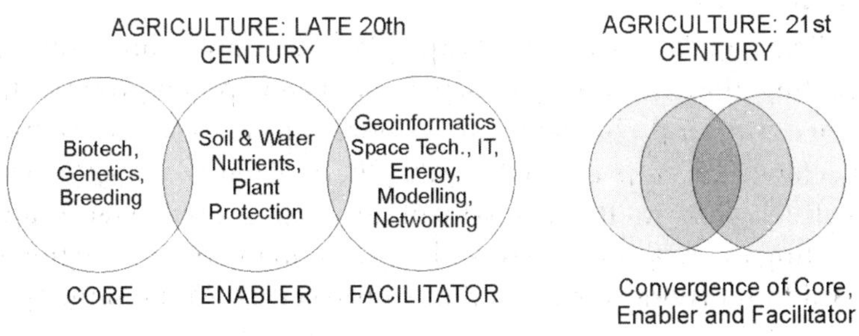

convergence of biotechnology with space and information technologies. While biotechnology provides a molecular understanding of the genetic basis of living organisms as well as the ability to use this understanding to develop genetically improved processes and products, space and information technologies allow quick propagation and assimilation of genetic materials. Information rich pulse agro-ecosystem needs to have gene banks, data warehouse, spectral libraries and appropriate informatics base for decision support. Total quality management which includes the optimal management of land, water and agricultural inputs, viz., fertilizer, seeds, integrated pest management, etc., is critical element in view of realizing the genetic potential of these crops with ecological integrity.

### *Agro-climatic Suitability for Pulse Crop*

The pulse crops require certain cardinal levels of environment like rainfall, air, soil temperature, length, quality and intensity of light, humidity conditions, etc., for optimum physiological functioning. These requirements not only vary from species to species but may also be different for the various stages of crop development. Deviations from these, resulting for either in situ climate or from weather, result in abnormal crop growth and performance (Venkataraman and Krishnan, 1992). Besides the above direct influences, weather may exert an indirect influence through the prevailing of conditions congenial for outbreak of pest and diseases. Therefore, crop growth is the resultant of many physical and physiological processes, each of which are affected separately or jointly, to various degree, by environmental factors.

The space technology can play an important role in revealing the dynamic interaction between the environmental factors and crop characters. The thorough analysis of spectral reflectance of crop as well as the thermal environmental can effectively use in the assessment of crop growth status and detecting the crop stress. The radar return in microwave region is useful for deriving surface soil moisture status. At the same time the reliability of rainfall/irrigation water can be worked out from the assessment of soil moisture during the crop-growing season.

### *Watershed Management*

The emphasis on sustainable agriculture through optimum management of natural resources for increasing the productivity of per unit of land has been the key criteria for development programme based on watershed approach. This approach will go a long way in restoration of ecological balance, maintenance of biological diversity, and protection of catchments for soil and water conservation and thereby enhancing the productivity (Table 1).

**Table 1**

*Watershed Development Capacity Building*

| *Human Dimensions* | *Natural Resources* | *Techno-Convergence* |
|---|---|---|
| • Reducing poverty/inequity | • Soil and water conservation | • Geoinformatics: Remote sensing/GIS/Info technology |
| • Rural livelihood support system | • Land-use planning | • Communications Technology |
| • Community participation awareness/training | • Ridge to valley treatment | • Contemporary technologies |
| | • Focus on wastelands | • Local wisdom |

Remote sensing based Integrated Mission for Sustainable Development (IMSD) experiment was conducted in the terrain of low productivity, mainly drylands/rainfed areas predominantly wastelands, where the scope of extensive pulse production is the main issue. Introduction of land shaping, graded bunds (ridges), and grass waterways significantly helped in reducing soil erosion on arable lands, increase on-site moisture conservation together with improved cropping techniques and inputs (such as high-quality seeds and fertilizers); stimulated agricultural production leads to boost in farm income.

The synergistic interfaces among grass-root level farmers, administrator/policy makers and multidisciplinary scientific communities along with the spatial information from space inputs forms the basis of decision-making process. With the availability of high-resolution data from IRS – P6 and planned future missions, such assessment can be carried out to generate finer level information to monitor and manage pulse growing areas more effectively.

## Future Perspective

The demand on pulses to meet the requirements of the projected population of 1.25 billion by 2010, even at the present level of pulses consumption, is 21.6 million. This means there would be a shortfall of about 19 million tonnes of pulses by 2010 over the production achieved during 1994-95. Implication of realizing the projected production of 21.6 million tonnes by 2010 is that average pulse productivity has to increase from the present 0.61 tonnes/ha to at least 0.96 tonnes/ha by 2010. It is, however, well within the technology capability to achieve the required pulse production by 2010, provided the developmental plans are oriented towards: (i) increasing the net irrigation potential based on in situ watershed management, (ii) reclamation of wastelands and salt affected lands by appropriate agricultural practices, including cultivation of short duration varieties of pigeonpea (arhar), moong, urd, chickpea, etc. (iii) using of biological techniques to control soil erosion (iv) development of genetically engineered varieties of pulses which are resistant to major biotic stress like disease, pest, nematode etc., flood resistant, water stress resistant and short duration (v) providing better plant protection measures, improved nutrition, particularly through phosphorus application and the use of effective rhizobium cultures, and (vi) adopting improved post harvest technology.

Moreover, to encourage further social and economic development in dryland/rainfed areas as a whole, future development projects might include investments in complementary infrastructure such as rural roads and drinking water supply and, depending on local potential, might encourage agro-processing and marketing development.

## Conclusion

Integrated information on agro-ecological distribution of pulse crop in the country is essential for intervening the production system with the possible package of practices. The satellite images, through preparation of thematic maps, provide ample opportunity towards mapping and monitoring of variable natural resources. As more than 75 per cent of pulses are grown in the stored soil moisture as rainfed crops, integrating and analysing the information generated through the research works in various rainfed areas across the variable natural resource environments throughout the country, optimum land-use plans at local, regional and national level can be derived. Once these spatial database on climate, soils, land use and water resources are created accurate characterization of rainfed areas along with different crop growing environments is possible. Thus, the sustainable management of the pulse agro-ecosystems has the potential to achieve a common ground between environmental and economic goals.

## References

Directorate of Economics and Statistics. 1999. "Agricultural Statistics at a Glance", Department of Agriculture and Cooperation, Ministry of Agriculture, Government of India, New Delhi.

Economic Survey of India. 1999-2000. Ministry of Finance, Government of India.

Navalgund, R.R., Parihar, J.S., Venkataratnam, L., Krishna Rao, M.V., Panigrahi, S. Chakroborty, M.C., Hebber, K.R., Oza, M.P, Sharma, S.A. and Bhagia, N., Dadhal V.K. 1996. Early results from crop studies using IRS-1C data. *Current Science,* 70(7): 568-574.

Singh, H.P., Venkateswaralu, B., Vittal, KP.R. and Ramachandran, K. 2000. "Management of rainfed agro-ecosystem", In, *Natural Resource*

*Management for Agricultural Productions,* (Eds.) Yadav, J.S.P. and Singh, G.B., pp. 669-774.

Venkataraman, S. 1992. "Weather Relation of Crops", In, *Crops and Weather* (by Venkataraman, S. and Krishnan, A.). Indian Council of Agriculture Research, Pusa, New Delhi.

Yadav, R.L., Singh, S.R., Prasad, K., Dwivedi, B.S., Batta, R.K., Singh, A.K., Patil, N.G. and Chaudhary, S.K. 2000. "Management of Irrigated Agro-ecosystem", in *Natural Resource Management for Agricultural Productions,* (Eds.) Yadav, J.S.P. and Singh, G.B., pp. 775-870.

# 3

# Use of High Resolution LISS III Images for Capacity Assessment of Reservoirs in Maharashtra

A.M. Deshmukh, S.R. Vaijapurkar and S.K. Kalvit

The State of Maharashtra has the highest number of large dams in the country. Most of these dams have been constructed across the rivers, which carry water only during monsoon period from June to October. The post-monsoon flow in the river is hardly about 3 per cent to 5 per cent of the monsoon flow in most of the rivers which lasts till the month of February and then the rivers remain mostly dry for remaining part of the year. The rainfall in the state is also unevenly distributed. The maximum annual rainfall occurs in the Western Ghats area of the state covering the Sahyadri range, which is about 3500-4000 mm, and the minimum rainfall occurs in some parts of North Maharashtra and Marathwada region, which is 300 mm only. Due to inconsistency of sufficient rains in various parts of the state and over-exploitation of groundwater through bore-wells and tube-wells, the water table in many areas has gone down very deep. The situation in many areas, especially, in the Orange belt of the state, i.e., the eastern part of Maharashtra is worsening every year. The best possible solution to this problem is to store the flowing river water during the monsoon period, as much as possible, by constructing dams and barrages across the rivers to cater the needs for different purposes, viz. irrigation, domestic and industrial water supply as also for power generation. So far, number of dams have been constructed in the state and many are under construction. Huge capital expenditure has been incurred on construction of these dams. These dams have been designed for some definite storage capacities for a definite lifespan. But in course of time, these dams are getting silted at different rates depending upon the type of the catchment areas, different rainfall pattern and various different hydrological parameters, etc., defeating gradually the very purpose for which they have been constructed.

The reservoirs are normally operated in the live storage zone, which is storage between F.R.L. i.e. full reservoir level and M.D.D.L., i.e., maximum draw down level. In general, the planning of the storage between F.R.L. and M.D.D.L. is done in advance depending upon the availability of water at the end of monsoon period. The storage below M.D.D.L. is occasionally used by letting the water into the river or by lifting it from the reservoir mainly for meeting the drinking water requirement and for saving the standing crops, if any. But it is the live storage zone of the reservoir, which is important for the management authority to plan optimal use of the storage.

In the earlier period there was a practice to fix the dead storage level of the reservoir at a level, which has a capacity to accommodate the total silt estimated and likely to be deposited in the reservoir based on observed silt rate in the catchment or based on some rational or empirical formula for certain assumed life period. It was assumed that the entire silt would be deposited below this dead storage level and there would not be any encroachment over the live storage part and the reservoir would continue to serve its planned benefits throughout its lifespan. But in practice, it is observed that the sedimentation takes place in the reservoir throughout its full range of depth at different rates, encroaching the live storage which results in reduction of the useful storage capacity, causing an adverse impact on the planned irrigation potential, power generation and other uses.

## Necessity of Resurvey of Reservoirs

Sedimentation in reservoir is a natural phenomenon and it occurs because of soil erosion, its transportation and subsequent deposition in reservoirs. The factors responsible for sedimentation are uncontrolled deforestation, overgrazing, improper method of tillage, faulty agriculture practices and other human activities, etc. The eroded soil is carried away by the rivers/nalas from the catchment and is deposited in the reservoir. Quantum of sediment transported in the reservoir depends upon rate of erosion of soil in the watershed and the ability of the streams/nallas to transport the eroded material into the reservoir which in turn depends on hydrological characteristics of the watershed. The rate of erosion also depends on climatic conditions, nature of soil, topography, land use, etc. The eroded material transported into the reservoir is deposited in full range of its depth, resulting in reduction in storage capacity. As huge capital cost is incurred on construction of reservoirs, it is very essential that these reservoirs sustain for a longer period to give economic benefit as envisaged at the planning stage. But the reservoirs progressively lose their storage capacity due to sedimentation. This sedimentation has an adverse effect on (a) irrigation potential of the reservoir; (b) contemplated hydropower generation, and (c) quantum of water for domestic and industrial supply. The sedimentation in a reservoir can cause excessive growth of unwanted vegetation like aquatic weeds and Besharam, which are harmful from environmental point of view. These aquatic weeds obstruct the flow courses and provide platform for silt deposition. These waterweeds multiply and spread profusely. Thus, sedimentation process and growth of water-weeds runs in a cyclic manner, thereby causing progressive reduction in storage capacity of a reservoir. It also causes degradation of water quality and also aquatic life.

This sedimentation in reservoirs seriously affects the designed storage capacities. There is a reduction in the storage capacity of the reservoirs on one side and increase in the demand of water due to increasing population and expanding industrial sectors on the other. For better management of the available storage in the dams, it is very important for the authorities of the reservoirs to know the up to date content in the reservoirs at different levels. For this purpose the recurrent capacity assessment studies at regular interval are very essential. The interval may be flexible depending upon the risk of the floods in the reservoirs. Heavy floods every year cause more sedimentation in the reservoirs. Hence, this interval may be of 3 to 5 years for the reservoirs which normally receive heavy flood every year. For other reservoirs, which do not receive heavy floods every year, the interval may be 5 to 10 years.

Maharashtra Engineering Research Institute (MERI), Nashik, working under the control of the State Irrigation Department, has done considerable work in the field of reservoir capacity assessment studies using both hydrographic survey method as well as satellite remote sensing technique. No doubt the hydrographic survey method gives a total picture of sedimentation throughout the full range of depth covering a dead storage portion also. But since this method requires more time, greater manpower and higher cost, it is unsuitable for frequent repeat surveys. Capacity assessment of one major reservoir by hydrographic survey method requires about 6 months period. On an average at the most three reservoirs can be covered under hydrographic survey every year. With this rate it would take a very long period of about 25 to 30 years to complete one hydrographic survey for 100 completed major and medium reservoirs in the state. Cost of such survey per reservoir ranges between Rs 3 lakh to 9 lakh depending up on the size of the reservoir, i.e., area to be surveyed.

Satellite remote sensing technique is found to be useful, which provides a faster and cost effective solution to this problem. It has been experienced that for reservoirs having large water spread area, remote sensing technique is faster and cheaper as compared to hydrographic survey method. The Institute has so far conducted capacity assessment studies by remote sensing technique for about 20 major reservoirs including Jayakwadi, Ujjani and Koyna reservoirs which are the three largest reservoirs in the state both in terms of water spread and storage capacity. All these reservoirs are having storage capacity of more than 100 TMC. The studies were found to be very economical. Manpower requirement for remote sensing based study is also less. Another advantage of this technique is that one LISS III image has a large ground coverage of 141 km by 141 km, which can accommodate number of adjacent reservoirs in one scene. The satellite data selection for the capacity assessment study of all the reservoirs covered in one image, if carefully done with due consideration to cover maximum fluctuation in the water levels in all these reservoirs, shall help reduce the cost of capacity assessment work per reservoir substantially. Only few additional images may be required to cover the total live storage under the study for all the reservoirs, as the fluctuation pattern in all the reservoirs may not be same for the same period. The cost of the satellite data is the main expenditure in such studies. This cost thus gets distributed over the number of reservoirs and is reduced substantially per reservoir study. Considering this aspect, the Institute has planned to undertake capacity assessment studies of nearly 100 major and medium reservoirs in the State using high resolution LISS III images. The entire State is covered in total 21 LISS III scenes. With proper selection of shift along track (SAT) product of the LISS III data to cover maximum number of reservoirs (100 reservoirs), the number of total LISS III scenes can be reduced to 18 only.

## Cost Estimate of Capacity Assessment Work for 100 Reservoirs

Normally on an average, 10 multidate images of different water levels in the reservoir covering the entire live storage are felt necessary for one reservoir study.

Therefore to cover 100 reservoirs in the state as mentioned above, 18*10 = 180 scenes would be required. As per the current price list of N.R.S.A., the cost of one Resourcesat (IRS-P6) LISS III image is Rs 20,000. Thus:

| | | |
|---|---|---|
| The total cost of the satellite data would be about (20,000 × 180) | = | Rs 36,00,000 |
| Adding 20 per cent for the analysis cost | = | Rs 7,20,000 |
| Adding 10 per cent for maintenance of the computer system | = | Rs 3,60,000 |
| Adding 5 per cent for misc. expenses including preparation of Reports, stationary charges, etc. | = | Rs 1,80,000 |
| Total Rs | = | Rs 48,60,000 |
| | | Say about Rs 50,00,000 |

Thus, the cost of the study per reservoir works out to Rs (50,00,000/100) i.e., Rs 50,000 only.

It is also felt that the comparison of results of the capacity assessment survey by three different methods, viz., hydrographic survey method, remote sensing technique and the original capacity survey using the conventional theodolite and level instruments at the time of planning and construction may differ by 3 per cent to 5 per cent because of some inherent limitations in all the three techniques. So the comparison of results obtained by two different techniques is also not proper for correct assessment. Moreover, in some cases it has been found that the original survey itself was not correct and drawing any conclusion about reduction or increase in capacity and the volume of silt deposited would be a misleading one. Hence it is always better to compare the results of the study repeated by the same method so that the possible errors due to inherent limitations get nullified. As mentioned earlier, hydrographic surveys are costly and cannot be carried out repeatedly at regular intervals for the same reservoir. Application of remote sensing technique for repeat survey should have first priority. Moreover, the current discount facility provided by N.R.S.A. for the satellite data which is three years old can be effectively availed for the studies to form a data base of the sedimentation pattern in the storage reservoirs for the last few years which would certainly help draw some important conclusions, like trend of annual percentage loss in the capacity of reservoirs, increase or decrease in the rate of siltation over a period of time, etc.

## Methodology and Data Used

### *Field Data*

The data of water levels in the reservoir on different dates of satellite pass, their respective water spreads and the volumes as per the project report are required for the study, which can be obtained from the field authorities. A submergence plan showing F.R.L. and M.D.D.L. contour is also desirable to know the extent of submergence at F.R.L. and M.D.D.L.

### *Satellite Data*

IRS 1C/1D or P6 (Resourcesat) Satellite data of LISS III sensor having a spatial resolution of 23.5 m has been used for the study of reservoirs in Maharashtra. Satellite data on compact disks, (CDs) is procured from National Remote Sensing Agency (N.R.S.A.), Hyderabad. Although the spatial resolution of LISS III sensor is 23.5 metres, N.R.S.A., provides LISS III data after re-sampling to 24 metres. Satellite images of the reservoir were purchased for different water levels.

## Methodology

Any standard image analysis software can be used for the digital analysis of the satellite image. At MERI, ERDAS/IMAGINE (Version 8.4), a standard image analysis software, is used for analysis of satellite data. All the satellite scenes are first downloaded on PC Pentium IV. These digital images are to be rectified first with corresponding Survey of India toposheets of 1:50, 000 scale. The temporal satellite images of the same path and row do not cover exactly the same area. There is a little shift in lateral and vertical direction. In the process of rectification, some common points in the image to be rectified are tied up with corresponding control points in the Survey of India maps, i.e., toposheets and the image is oriented to true north by the process of transformation. After rectification, the satellite image gets north orientation and the geographical locations in the two rectified images exactly coincide when superimposed over one another.

One full scene of IRS -IC LISS III image covers a large area of 141 km × 141 km. As compared to it, reservoir area is small. Hence sub-scenes covering full reservoir area are extracted from full scenes. The digital images are classified in different classes such as land, water, vegetation, etc., by taking appropriate training samples of the respective classes. For classification of the images, Supervised Classification technique with Maximum Likelihood Classifier is generally adopted. After classification, areas of different classes are measured from histogram of pixels belonging to respective classes, by multiplying pixel counts of each class by the size of the pixel which is 24 × 24 m. Water-spread areas of the reservoir in all satellite sub-images are thus measured. From the water-spread area obtained as above for different levels, the storage capacity between successive levels under study is computed using standard prismoidal formula as below.

$$V = H/3 \times [A_1 + A_2 + \text{SQ.RT.}(A_1 \times A_2)]$$

where V is the volume between two levels.

H is difference between two levels.

$A_1$, and $A_2$ are areas at successive levels.

The volume of water, i.e., the storage capacity calculated at various levels is then compared with the designed capacity, i.e., storage capacity between the same levels, as per the project report. Any change, either increase or decrease is broadly attributed to erosion or siltation respectively. For correct assessment it is better to compare the results on a common platform, i.e., results of the surveys carried out by the same method repeatedly.

So far, about 20 major reservoirs have been studied by satellite remote sensing technique at MERI, Nashik, in last 3 years and about 10 reservoirs are under study at present. On an average about 75 per cent to 80 per cent live storage of the reservoirs has been covered under study. The area-capacity curves and the tables have been updated for these reservoirs. A statement giving the list of reservoirs proposed to be covered under study on priority and the results of the studies completed so far are also enclosed at the end.

## Some Other Important Findings of the Studies

While studying the reservoirs, it was observed that around and within the submergence area, growth of vegetation like Besharam and other aquatic weeds like Pan-kanis has been increasing, especially in the reservoirs in the flat areas. Ujjani, Jayakadi, Manjara, Lower Terna, and Lower Wunna reservoirs are some of the reservoirs where such growth is predominant. But for the reservoirs located at or near the foot of the hills such harmful vegetation growth has not been observed. Koyna, Dhom, Kanher, Warasgaon, Panshet and Bhandardara are some of these reservoirs. As stated earlier such vegetation grows and spreads profusely in the water, consuming the dissolved oxygen in the water and prevents sunlight from reaching the water surface. Such areas can provide suitable places as breeding zones for mosquitoes. It is therefore necessary to eradicate such vegetation periodically.

# 4

# High Resolution Satellite Imagery in Site Identification for Telecommunication Networks

Shalini Ajayan, Deva Pratap, Siva Sankara Reddy and P. Sreedhar Reddy

The haphazard urbanization pattern in the country has led to the telecommunication network being concentrated only in and around the urban areas owing to increased market values; making the system a mess, due to poor planning or the lack of it. Vast geographical expanse and a large number of features characterize telecommunication networks. But, localized concentration of these networks has led to a congestion of facilities. A strong geographical inventory is required to support several business functions in telecom right from customer contact, service order, network planning and engineering. Many telecom service providers are facing tough competition and challenges in terms of availability of land and optimum usage of available land resources.

The skyrocketing development of satellite imagery and its increasing use as an efficient source of spatial, temporal and spectral information is prompting the telecom giants to explore its burgeoning potential as an effective means for siting and location identification problems in establishing mobile telecommunication networks. High resolution satellite imagery will add an entirely new level of geographic knowledge and detail to the intelligent maps that can be created from the imagery. They provide the most up-to-date source of data available helping to ensure accurate and reliable decisions. The potential of using high resolution satellite imagery in the identification of site locations for telecommunication networks is presented as a case study on "Cellular Tower Siting for Mobile Communication Network" for an area within Ranga Reddy – Hyderabad revenue districts, Andhra Pradesh, India. The IRS – 1C, LISS III and IRS – 1C PAN images of the area are used for the study which is aimed at planning a suitable network for mobile communication with high resolution satellite imagery to have continuous signal coverage for the cellular phone users.

## Use of High Resolution Satellite Imagery

Indian metropolis is witnessing a fast paced technological development, which has made communication with the entire world possible in a matter of minutes. There are rapid structural changes in urban and rural areas. However, there are very few base maps and high resolution satellite images available to reflect these rapid changes. Besides, local survey mapping and subdivision activities lack any acceptable standard or format. These systems are far from representing most parts of the areas.

This is where high resolution satellite imagery scores over data collected through conventional techniques. Incredibly sharp earth photos from space platforms reveal spatial, spectral and temporal information about our world in ways never before possible or even imagined. The importance of high resolution satellite imagery is continuously expanding because of the rapid increase in ground resolution of successive systems. The commercial capability of high resolution space borne imagery is now generating a compelling interest in its potential use in siting and location identification problems. Satellite imagery is ideally suited for applications requiring large area coverage, such as agricultural monitoring, regional mapping, environmental assessment and infrastructural planning, and telecommunication network planning; emergency response requirements like post-disaster relief management, etc.

High resolution satellite images, having spatial resolution of the order of 1 to 4 metres, obtained from IKONOS and IRS satellites can be used to generate maps of scales ranging from 1: 25,000 to 1: 10,000, which can be suitably used for site identification problems These images have a definitive advantage over the conventional land-based techniques owing to their superior spectral as well as spatial capabilities, which render them helpful in clearly differentiating various physical, natural and cultural features present at the site.

Some of the high resolution satellites currently in use and their spatial resolutions are given below.

| *System* | *Ground Resolution (metres)* | |
|---|---|---|
| | *Panchromatic* | *Multispectral* |
| Radarsat | 10 | 30 |
| IKONOS | 1 | 4 |
| Quick Bird 2 | 0.61 | 2.44 |
| SPOT 5 | 2.5 | 10 |
| IRS – 1C and IRS – 1D | 5.8 | 23.5 (LISS III) |
| IRS – P5 CARTOSAT – 1 | 2.5 | – |
| IRS – P6 | 5.8 (LISS IV) | 5.8 (LISS IV)<br>23.5 (LISS III) |

However, high-resolution imagery from both aerial and space-borne sensors provides a challenge to the user community in terms of information extraction. Low resolution and high resolution satellite images greatly differ in terms of interpretability. While low resolution imagery smoothes out troubles due to shadows or variations across individual pixels, linear feature extraction, etc., high resolution images can prove to be a handful with these problems. Thus, high resolution satellite imagery requires a very different feature extraction and classification technique.

## Site Identification for Telecommunication Networks

Site identification is the prima facie requirement of any human related activity. The signal for all the telecommunication systems originate from a network of antennae situated at a number of strategic locations across the landscape. For a better planning of telecommunication networks, the sites for the establishment of these antennae towers should be carefully selected keeping in mind topography, land use/land cover pattern, population density, cost of the system, user density, etc.

The site selection can be carried out through conventional methods like land surveying, census data collection for population, cost studies, etc. However, trying to work out the best network of antennae over a large area is difficult and is dependent on numerous factors; including land cover, height variation, building heights, composition and morphology. For many areas in the world, this information, together with good base mapping is not readily available and it would be very expensive and time consuming to measure using standard ground-based techniques. Moreover, this can create a very big problem in inaccessible areas.

The stereo satellite or aerial images can be used to generate the most up-to-date height and land use information. Using this information, telecom companies can make use of utilities like Geographic Information Systems (GIS) to calculate and model the likely signal strength for any given theoretical antenna site, without physically having to visit the site. This can be done by developing Digital Elevation Models (DEMs) including different parameters, such as the topography, number of antennae, position, antenna heights, etc. After that, the commercial and topographical viability of the area as a site for establishing communication tower can be studied using information obtained through land use/land cover pattern interpretation of the satellite imagery of the area.

Thus, through the software model, the planner can identify sites, which will produce the best coverage in view of economy and utility. This has the potential to eliminate the undisciplined and haphazard growth of telecommunication networks in the country.

## Case Study: Cellular Tower Siting for Mobile Communication Network

| | | | |
|---|---|---|---|
| Study Area | Ranga Reddy | Hyderabad Revenue Districts, Andhra Pradesh, India | |
| Data used | | 1. Survey of India Toposheet No. 56K/7 between | |
| | | Latitude | : 17° 30′ 5.73″ N and 17° 22′ 24.01″ |
| | | Longitude | : 78° 30′ 9.28″ E and 78° 22′ 21.63″ E |
| | | Relative Height | : 458.502 m to 621.058 m |
| | | 2. High Resolution Satellite images used for the study | |
| | | IRS – 1C, LISS III dated 5 February 2002. | |
| | | IRS – PAN data dated 5 March 2002 | |
| Projection System | | Polyconic (units: metres) | |
| Spheroid | | Everest | |

The study is aimed at planning a suitable network for mobile communication with high resolution satellite imagery and raster GIS techniques to have a continuous signal coverage for the cellular phone users. The selection of site is based on local peaks covered under all the tower points to achieve total coverage based on ancillary data.

The remote sensing data acquired from IRS 1C LISS III has been classified using unsupervised classification technique, based on ISODATA algorithm with 16 classes to generate land use/land cover thematic maps. IRS 1C LISS III and 1C PAN data sets were merged using Brovey method for improving the interpretability of the multispectral data to digitize roads, unsuitable and protected areas for tower placement.

**Figure 1**
*IRS 1 C PAN Data*

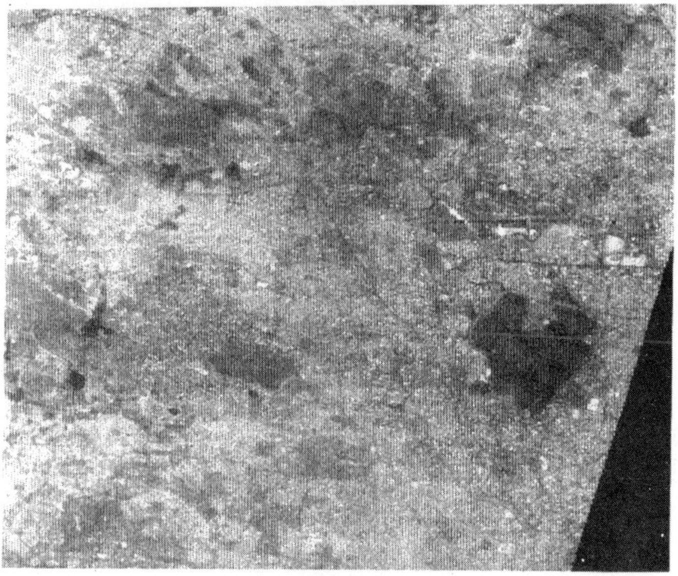

**Figure 2**
*IRS 1C LISS III Image*

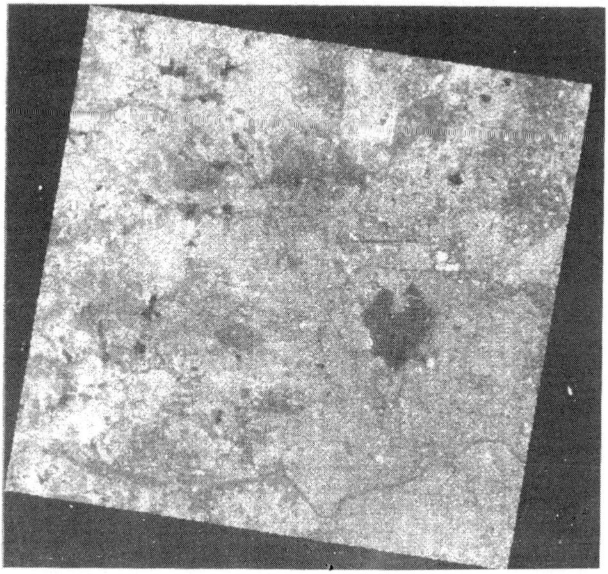

**Figure 3**
*LISS and Merged Data using BROVEY Method*

**Figure 4**
*Land Use/Land Cover Grids*

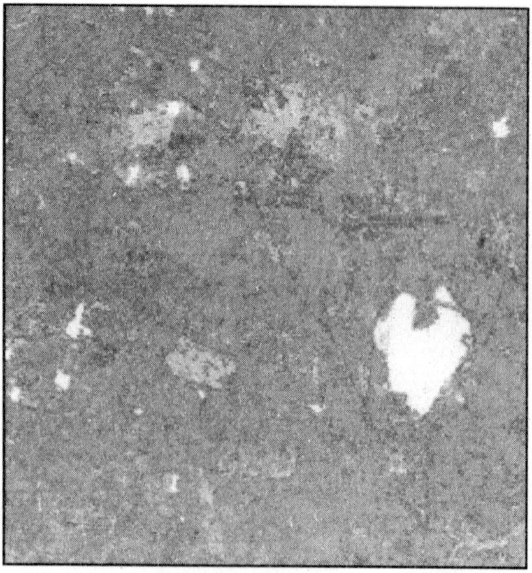

**Figure 5**

*Point Coverage Generated Using Land Use/Land Cover Grid*

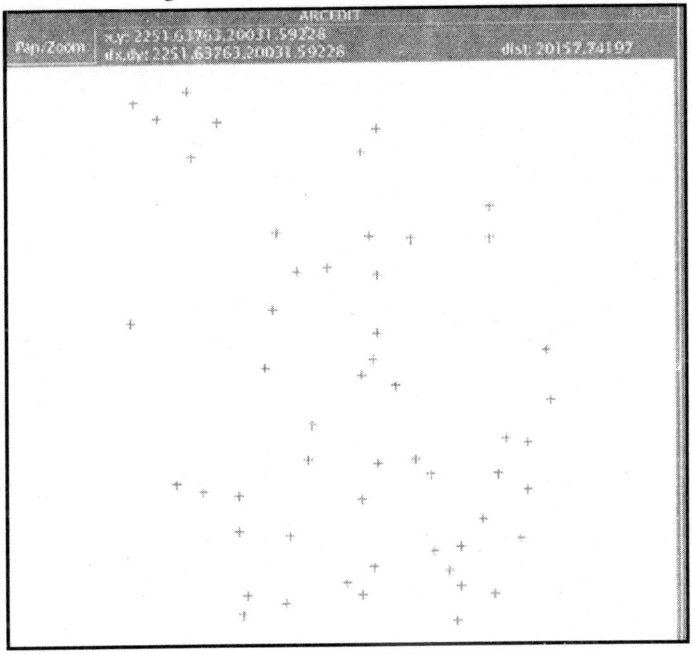


The Digital Elevation Model (DEM), which is prepared using the digitized contours from Survey of India toposheet, was used for calculation of slope map and later in the study for getting the elevation values to the selected sites in the planning mode. The methodology considered for this is based on raster GIS modelling techniques available in ARCINFO software. Also, a land use/land cover grid was created from the merged data.

The local peaks from the created DEM were found out and the points from the land use/land cover grid falling in the built-up area were identified. The sites with slopes less than 3 per cent were taken into consideration due to engineering aspects and the points were refined for their accessibility with roads with a buffer of 250 metres on either side of the road.

## Result of the Study

The points were then analysed for view shed and points with maximum coverage were selected using GRID modeller in Workstation ARCGIS 8.1 software. The result shows the sites identified from the DEMs overlaid on the merged IRS LISS III and PAN data.

**Figure 6**

*Points Identified for Cellular Tower Network over the Study Area*

The study has given a satisfactory output with the available data. Using the basic parameters, i.e., height, land use and road network coverage, this study has given 54 number of points along various strategic points across the study area.

The other factors that were considered for this study are the environmental factors, attenuation due to various land use in typical urban area, buildings, radio frequency emission factor of the cell tower along with the factor for economy of the land acquisition or lease. These studies could be taken up at the initial stage of planning and in the end to have a final analysis including all parameters using various GRID modelling techniques.

## Conclusion

It has been observed from the result that though this model has given a reasonable output, it would have given a better output when used with satellite imagery of very high resolution and other ancillary information like building heights, RF signal strength and propagation values. The integration of these parameters could be an advantage for the planners, to have a realistic approach towards a good planning for the economy of network tower installation and maintenance.

Thus, by employing higher resolution satellite imagery and using more up-to-date ancillary information it is possible to maintain the competitive edge that allows succeeding in the highly competitive and constantly shifting telecom industry. By doing so, users can analyse the utilization, network and infrastructure investments; plan for future growth by analysing potential growth zones and their direction of growth and accurately budget the infrastructure expenditure required to effectively meet the growth.

The power of high resolution imagery coupled with GIS technology can prove to be a good match to the needs for telecommunication industry, where fierce competition and rapid change of innovation require operators to quickly adapt to the changing technological and economic market place.

*Acknowledgements*

We sincerely thank Dr. Deva Pratap, for his constant guidance rendered during the preparation of this paper. We also thank our lecturer, Ms. Vindhya for her valuable suggestions.

## Select References

Deshpande, Milind and K.M. Jagadessh, *Telecom GIS: An Integrated Approach.*

Jamaludin, Syed, S. Ali and Alikin Ahmad, *Facilities Management in Telecommunication Industry using GIS Technology.*

John, W. Allen, *High Resloution Geographic Imagery and its Impact on GIS.*

Lillesand and Kiefer, *Remote Sensing and Image Interpretation,* Fourth Edition.

Misra, P., *Town Base Mapping through Very High Resolution Satellite Imagery.*

# 5

# Estimation of Crop Evapotranspiration Using Remote Sensing and GIS: A Case Study from Tarafeni Command

P.K. Mishra and R. Saha

The knowledge of evapotranspiration (ET) is of great importance for water use assessment of a large area. Estimation of crop evapotranspiration is essential for computing the soil water balance and irrigation scheduling. Conventional ground-based methods for estimating evapotranspiration provide accurate measurements over a homogeneous area surrounding the instruments, but the results are not applicable to large heterogeneous area (Caselles, et al., 1992). However, in catchments with reasonable topographic variation, even if the regional potential ET based on atmospheric parameters and intensity of solar radiation is estimated accurately, the actual ET will not be uniform spatially. The only viable means of mapping evapotranspiration at large scale is the use of remotely sensed satellite data.

Crop evapotranspiration ($ET_c$) refers to evapotranspiration of a disease-free crop, grown in very large fields, not short of water and fertilizer (Doorenbos and Pruitt, 1977). $ET_c$ is governed by weather and crop condition. Mathematically, $ET_c$ can be expressed as:

$$ET_c = K_c \times ET_o \qquad ...(1)$$

where $K_c$ is crop coefficient, which varies for different crops and their growth stages, and $ET_o$ is the reference crop evapotranspiration.

Crop evapotranspiration is governed by weather and crop condition. Most of the current water demand models are non-spatial, which use point data of reference evapotranspiration and the crop coefficient values from available literature (ibid.). However, the climatic data used to measure ET are highly spatially variable. Also the land use and the crop condition can vary from field to field, thus affecting the crop coefficients and thereby crop evapotranspiration rates. Hence, using crop coefficient values from available literature may provide a practical guide for scheduling irrigation, but considerable error in estimating crop water requirement can occur due to their empirical nature (Jagtap and Jones, 1989). Bausch and Neale (1987) have shown the usefulness of remotely sensed data to represent the crop coefficient. The feasibility of estimating crop coefficients from remote sensing data is studied in the present work as both crop coefficients and vegetation indices (derived from reflectance) are affected by leaf area index and fractional ground cover (Bausch, 1995). The advantage of remote sensing derived crop coefficient over traditional one lies in its representation as a real-time crop coefficient which represents to actual field crop conditions. Hence, crop evapotranspiration estimation based on remotely sensed crop coefficients can be a real boost as compared to traditional methods.

The objective of this study was to compute the crop evapotranspiration of Tarafeni command area in West Midnapur district of West Bengal, India using satellite data and geographic information system (GIS).

## Materials and Methods

### *Study Area*

The proposed study was conducted in the Tarafeni command area of West Midnapur district of West Bengal state, India. This command area lies between 86°45′ and 87°15′ East longitudes, 22°15′ and 22°45′ North latitudes. The geographical area was about 716 sq. km. The command area is covered in four Survey of India (SoI) 1:50,000-scale toposheets with index nos. 73J/14, 73J/15, 73N/2 and 73N/3. The location map of the study area is presented schematically in Figure 1.

**Figure 1**
*Location Map of Tarafeni Command Area*

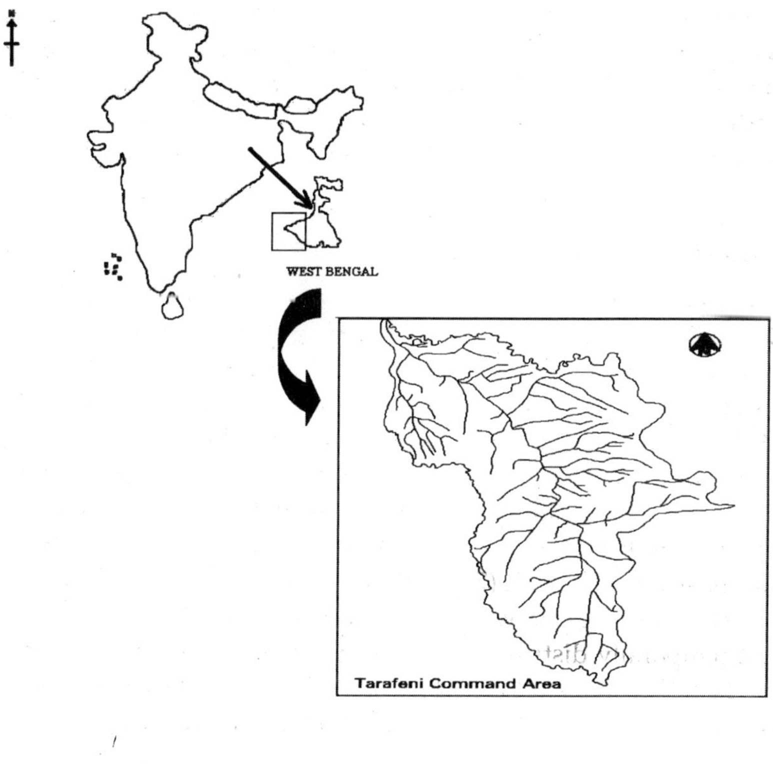

## *Physiography and Climate of the Study Area*

Tarafeni command area is one of the drought-prone areas of West Bengal. The topography of the major part of this command area is nearly flat, with an average slope of 3 to 4 per cent except a few places rolling. The predominant soil group in the command area is lateritic and red soils. The climate of the precinct is characterized by an oppressively hot summer with high humidity nearly throughout the year (except a brief spell of winter when it is relatively drier), and a well distributed rainfall during monsoon months. South-westerly monsoon prevails over the entire area for about four months. The seasons are generally distributed as hot summer (April-May), monsoon season (June-September) and winter (November-February) while October-November is generally regarded as post-monsoon months when temperature and humidity come down to a moderate level. The average rainfall for the command area is about 1428 mm. Almost 85 per cent occur during the four monsoon months from June to September. A considerable portion of the command area comprises forest coverages with deep and shallow water bodies and some patches of plantation of Eucalyptus.

## *Data Used*

The digital data of Indian Remote Sensing satellites IRS-1D LISS III (Path 107 Row 56) of 9th March 2000 and 14th November 2000 were used for digital classification of land use/land cover of Tarafeni command area. The spatial resolution of LISS III is 23 m. Multidate Remote sensing data of Wide Field Sensor (WiFS); on-board IRS-1C satellite was used for generation of vegetation index and computation of crop coefficients. WiFS has 188 m ground resolution and two spectral bands in red (620-680 nm) and near-infrared, NIR (770-860 nm) region. One scene for each month of Rabi season was used for the development of spatially and temporally distributed $ET_c$ map. The dates of satellite data are 14th December 1999, 19th January 2000, 19th February 2000, 9th March 2000 and 3rd April 2000. Daily rainfall data of five consecutive months (from 1st December 1999 to 30th April 2000) were collected from the Tarafeni Barrage. The Survey of India topographical sheets nos. 73J/14, 73J/15, 73N/2 and 73N/3 and the command area map supplied by Kangsabati Canal Division V, Govt. of West Bengal were used to prepare the base map and to finalize the different land use/land cover classes.

### *Concepts and Methods*

The following steps were carried out to estimate the crop evapotranspiration from the remotely sensed data:

1. Collection of Multidate Remote Sensing data
2. Image rectification and atmospheric correction
3. Vegetation index model development
4. $K_c$ map generation
5. $ET_c$ map generation

Multidate remote sensing data of WiFS were taken for the evaluation of the evapotranspiration. One scene for five consecutive months of Rabi season dated 14th December 1999, 19th January 2000, 19th February 2000, 9th March 2000 and 3rd April 2000 were taken for the study. The satellite data were geographically registered onto one another, with respect to LISS III data. From the classified scene representing the Rabi season, only the pixels representing the paddy crop were chosen by a process called masking. Normalized Difference Vegetation Index (NDVI) was generated for each pixel representing paddy crop.

Normalized Difference Vegetation Index (NDVI) which is related to the proportion of photosynthetically absorbed radiation, was calculated from atmospherically corrected reflectances from the visible and near infrared band as:

$$\text{NDVI} = [(\text{NIR} - \text{R})/(\text{NIR} + \text{R})] \qquad \ldots(2)$$

where, NIR and R are reflectances in near-infrared and red wavelengths, respectively.

For each scene six-sample areas, and from each area few pixels were chosen and corresponding NDVI values were noted down. The average of the NDVI values picked from each scene was considered to be as representative NDVI value for that scene.

Thus for five scenes, five NDVI values were obtained. Regression equations between the NDVI values and corresponding crop coefficient of paddy taken from literature for the five consecutive months were developed. Using these empirical equations, crop coefficient maps ($K_c$ map) were generated. Daily reference evapotranspiration (ETo) values for the five months were calculated using the FAO-56 modified Penmann-Monteith equation. Monthly average value of the reference evapotranspiration was multiplied with the crop coefficient maps developed, and thus corresponding crop evapotranspiration maps ($ET_c$ map) were generated for five different months. The flow chart for generation of $K_c$ and $ET_c$ maps is shown in Figure 2.

**Figure 2**

*Flow Chart for Evapotranspiration Estimation using RS and GIS*

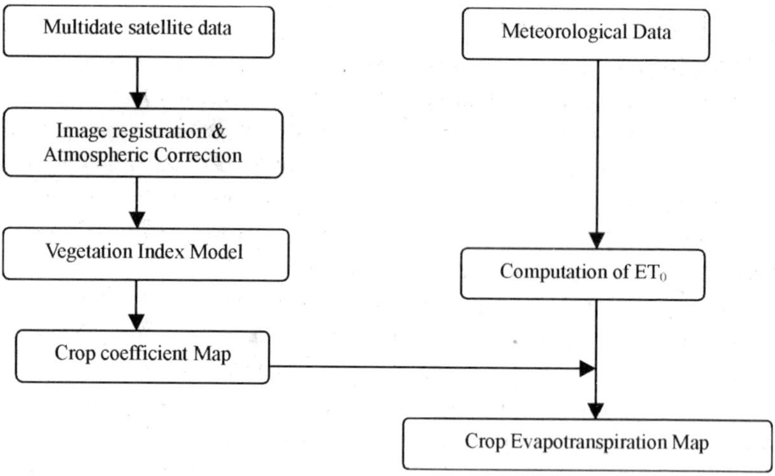


## Results and Discussion

False Colour Composite (FCC) of five different scenes used to generate $K_c$ and $ET_c$ map are given in Figure 4. The NDVI values generated for five scenes in five months (14th December 1999, 19th January 2000, 19th February 2000, 9th March 2000 and 3rd April 2000) were 0.147, 0.188, 0.395, 0.341, and 0.163 respectively. Regression equations between the NDVI values and corresponding crop coefficient of paddy as taken from literature (Lenka, 1999) for the five consecutive months were developed. Though the crop coefficients proposed by Lenka, 1999 were at higher side as compared to those of FAO, the former method was adopted in the present study as they were derived from the same climatic zone. The NDVI values and corresponding crop coefficients of paddy used in the development of regression equation have been presented in Table 1. The linear regression equation corresponding to $K_c$ and NDVI values has been given in Figure 3. The coefficient of determination ($R^2$) obtained between $K_c$-NDVI relation was 0.8351. The analysis of regression showed that the equation was highly significant with high $R^2$ values. Using this empirical equation, pixel-wise crop coefficients from the corresponding NDVI values for five months were computed. Crop coefficient maps ($K_c$ map) generated for different months based on NDVI has been given in Figure 5. Climatic data collected from nearest weather station at Indian Institute of Technology, Kharagpur were used to compute the reference evapotranspiration. Daily reference evapotranspiration ($ET_0$) values for the five months were calculated using the FAO-56 modified Penmann-Monteith equation. Monthly average value of the reference evapotranspiration was multiplied with the pixel-wise crop coefficient values, and thus corresponding crop evapotranspiration maps ($ET_c$ map) were generated for five different months. Crop evapotranspiration map ($ET_c$) generated for five months based on NDVI has been presented in Figure 6.

**Table 1**

*$K_c$ and NDVI Values for Different Scenes Used in Regression Relationship*

| *Month* | *$K_c$** | *NDVI* |
|---|---|---|
| December | 0.98 | 0.147 |
| January | 1.2 | 0.188 |
| February | 1.3 | 0.395 |
| March | 1.7 | 0.341 |
| April | 1.1 | 0.163 |

**Source:* Lenka, D., 1999

**Figure 3**

*Regression Curve Between Kc and NDVI*

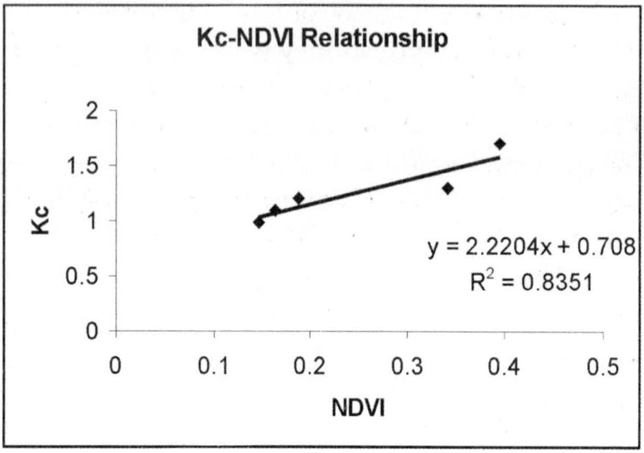


It was found that the $K_c$ values based on NDVI were found to be in the range of 0.13 to 2.23 for entire season and $ET_c$ values based on NDVI were found to be in the range of 0.38 to 8.54 mm/day. The maximum and minimum $K_c$ and $ET_c$ values obtained for five scenes based on NDVI have been presented in Table 2.

**Table 2**

*Scene-wise Kc and $ET_c$ Ranges based on NDVI and SAVI*

| *Month* | $K_c$ | $ET_c$ *(mm/day)* |
|---|---|---|
| 14th Dec'99 | 0.14 – 1.58 | 0.44 – 5.00 |
| 19th Jan'00 | 0.13 – 2.34 | 0.38 – 6.43 |
| 19th Feb'00 | 0.34 – 2.23 | 1.17 – 7.60 |
| 9th Mar'00 | 0.29 – 1.74 | 1.25 – 7.34 |
| 3rd Apr'00 | 0.25 – 1.73 | 1.23 – 8.54 |

**Figure 4**

*FCC of Tarafeni Command Used in $ET_c$ Estimation*

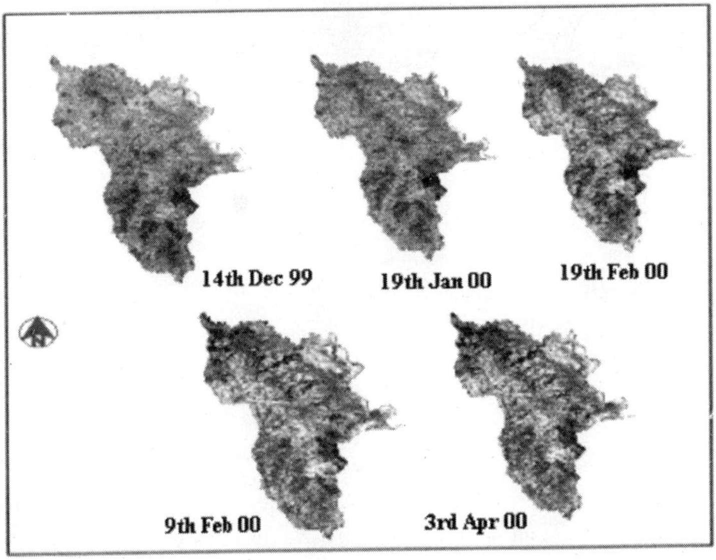


**Figure 5**

*$K_c$ Map (based on NDVI)*

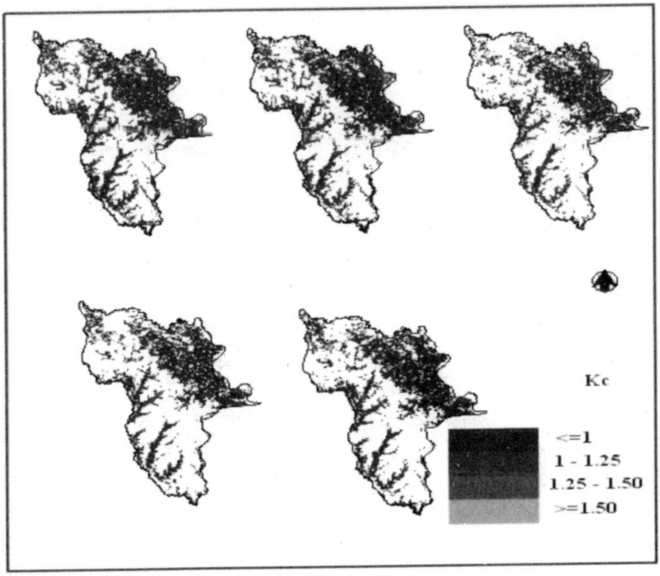

**Figure 6**

*$ET_c$ Map (based on NDVI)*

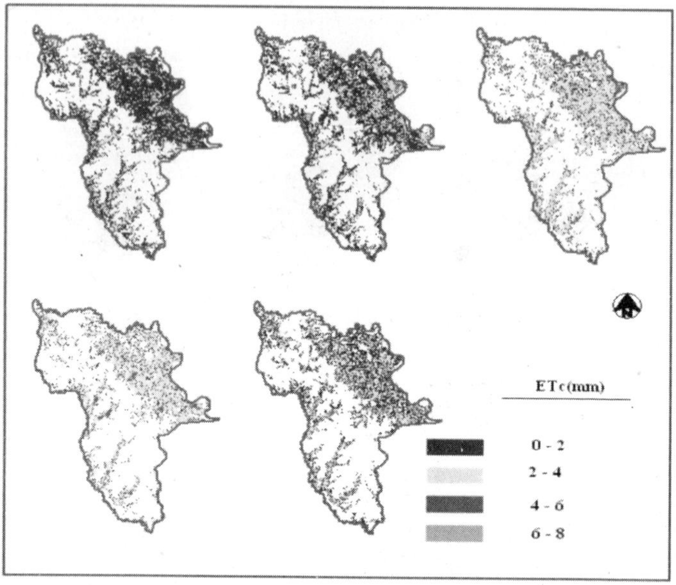


## Conclusions

In this study, a procedure was developed to estimate the realistic spatially and temporally distributed crop evapotranspiration using Remote Sensing and Geographic Information System. Crop coefficients for paddy were estimated from the remotely sensed crop reflectance values captured from different sensors. This study explains the use of satellite remote sensing for command water assessment of a large watershed. Because of high repetition and spatial resolution, satellite remote sensing is a viable mean to estimate crop water use at large scale. The crop water use estimated by this method is a better estimate as compared to other methods, since it is a direct representation of the actual crop growth conditions in the field. The proposed study estimates the monthly crop water requirement. The same procedure adopted for other areas and crops for better water management towards optimizing crop production.

## References

Bausch, W.C., 1995, Remote sensing of crop coefficients for improving the irrigation scheduling of corn. *Agricultural Water Management*, 27: 55-68.

Bausch, W.C., Neale, C.M.V., 1987, Crop coefficients derived from reflectance data. *Trans. Am. Soc. Agric. Eng*, 30: 703-709.

Caselles, V., Sobrino, J.A., and Coll, A., 1992, On the use of satellite thermal data for determining evapotranspiration in partially vegetated areas. *International Journal of Remote Sensing*, 13, 2669-2682.

Doorenbos, J., Pruitt, W.O., 1977, Guidelines for predicting crop water requirements. FAO Irrigation and Drainage Paper No. 24. Food and Agriculture Organization, Rome, p. 144 .

Huete, A.R., 1988, A soil-adjusted vegetation index (SAVI). *Remote Sensing Environment*, 25: 295-309.

Jagtap, S.S., Jones, J.W., 1989, Stability of crop coefficients under different climatic and irrigation management practices. *Irrigation Science*, 10: 231-244.

Lenka, D., 1999, *Irrigation and Drainage*. Kalyani Publisher, New Delhi, p. 136.

Ray, S.S., and Dadhwal, V.K., 2000, Estimation of evapotranspiration of irrigation command area using Remote Sensing and GIS. *Agricultural Water Management*, 49: 239-249.

Schultz, G.A., 1988, Remote Sensing in Hydrology, *Journal of Hydrology*, 100:239-265.

# 6

# Orthomap Accuracy Evaluation with Various Stereo Restitution Models and Matching Techniques

Hina Pande, Poonam Seth and Sushma Yadav

Digital Orthophoto Imagery (DOI) has become an affordable and highly effective component to the modern GIS. However, a variety of considerations must be addressed in order to ensure the DOI that one is investing in will meet expectations and intended applications. The quality (i.e., accuracy) and suitability of the product are the critical issues that should be considered. At present, although digital photogrammetric system (DPS) can efficiently produce digital orthophotos, its accuracy is doubted by most users. In order to make digital photogrammetric products accepted widely, it is necessary to analyse their accuracy. The best orthorectification software for any particular individual or organization is completely dependent on general considerations such as cost, ease of use, accuracy, and flexibility.

## Accuracy of an Orthophoto

There exist many factors that affect digital orthophoto accuracy, i.e., the accuracy of the ground control, the accuracy of image matching and the accuracy of DEM used in the orthorectification process.

The major error associated with photogrammetry is mislocation of ground control. This type of error results in offset of elevation features and miscalculation of elevation over a significant portion of imagery. Such errors are difficult to detect and correct after DEM collection and mosaicking and therefore clearly illustrates the need for exceptionally good ground control in photogrammetry to preserve accuracy in the output.

In photogrammetry and remote sensing, matching can be defined as the establishment of the correspondence between various data sets. The matching problem is also referred to as the correspondence problem. Many steps of the photogrammetric processing chain are linked to matching in one way or another for example in interior orientation the image of a fiducial is matched with a two-dimensional model of the fiducial, in relative orientation and aerial triangulation parts of one image are matched with parts of other images in order to generate tie points, in absolute orientation parts of the image are matched with a description of ground control points, in generation of digital terrain models (DTM) parts of an image are matched with parts of another image in order to generate three-dimensional object points.

If the algorithm matches two points on a stereopair incorrectly, the result is usually a 'spike' or a 'sump'. A spike is a single (or sometimes multiple) high value that is clearly incorrect. A sump is similar, except it is a low value. If the pattern-matching algorithm cannot find a match, however, the point is not left uncalculated; interpolation is used to estimate the elevation instead.

The matching strategy plays a decisive role for the success of any particular algorithm. A hierarchical approach combined with epipolar geometry and a highly redundant set of conjugate primitives is found to be very important in order to control the computational complexity and to ensure a high reliability for the results.

## Objective

Digital photogrammetry technique is now widely available, but many users may be unaware of the potential inaccuracies in the approach. Problems in Digital Orthophotography usually relate to cost, quality, accuracy and hardware/software (Addendum to *Manual of Photogrammetry,* Chapter 7). The primary objective of this study is to make a quantitative assessment of accuracy achieved by different stereo restitution methods and matching algorithms.

1. To perform Bundle Block Adjustment using GCPs, Automatic tie points and Manual tie points and to generate DEM for the same.
2. To compare the accuracy of rectified image and orthorectified image.
3. To compare the planimmetric accuracy of orthorectified image derived from different software.

## Data Used and Study Area

For the present study digital stereoscopic PAN data of IRS 1C of Dehradun area has been used (Plate I).

**Table 1**

| *Imagery* | *Path* | *Row* | *Date of Acquisition* | *Side Incidence Angle* |
|---|---|---|---|---|
| Imagery 1 | 98 | 50 | 21-Feb-97 | 9.6995 |
| Imagery 2 | 97 | 50 | 18-Oct-97 | -18.7282 |

## Software Used

A number of commercial off the shelf DPWS are available. For the present study following softwares have been used:

(1) Orthobase Pro 8.5.1 (ERDAS)
(2) OrthoEngine 8.2.0 (PCI)

## Methodology

Image and auxillary data was downloaded in S/W specific format. Ground Control was identified on image and collected

**Plate I**

*Stereo Pair of IRS 1C PAN of Dehradun Area*

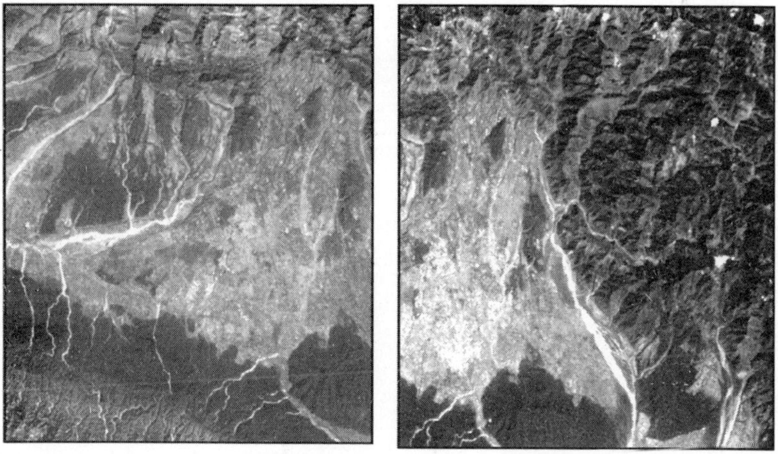

using Magellan ProMARK X Handheld GPS in differential mode. Interior Orientation (I.O.) i.e., defining the angular relationship between object space rays on the basis of the location of image points with respect to perspective center was carried out using geometric elements of I.O. from the auxillary data. Satellite orientation parameters were updated using stereo restitution model available in the respective s/w's and collected GCPs. Updation was carried out by using (i) GCPs, (ii) GCPs and automatically generated tie points, (iii) GCPs, automatically generated tie points and manual tie points collected in problematic areas. This was done to study the effect of increasing tie points on overall accuracy of restitution and orthoproduct. Automatic Image Matching was performed using different matching techniques and DEM was generated. The output DEM, and updated satellite orientation parameters were utilized to produce the orthorectified product (Plate II and III). Simple Polynomial rectification of the image was carried out using same GCPs to emphasize advantages of orthorectification over simple rectification. The outputs were statistically analyzed to make a quantitative assessment of accuracies achieved by different stereo restitution methods and matching algorithms.

**Figure 1**

*Functional Block Diagram of Orthoimage Generation Process*

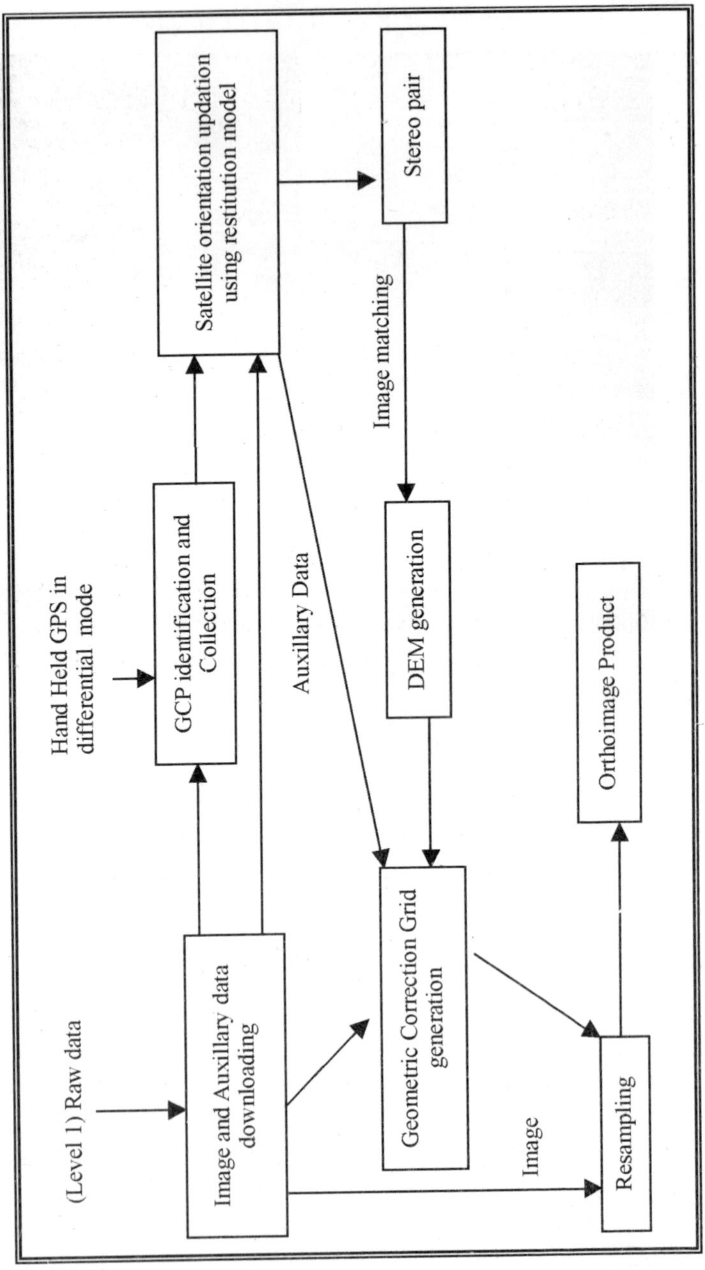

**Plate II**

*Outputs from OrthoEngine*

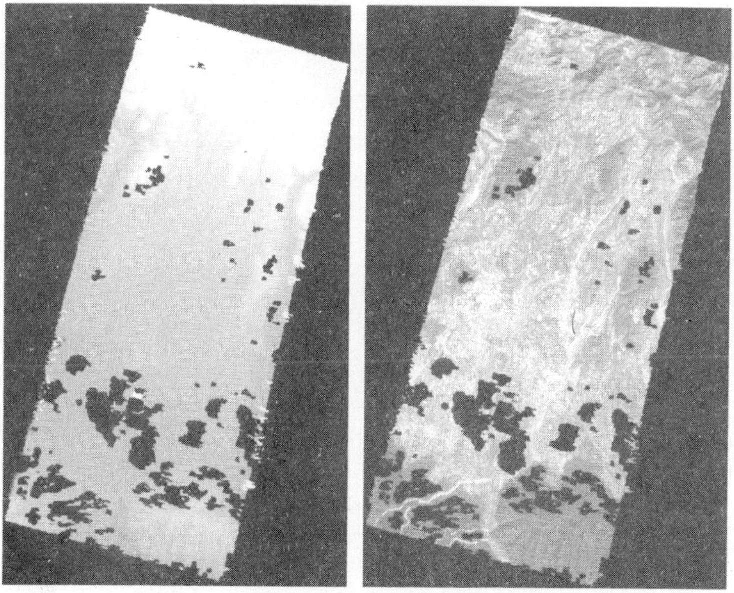


**Table 2**

*Ground Control Points*

| *No.* | *Point ID* | *X (UTM, WGS84)* | *Y (UTM, WGS84)* | *Height (Above ellip.)* |
|---|---|---|---|---|
| 1. | Wild Life Inst | 208915.44818 | 3354441.04151 | 557.86 |
| 2. | IIRS | 215855.87509 | 3360372.67486 | 667.07748 |
| 3. | Pump house Clementown | 213612.36335 | 3352161.88385 | 543.79729 |
| 4. | Kedarpur canal and road crossing | 215280.46411 | 3353584.71775 | 571.52364 |
| 5. | Kedarpur canal and bypass crossing | 215250.99036 | 3354775.09410 | 588.01013 |
| 6. | Sahastradhara Road near river | 222959.819 | 3364916.094 | 780.77776 |
| 7. | River crossing near village Postari | 222614.028 | 3362774.590 | 719.56402 |
| 8. | Bridge near IRDE colony | 219547.36054 | 3355754.37164 | 609.65336 |
| 9. | Bridge near IIRD colony Miyanwala | 219411.19578 | 3351884.33662 | 558.95277 |
| 10. | Hathibarkala bridge | 216054.46648 | 3361169.85336 | 653.97912 |

**Plate III**

*Outputs from OrthoBase*

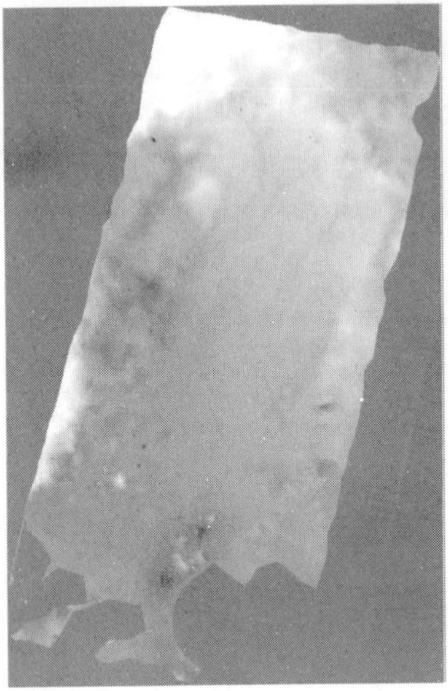

**DEM**

**Orthoimage**

## Results

Ground control was collected with an overall accuracy between 2-3 m in undulating areas and less than 2m in central flat terrain. The accuracy levels achieved were acceptable since the input image data is of the order of 6 m spatially.

In ERDAS IMAGINE OrthoBASE, Block Triangulation was performed with: (i) GCPs, (ii) GCPs and automatically generated tie points, (iii) GCPs, automatically generated tie points and manual tie points collected in problematic areas. DTMs were extracted and orthoimages were created for the same. The triangulation results are:

**Table 3**

| | |
|---|---|
| (Block Adjustment using GCPs only) | |
| Ground Control Points | |
| Ground X: | 0.1081 (5) |
| Ground Y: | 0.4955 (5) |
| Ground Z: | 0.2514 (5) |
| Block Adjustment using GCPs automatic tie points.) | |
| Ground Control Points | |
| Ground X: | 0.0596 (5) |
| Ground Y: | 0.3080 (5) |
| Ground Z: | 0.2645 (5) |
| (Block Adjustment using GCPs and automatic tie points and manual tie points.) | |
| Ground Control Points | |
| Ground X: | 0.0665 (6) |
| Ground Y: | 0.3163 (6) |
| Ground Z: | 0.2216 (6) |

In OrthoEngine(PCI) Inner orientation was performed while defining the sensor model during project setup. After completing project setup GCPs were collected in both images. Once sufficient GCPs were collected, bundle adjustment was performed. Epipolar Images were created using which DEM was generated. Triangulation, DEM extraction and Orthophoto generation was carried out as in previous method. The triangulation results are shown below:

**Table 4**

| | | | | | |
|---|---|---|---|---|---|
| (Block Adjustment using GCPs only) | | | | | |
| GCPs: | | | | | |
| X RMS | 0.52 | Y RMS | 0.58 | Z RMS | 0.64 |
| (Block Adjustment using GCPs and automatic tie points.) | | | | | |
| GCPs: | | | | | |
| X RMS | 0.28 | Y RMS | 0.35 | Z RMS | 0.43 |
| (Block Adjustment using GCPs automatic tie points and manual tie points.) | | | | | |
| GCPs: | | | | | |
| X RMS | 0.26 | Y RMS | 0.33 | Z RMS | 0.37 |

## Analysis and Conclusion

When triangulation was performed using GCPs only, GCPs and automatic tie points and then using GCPs along with automatic tie points and manual tie points it was observed that RMS error in triangulation decreased as number of automatic tie points was increased. It has however been observed that significant difference in accuracy was not achieved after adding manual tie points in problematic areas.

### *Comparison Between Rectified Image and Orthorectified Image*

For the comparison of simple polynomial rectified image and the orthorectified images obtained from both the software, graph is plotted for X residuals and Y residuals of GCPs

The mean error and standard deviation for X residuals and Y residuals in GCPs for both rectified and orthorectified images obtained from OrthoBase are as follows (Table 5, Figure 2):

**Table 5**

| | *Mean Error* | | *Standard Deviation* | |
|---|---|---|---|---|
| | *X residual* | *Y residual* | *X residual* | *Y residual* |
| Rectified Image | 2.522348 | 3.022606 | 1.583797 | 4.260844 |
| Orthorectified image | 0.351396 | 0.375695 | 0.299824 | 0.61377 |

The mean error and standard deviation for X residuals and Y residuals in GCPs for both rectified and orthorectified images obtained from OrthoEngine are as follows (Table 6, Figure 3):

**Table 6**

| | *Mean Error* | | *Standard Deviation* | |
|---|---|---|---|---|
| | *X residual* | *Y residual* | *X residual* | *Y residual* |
| Rectified Image | 2.522348 | 3.022606 | 1.583797 | 4.260844 |
| Orthorectified image | 0.749346 | 0.527861 | 0.745001 | 0.407683 |

**Figure 2**

*Plotted Residuals in GCPs (X axis: GCP no. Y axis: Residual in m)*

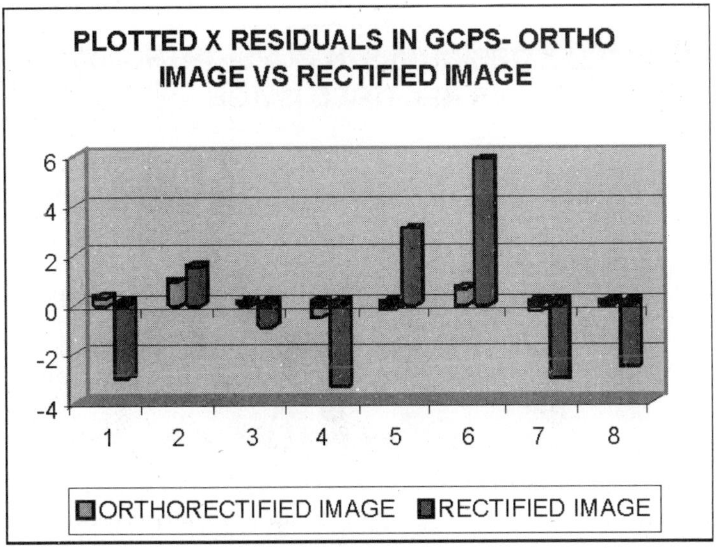


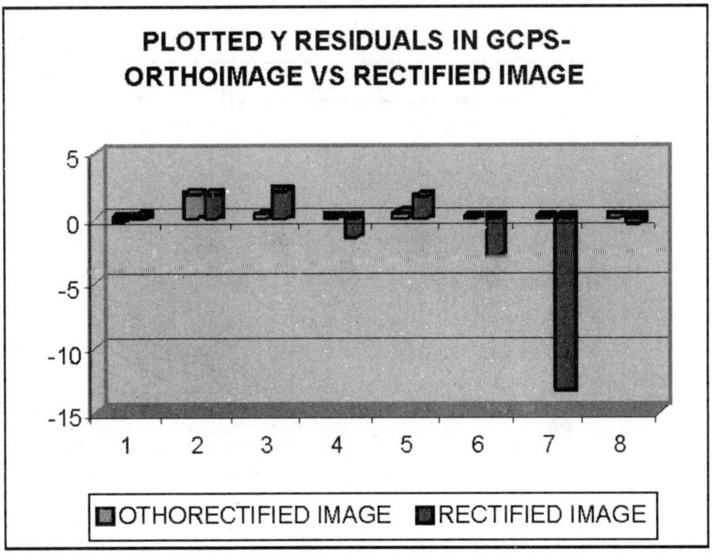

**Figure 3**

*Plotted Residuals in GCPs (X axis: GCP no. Y axis: Residual in m)*

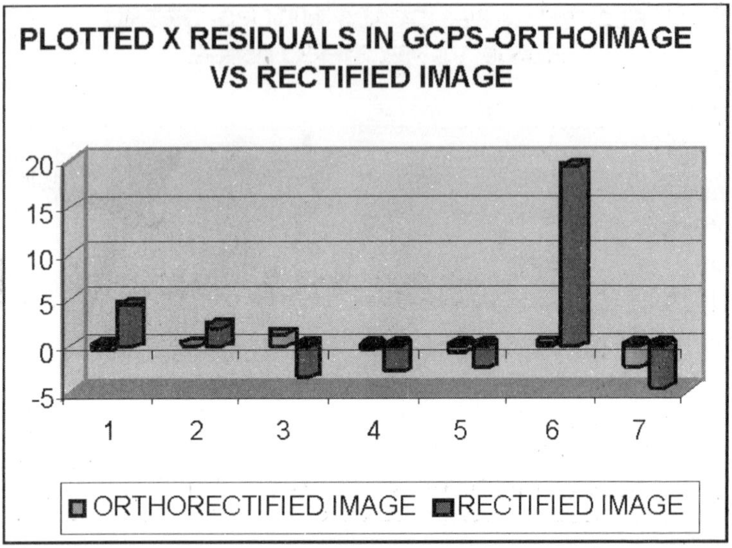


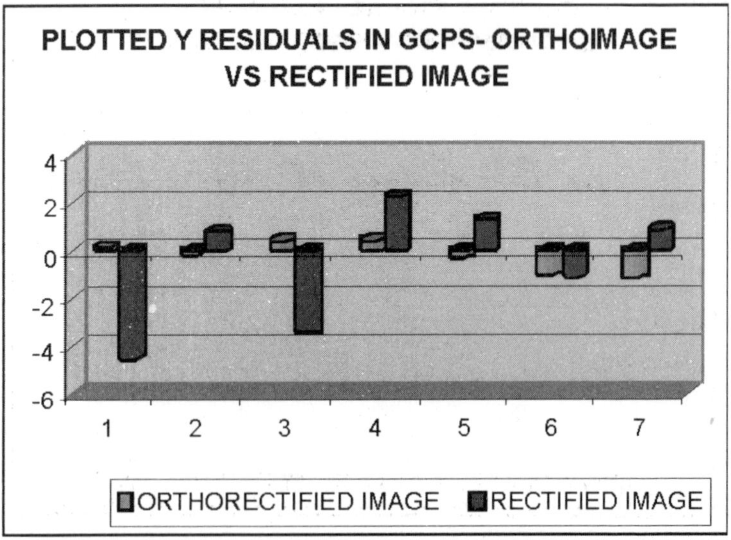

The comparison of rectified image and orthorectified image makes it obvious that orthorectified images are more reliable than rectified images (Figure 2 and 3).

### *Comparison Between Ortho-images Generated Through OrthoBASE and OrthoEngine*

For the comparison of planimetric accuracy between Othophoto obtained from OrthoBASE and OrthoEngine, graphs were plotted between X residuals and Y residuals for GCPs and tie points (Figure 4 and 5). Although the accuracy achieved from both the software is acceptable (within 2-3 metres) but OrthoBASE is giving better results than OrthoEngine.

OrthoBASE (ERDAS) has given better results with the data used in the study. Although both the softwares have given acceptable accuracy, but to validate the accuracy of any software one needs to do the same exercise with data sets of different sensors, terrain conditions and spatial resolution.

The ultimate usability of the digitally generated orthoproduct is dependent upon it being created to a known accuracy. The study demonstrates that though digital photogrammetric system (DPS) can efficiently produce digital orthophotos, it is recommended to obtain information regarding accuracy standards prior to usage.

### **Limitations**

DEM generated by both matching techniques reflected problems in certain areas. The DEM generated from OrthoBASE (ERDAS) shows spike in the flat region (Plates IV) and in the South-East region where there is dense forest DEM extaction has failed, whereas the DEM generated from OrthoEngine (PCI) displays some black patches (failed areas) (Plate V) in the same regions. This may be due to the radiometric inconsistency as there is an eight-month gap between the acquisitions of both the scenes of stereo pair. Such data is not conducive for automatic image matching. The drastic change in the radiometry in above-mentioned areas (Plates VI) may have resulted in failed

**Figure 4**

*X residual and Y residuals in GCPs*

*(X axis: GCP no. Y axis: Residual in m)*

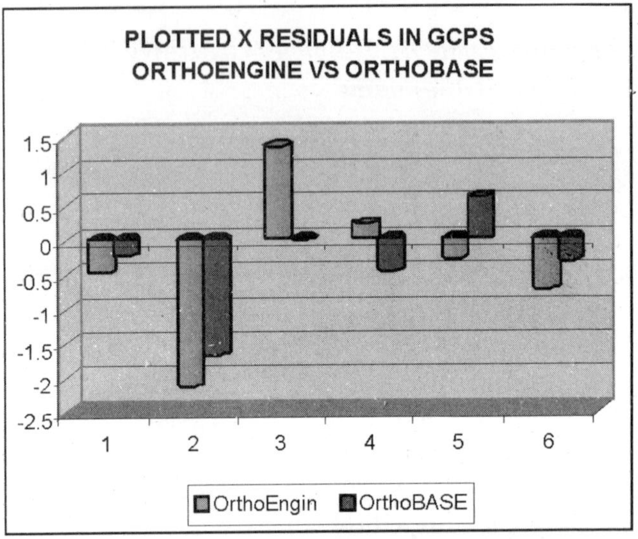


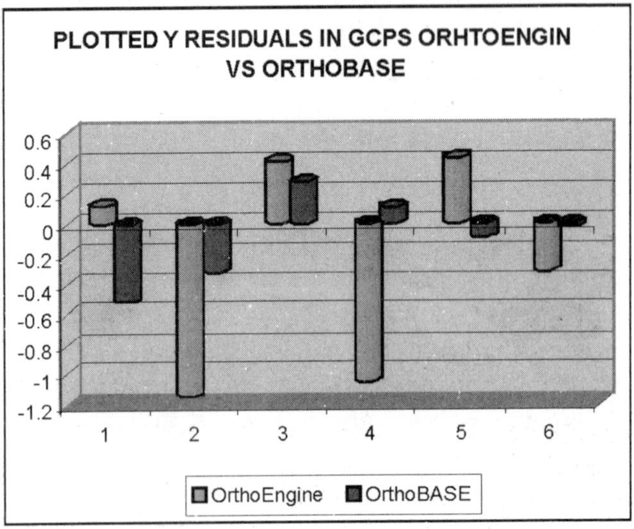

**Figure 5**

*X residual and Y residuals in Tie points*
*(X axis: Tie pt. no. Y axis: Residual in m).*

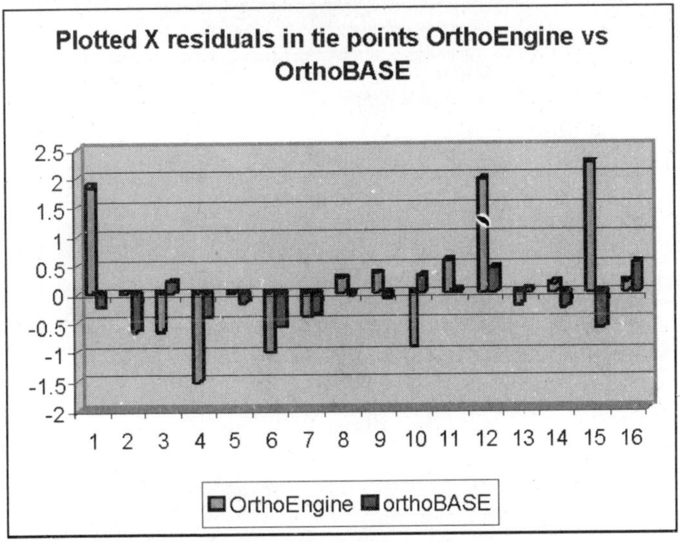


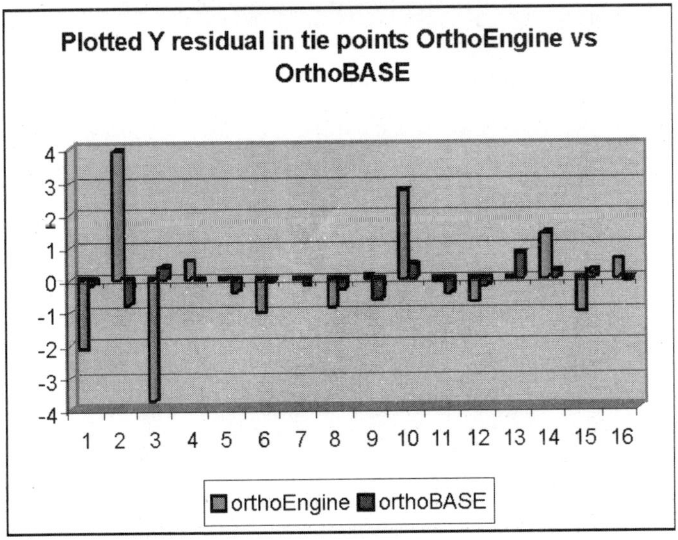

**Plate IV**

*Spike in DEM (3D view, OrthoBASE)*

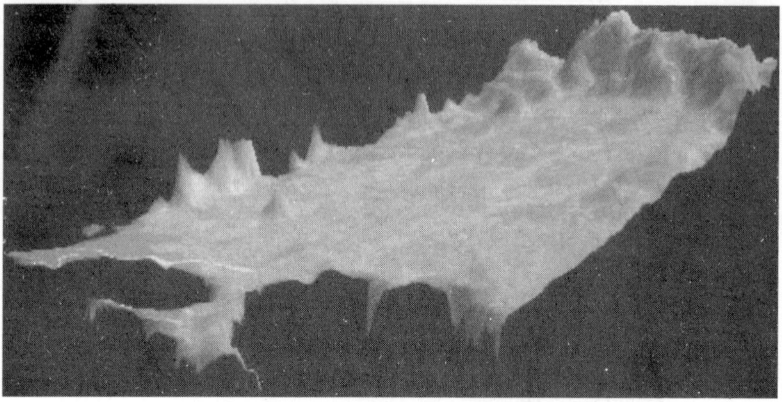

**Plate V**

*Failed Areas (3D view, OrthoEngine)*

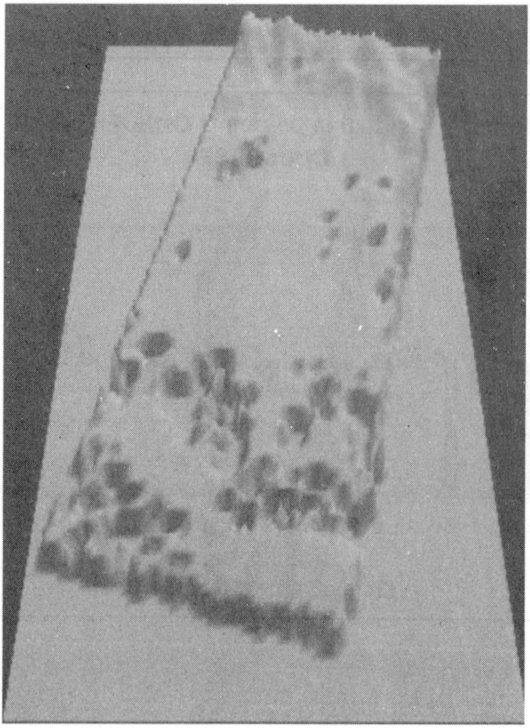

**Plate VI**

*Radiometric Inconsistency*

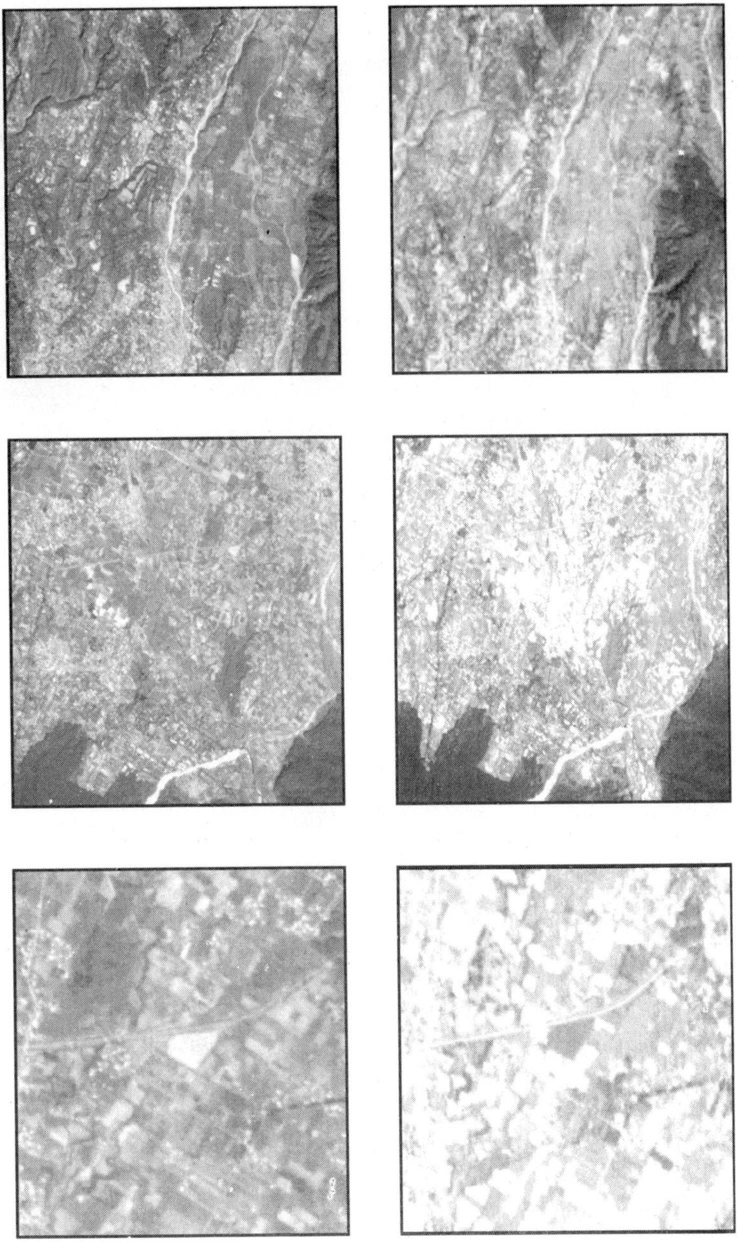

areas/erroneous heights in the DEM, which will ultimately affect the accuracy of output Ortho-image.

If one extracts DEM from automated stereo matching following points should be taken into consideration (Gugan and Dowman, 1988):

(1) Availability of a high Baseline to Height ratio (B/H >= 0.6) stereo pair,
(2) Acquisition of either one or both scenes during high-sun angle conditions,
(3) Acquisition interval between the stereo pair scenes should be <=30 days to assure radiometric stability and in fact, the same-date along-track stereo-data acquisition gives a strong advantage versus the multi-date across-track stereo-data acquisition. It reduces the radiometric image variations (refractive effects, sun illumination, temporal changes) and thus increases the correlation success rate in any image matching.
(4) Clear atmospheric conditions during image acquisitions.

## Select References

A New Epipolarity Model Based on the Simplified Pushbroom Sensor Model, Hae-Yeoun Lee a, Wonkyu Park a

Acquisition of Digital Elevation Models at High Spatial and Temporal Resolutions Using Automated Digital Photogrammetry, Jim Chandler, Stuart Lane and Koji Shiono, http://www-staff.lboro.ac.uk/~cvjhc/HRWallRep.html

Digital Aerial Triangulation for DTM and Orthophoto Generation, Romuald Kaczynski, Jan Ziobro Institute of Geodesy and Cartography, http://www.ifp.uni-stuttgart.de/publications/commIV/kaczy115.pdf

Digital Orthophotography – Principles, Project Design Issues, http://www.gisqatar.org.qa/conf97/links/h1.html

Digital Orthophotos from IRS – Production and Utilization Dan Rosenholm and Daniel Åkerman, http://www.ifp.uni-stuttgart.de/publications/ commIV/rosenholm66.pdf

Digital Photogrammetry: The Future of Spatial Data Collection, C. Vincent Tao, http://www.geoplace.com/gw/2002/0205/0205dp.asp

Karsten, Jacobsen, Status and Tendency of Sensors for Mapping, University of Hannover.

Mapping With IRS-1C-Images, Dr. Karsten Jacobsen, Institute For Photogrammetry and Engineering Surveys, University of Hannover, Germany, ASPRS Annual Convention Tampa, 1998, http://www.ipi.uni-hannover.de/html/publikationen/1998/jacobsen/jac_98_mapping_irs1c.pdf

Mikhail, Edward M., James S. Bethel and J. Chris McGlone, Introduction to Modern Photogrammetry.

New Developments in Digital Elevation Modelling, Karsten Jacobsen, Geoinformatics, June 2001. http://www.geoinformatics.com/issueonline/issues/2001/06_2001/pdf_06_2001/18_21_joc_dem.pdf

Orthophotograph Production in Urban Areas, D. SKARLATOS, University College London Photogrammetric Record, 16(94): 643–650 (October 1999), http://snap.lut.ac.uk/PhotSoc/abs/phroct99/Skarlatos.pdf

Orthorectification and Dem Generation from High Resolution, http://www.crisp.nus.edu.sg/~acrs2001/pdf/070cheng.pdf

Quality Assessment of DTM and Orthophoto Generated by Airborne Laser Scanning System Using Automated Digital Photogrammetry, O.A. Al-Bayari a, N.N. Al-Hanbali b, M. Barbarella c, A. Nashwan d, http://www.isprs.org/commission3/proceedings/papers/paper097.pdf

Rigorous Photogrammetric Processing of High Resolution Satellite Imagery, Dieter FRITSCH, Dirk STALLMANN, Institute for Photogrammetry, Stuttgart University, Germany, ISPRS, Vol. XXXIII, Amsterdam, 2000y, http://www.ifp.uni-stuttgart.de/publications/2000/Stallmann_1576.pdf

# 7

# GIS and Health Management

Vijay Kumar Baraik

Health is not merely the absence of disease or infirmity but is a state of complete physical, mental and social well-being. It is greatly affected by some non-medical factors (social, cultural, economic, educational, political, bureaucratic, etc.) too. There are two kinds of determinants of health status: (1) internal/individual and (2) external or environmental. Internal determinants are hereditary while external determinants include physical and social environment, economy, occupation, education, lifestyle, access to infrastructure facilities, amenities, services, and environment inside and outside the house structure, government policies, and levels of community participation in overall terms 'development'. Some of these are the part of preventive health care, without which the success of curative health care is not at all possible. This work with an example of a case study on Jharkhand is an attempt to provide a theoretical framework for the true integrated kind of development leading to the higher level of health status at this era of microchip technology.

However, in India, health is seen in isolation, especially in rural India. It has hardly been integrated with other aspects of development leading to the higher level of health status, which is inevitable now. All above factors associated are intricately related to health. And so the management of correlates is also an earnest need for the health management. Management with quick diagnosis and decision-making system will give an impetus or will improve the health status of the people with much faster rate. It will also accelerate the development in all associated aspects at even more pace.

GIS (Geographical Information System) is in infancy in health sector in India as only few southern states have started mapping health indicators over physical features and infra-structure. There are some isolated works on mapping diseases (Balaji, 2000; Sastry, et al., 2000; Nagpal and Srivastava, 2000). NATMO (National Thematic Mapping Organization) has also prepared a health atlas (Nag, 2000). At the same time the western model is blindly followed in conceptualizing the use of GIS in health only to locate a hospital/health centre or doctor specialized. This model, however, at present, is applicable only to the metro or urban environment for insignificant or handful of people because of much limitations. The alternate model being presented here is the future of health management in India covering the entire population once database is created.

In this work a small example has been cited through the primary data collected from the field. It uses the spaghetti data for the superimposition of the health status over the map of the levels of development of the sample districts, blocks and villages. From this example one can easily find out the relation and gap between the health status within moments in the GIS environment. In this exercise, health status has been measured condensing the indicators of morbidity, nutritional status and health care utilization. Similarly, levels of socio-economic devel-opment have been derived by constructing composite index combining the indicators of demography, economic and social development, and amenities and infrastructure.

Since health and diseases are not essentially biomedical elements and are related to various spheres of physical environment and socio-economic, cultural, political, and bureau-cratic aspects of the region, the information on all these aspects should be readily available with the concerned departments for the overall meaningful and effective planning. It is not a difficult task as now certain levels of awareness has come to the people through various agents like, village Panchayats, social, cultural and religious gatherings, NGOs, and through other kinds of diffusion. Moreover, the data should be electronic as National Informatics Centre (NIC) is doing at the districts headquarters in almost all the states of India and the Census of India has been doing it since the 1991 Census. When all the information in structured format come in an integrated manner, it can be utilized for various kinds of analysis based on present status and future modelling using the most efficient tools – Remote Sensing and GIS.

Remote Sensing is a very good cost-effective source to get the temporal information at regular interval about the phenomena on the ground without any physical contact. GIS takes account of spatial variation in any attributes and here it can be health investment as well as outcomes. Through the integrated data one can go for any kind of query using all sorts of data related to physical environment including topographic features, climate, vegetation, land use, water quality; settlement, various infrastructure, amenities and utilities, demographic including literacy and levels of education, socio-cultural, economic including employment, and health. Health should include all information ranging from available health centres at various levels, facilities and services offered at each centre, health personnel, flow of in- and out-patients, and the entire record of various diseases, record of the status of treatment. Once structured digital data is prepared, it can be easily updated within certain intervals. This way, analysis of all aspects of health can be done integrating other associated factors. For example, if malaria outbreak takes place in certain area, that may be seen in relation to the surroundings of the settlements, like vegetations and water-logging areas and then solution may be found out immediately. Similarly, if a region is affected by heavy morbidity, that may be studied in relation to the education, economy, employment, food and nutrition, the distance at which health facilities are available, whether approachable road and safe drinking water facilities are available, etc. Suitable sites for the new health centres at various levels can be located on the basis of all associated factors, especially the approachability of the place by all the hamlets and settlements falling on the roughly defined service area of that particular health centre. All these informations can be upgraded from time to time with little effort. And finally, one may also find out suitable models or projections or information for more efficient and effective planning based on the query performed, collating all the required information.

## The Management System

The diagramme (Figure 1) titled "Integrated Rural Health Management System (IRHMS)" may be helpful in understanding the function of GIS in rural health management clearly. There will be a common hub of information where every department can receive and share the necessary information of other departments through network directly like the Census of India is presently sharing its database through the Internet. This way the whole country can be connected through network and the vertical integration in planning and management is possible. This kind of management will enable the respective departments work much more efficiently and within much shorter time taken by the traditional methods.

The diagramme shows that every department shares its digital information for planning. For example, most of the village data are available with the Census of India in digital format, health department, Panchayat and block offices. Few more departments will have to generate this kind of data, especially for the socio-economic development, physical and environmental information. For many informations the state government's Remote Sensing Agency will be of a great use, for instance the maps on physiography or topography, settlements, forest areas, damp areas prone to malaria breeding, water quality, etc. Once the data is generated, it can be integrated, queried and mapped. Finally, the superimposition of health indicators on the various determinants of health related indicators like the Figure 2, where

# Figure 1

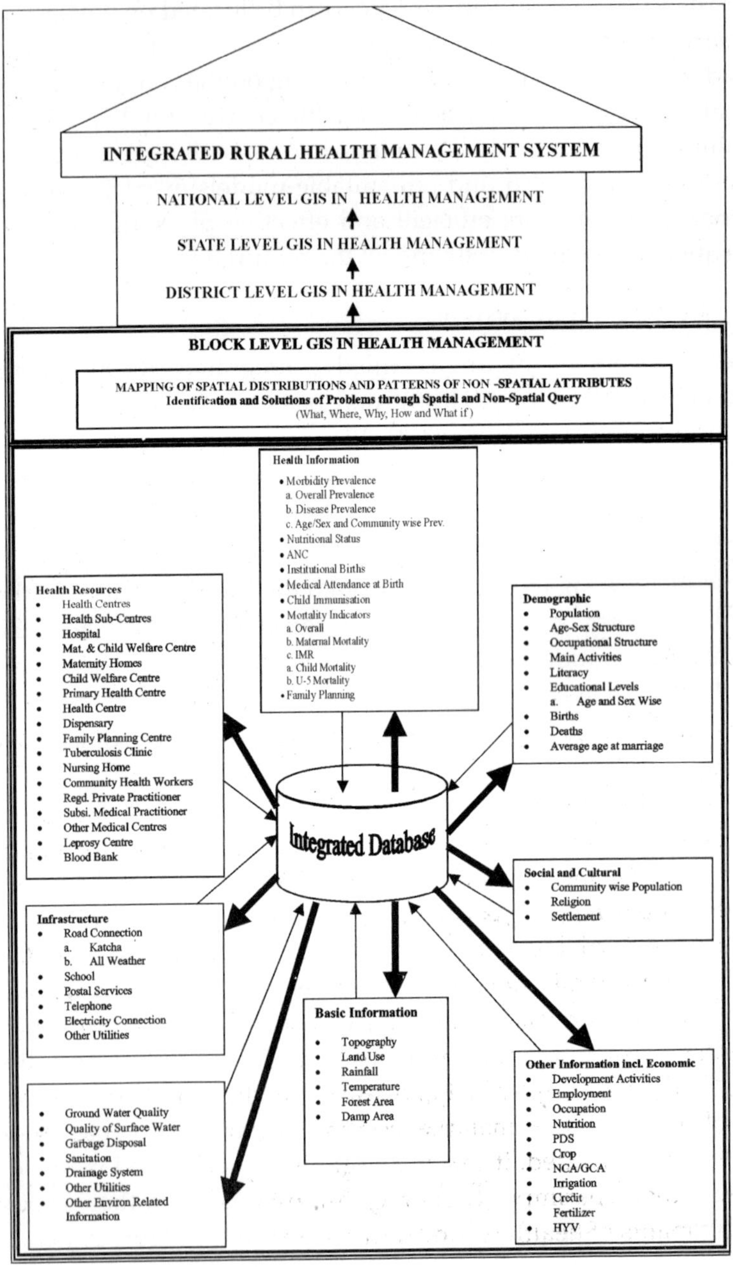
INTEGRATED RURAL HEALTH MANAGEMENT SYSTEM
NATIONAL LEVEL GIS IN HEALTH MANAGEMENT
STATE LEVEL GIS IN HEALTH MANAGEMENT
DISTRICT LEVEL GIS IN HEALTH MANAGEMENT
BLOCK LEVEL GIS IN HEALTH MANAGEMENT
MAPPING OF SPATIAL DISTRIBUTIONS AND PATTERNS OF NON -SPATIAL ATTRIBUTES
Identification and Solutions of Problems through Spatial and Non-Spatial Query
(What, Where, Why, How and What if)
Health Information
• Morbidity Prevalence
a. Overall Prevalence
b. Disease Prevalence
c. Age/Sex and Community wise Prev.
• Nutritional Status
• ANC
• Institutional Births
• Medical Attendance at Birth
• Child Immunisation
• Mortality Indicators
a. Overall
b. Maternal Mortality
c. IMR
a. Child Mortality
b. U-5 Mortality
• Family Planning
Health Resources
• Health Centres
• Health Sub-Centres
• Hospital
• Mat. & Child Welfare Centre
• Maternity Homes
• Child Welfare Centre
• Primary Health Centre
• Health Centre
• Dispensary
• Family Planning Centre
• Tuberculosis Clinic
• Nursing Home
• Community Health Workers
• Regd. Private Practitioner
• Subsi. Medical Practitioner
• Other Medical Centres
• Leprosy Centre
• Blood Bank
Demographic
• Population
• Age-Sex Structure
• Occupational Structure
• Main Activities
• Literacy
• Educational Levels
a. Age and Sex Wise
• Births
• Deaths
• Average age at marriage
Integrated Database
Social and Cultural
• Community wise Population
• Religion
• Settlement
Infrastructure
• Road Connection
a. Katcha
b. All Weather
• School
• Postal Services
• Telephone
• Electricity Connection
• Other Utilities
Basic Information
• Topography
• Land Use
• Rainfall
• Temperature
• Forest Area
• Damp Area
Other Information incl. Economic
• Development Activities
• Employment
• Occupation
• Nutrition
• PDS
• Crop
• NCA/GCA
• Irrigation
• Credit
• Fertilizer
• HYV
• Ground Water Quality
• Quality of Surface Water
• Garbage Disposal
• Sanitation
• Drainage System
• Other Utilities
• Other Environ Related Information

map of the health condition has been superimposed on the map of the levels of socio-economic development, will roughly or clearly reveal the problem through instant univariate, bivariate and multivariate query, analysis and mapping. For any kind of spatial query to detect or identify the problem area and quick solution through new models, one may go for end number of queries, which GIS will answer. The queries will be related to what (existence, patterns and trends), where (space), why (reasons) and what if (modelling based on certain conditions). For example, the villages beyond a particular radius of the health centre with low literacy, low ANC (Ante Natal Care) and low immunization can be had in tabular or visual forms (maps and graphs) through one click on customized button. Likewise, it will help in quick detection of catchment area, problem or affected area, neglected area, linear distance and future risk area and so on. It will also help in surveillance, environmental health assessment, and geographic allocation or equitable, logical and effective distribution of health resources and associated infra-structure. Finally, with the help of GIS the health department may consult the respective department due to which there is any hindrance in health development. And improvement can be done by overall development process in integrated manner that is intricately associated with the health, like without agricultural development and perennial and sufficient income normal nutritional status and health cannot be maintained in the rural society. Therefore, the GIS will increase the public health capability for higher levels of human development.

## Application and Problems

The Figure 2 shows that the levels of health status had largely followed the levels of socio-economic development in the tribal villages of Jharkhand. The micro-level study (Baraik, 2002) also finds that the health care facilities are positively associated with the preventive and curative health care. Availability of health care facilities, literacy and educational levels, occupation in non-primary sector, and income have also positive associations

Figure 2

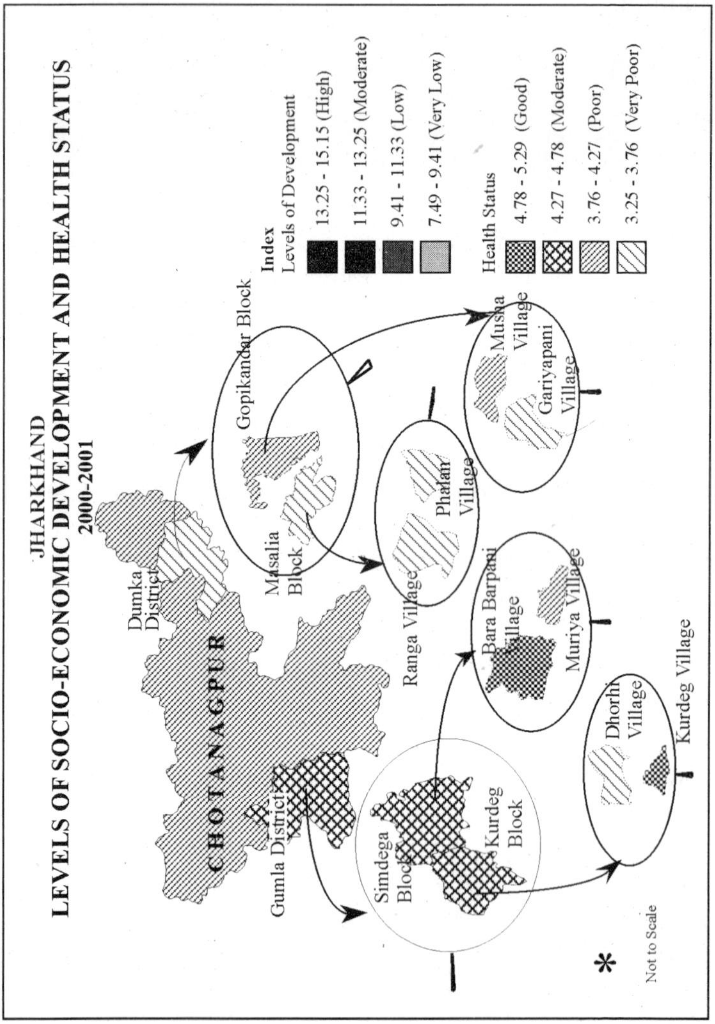
JHARKHAND
LEVELS OF SOCIO-ECONOMIC DEVELOPMENT AND HEALTH STATUS
2000-2001
CHOTANAGPUR
Dumka District
Gumla District
Gopikandar Block
Masalia Block
Simdega Block
Kurdeg Block
Ranga Village
Phalan Village
Musha Village
Gariyapani Village
Bara Barpani Village
Muriya Village
Dhorhi Village
Kurdeg Village
Index
Levels of Development
13.25 - 15.15 (High)
11.33 - 13.25 (Moderate)
9.41 - 11.33 (Low)
7.49 - 9.41 (Very Low)
Health Status
4.78 - 5.29 (Good)
4.27 - 4.78 (Moderate)
3.76 - 4.27 (Poor)
3.25 - 3.76 (Very Poor)
Not to Scale

with the indicators of health status. The other major factors for the existing health status are related to geographical, unavailability and inaccessibility of health care facilities, administrative hindrances, poor infrastructure, economic backwardness and lack of integration with other developmental activities. There is a need to have an overall development of the rural people of Indian in general and the tribal people of Jharkhand in particular with proper inter-sectoral integration along with revamping public health care services through all above measures to achieve the higher level of health status.

Use of the latest techniques for the effective planning and quick decision-making may prove to be an efficient step in the theoretically focused but practically neglected rural areas in general and tribal areas in particular. The big problem in the solution to the health problems is the lack of database or information on various aspects of health. If available with the concerned departments, it is not accessible to others. And even if available and accessible, it is highly unreliable because it is cooked up during any special requirements or otherwise no records are maintained.

## Conclusion

Computerization and integrated database management is not difficult and costly as every block is given computers and Internet connections and the districts already have a computer section. The only need is of some specialized persons to train the government officials, and for analysis, specialized persons may be employed at either district level or zonewise basis. Many governments in the world are using GIS for public health and in all the departments. It may be started in the form of small projects like block levels and will automatically expand with the success.

## References

Balaji, L.N. (2000): "GIS in Health", *GIS@development,* June, Vol. IV, Issue 6, pp. 17-20.

Baraik, V.K. (2002): Socio-Economic Development and Changing Health Condition among the Scheduled Tribes of Chotanagpur, (Ph.D. unpublished thesis), New Delhi: CSRD, JNU.

Burrogh, P.A. and R.A. McDowell (1998): *Principles of GIS,* New York: Oxford University Press.

Clarke, K.C. (2001): *Getting Started With Geographic Information Systems,* New Jersey: Prentice Hall.

ESRI (1995): Understanding GIS: The ARC/INFO, Redlands: ESRI. inc.

Heywood, I., S. Cornelius and S. Carver (2000): *An Introduction to Geographical Information Systems,* New York: Longman.

Gatrell, A.C. and M. Logtonen (1998): *GIS and Health,* London: Taylor and Francis.

Maguire, D.J., M.F. Goodchilde and D.W. Rhind (1991): *Geographic Information Systems: Principles and Applications,* New York: Longman Scientific and Technical.

Nag, P. (2002): "Health and Disease Atlas of India", *GIS@development,* June, Vol. IV, Issue 6, pp. 26-27.

Nagpal, B.N. and A. Srivastava (2000): "Mapping Malaria" *GIS@development,* June, Vol. IV, Issue 6, pp. 28-31.

Sastry, J.S.R.K., B.F.A.M. Peters and N. Paithankar (2000): Health GIS in India, *GIS@development,* June, Vol. IV, Issue 6, pp. 24-25.

# 8

# LISS III Data of IRS-1D and IRS-P6: A Comparative Appraisal

R.K. Sinha and Arabinda Sharma

Remote sensing technology has made a significant progress in last two decades in India. The country has unique constellation of five EO (Earth Observatory) satellites of IRS series, viz., IRS-1B, IRS-P3, IRS-1C, IRS-1D, IRS-P4. Being cost effective, accurate and fast the technology has high appreciation in resource planning and management (Stephen, et al., 1986; Roy, et al., 1988, 1990, 1991; Madhavan Unni, et al., 1991). Presently EO data is used for thematic mapping, generation of statistics/inventory, change detection/monitoring, spatial analysis/modelling, generation of decision support options/development plans, etc. One of the unique applications of IRS has been Integrated Mission for Sustainable Development (IMSD) where information from satellite remote sensing data and GIS concept are used for enriching the developmental planning processes in the country. EO data has now a wider acceptance as an operational tool in India. To cater the demands of users of various disciplines, ISRO has recently launched the latest version of IRS Resourcesat-1; popularly known as IRS-P6, in October, 2003. The sensors of Resourcesat 1 have been conceived to meet the objectives of providing continuity to existing data and providing data with improved spatial, spectral and temporal resolution. Prior to this IRS-1D was very much in use for resources management and monitoring. The coverage details of both the satellites are presented in Table 1. Both the IRS-P6 and IRS-1D satellites were planned to have multiple sensors of almost identical specifications for continued data service and multi-temporal analysis to cater the user requirements. LISS-III operating in visible and IR region of EMR has been extensively used in the field of agriculture, forestry, land use mapping, hydrology, geology and urban planning, etc. The LISS-III sensor of Resourcesat-1 is identical to IRS-1C/1D except that the SWIR band (1.55 to 1.7 microns) provides data with spatial resolution of 23.5 m unlike IRS -1C/1D where the spatial resolution is 70 m. However, their radiometric behaviour may differ due to any of the reasons, viz., the detector ageing, satellite injection vibration, on-board

**Table 1**

*Characteristics of the Two Sensors*

| *Characteristics* | *IRS- P 6* | *IRS- 1D* |
|---|---|---|
| Altitude | 817 km | 817 km |
| No. of orbit | 341 | 341 |
| Inclination | 98.69 | NA |
| No. of orbit/day | 14.203 | 14 |
| Orbit period | 101.35 min | 101.35 min |
| Repetivity | 24 days | 24 days |
| Equatorial crossing time | 10:30 ± 5 mins | 10:30 |
| Spectral Resolution LISS III | B2 0.52 – 0.59 | B2 0.52 – 0.59 |
| | B3 0.62 – 0.68 | B3 0.62 – 0.68 |
| | B4 0.77 – 0.86 | B4 0.77 – 0.86 |
| | B5 1. 55 – 1.70 | B5 1.55 – 1.70 |
| Spatial resolution LISS III | 23.5 m | 23.5 m for B2, B3, B4 and 70 m for band B5 |
| Radiometric resolution LISS III | 7 bits | 7 bits |

electronic system anomalies, orbital characteristics etc (Jayabharathi, et al., 2003). It is, thus, imperative to carry out a comparative study of these dual sensors to ascertain the feasibility of data use in conjunction to get improved image interpretation. In the present communication a comparative appraisal of radiometric behaviour of the LISS-III data of IRS-P6 and IRS-1D has been presented to analyse their significance in spectral clustering.

## Objective

The present research work are aimed at following prime objectives:

- to compare the radiometric behavior of the LISS III of the IRS 1D and IRS-P6
- to ascertain the acceptability of the IRS-P6
- to ascertain the feasibility of use of LISS III of the IRS-1D and IRS-P6 in continuity/conjunction with each other.

## Methodology

For the present study we have used the georeferenced LISS-III data of IRS-1D and sub-scene of IRS-P6 LISS-3 of Daltonganj (SoI 72-D4, 24°00′ – 24°15′N and 84°15′-84°30′ E) of Jharkhand state of 24 March, 2004 procured from NRSA, Hyderabad.

The sub-scene of IRS-P6 LISS-III was georeferenced with respect to IRS-1D LISS-3 data and SoI toposheet. Unsupervised classification (ISODATA clustering technique) was performed on both the images having all the four bands 1234 and band 321 to obtain spectral classes. The clustering difference was studied in context to relative difference in radiometric behaviour of the two data. The radiometric behaviour was studied by comparison of images based on their histograms. The histogram describes the grey level or colour distribution for a given image (Srikrishna, et al., 2003). Radiometric correlation analysis was carried out between corresponding bands of the two data. For this, altogether 38 GCPs were selected and their DN values were collected to find out the correlation coefficient and covariance values. GCP were taken for correlation study because these are the points which are considered to be relatively stable temporally and radiometrically as well.

## Results and Discussion

Histogram statistics of the two imageries, classification, and band-wise spectral response of 38 GCP are presented in Tables 1-3. The spectral reflectance value of IRS-1D of four bands were 117.37± 14.36, 104.39 ± 16.87, 100.49 ± 11.98 and 213.77 ± 22.43 respectively whereas in IRS-P6 the corresponding values were 152.71 ± 18.45, 115.72 ± 19.30, 118.68 ± 14.49 and 171. 01 ± 20.43 respectively. Table 1 reveals that band 4 of IRS-1D has better spectral reflectance than that of IRS-P6.The contrast characteristics of all the four bands were enhanced for IRS-P6 which exhibited better quality for data display and interpretation.

As planned, spectral clustering was performed on the LISS III data of IRS-1D and IRS-P6 obtained on the same date and identical situations with the concept that the similar output would be obtained. However, in classification output some anomalies were observed. There is a significant difference in classification output of the two data for the class 1 representing 9.25 per cent pixel values in IRS-1D and 7.775 per cent in IRS-P6, class 2 representing 10.817 per cent in IRS-1D and 13.909 per cent in IRS-P6 and class 5 representing 23.861 in IRS 1D and 22.203 in IRS P6. This may be assigned to the fact that these are the classes which are most affected by the different spectral behaviour of the two data. Earlier Thome, et al. (1997) as well as Teillet, et al (2000) had also performed radiometric calibration and cross calibration of Landsat data and found almost similar results. The comparative chart presented in Figure 4 reveals that band 4 of IRS-P6 has distinct behavior of reflectance in different GCP points than that of IRS-1D. It is supported by the clustering output of band 123 of the two data (Table 2). The significant difference in spectral clustering is mainly accounted for the different spectral behaviour of the sensor due to any of the reason, viz., detector ageing, satellite injection vibration, on board electronic system anomalies, etc. Moreover, IRS-ID has almost completed its tenure successfully serving the purpose of resource monitoring. Now it is developing snags as evidenced by appearance of stripes on the imageries and may not be of much use in near future. This has to be upgraded or supplemented and as such launching of Resourcesat-1 may be treated as desirable endeavour.

**Table 2**

*Band-wise Histogram Statistics of the Two Images*

| *Histogram statistics* | *IRS- 1D LISS III* | | | | *IRS- P 6 LISS III* | | | |
|---|---|---|---|---|---|---|---|---|
| | *Band 1* | *Band 2* | *Band 3* | *Band 4* | *Band 1* | *Band 2* | *Band 3* | *Band 4* |
| Min | 69 | 46 | 28 | 54 | 87 | 47 | 28 | 57 |
| Max | 203 | 189 | 160 | 255 | 255 | 220 | 186 | 255 |
| Mean | 117.37 | 104.39 | 100.49 | 213.77 | 152.71 | 115.72 | 118.68 | 171.01 |
| Median | 119 | 106 | 102 | 216 | 154 | 117 | 121 | 173 |
| Mode | 122 | 107 | 104 | 255 | 156 | 122 | 124 | 173 |
| S.D. | 14.36 | 16.87 | 11.98 | 22.43 | 18.45 | 19.30 | 14.49 | 20.43 |
| Contrast | 134 | 143 | 132 | 201 | 168 | 173 | 158 | 198 |

**Table 3**

*No of Pixel in Each Spectral Class of the Two Images*

| *Class code* | *IRS- 1D LISS III* | | | | *IRS- P6 LISS III* | | | |
|---|---|---|---|---|---|---|---|---|
| | *No of pixel* | | *%* | | *No of pixel* | | *%* | |
| | *1234* | | *1234* | | *1234* | | *1234* | |
| 1 | 132774 | 168729 | 09.259 | 11.767 | 111462 | 151222 | 07.775 | 10.546 |
| 2 | 155113 | 149045 | 10.817 | 10.394 | 223544 | 177330 | 13.909 | 12.367 |
| 3 | 228426 | 211789 | 15.930 | 14.770 | 273211 | 216511 | 15.589 | 15.100 |
| 4 | 275883 | 285786 | 19.239 | 19.930 | 318409 | 264375 | 19.054 | 18.437 |
| 5 | 342157 | 341805 | 23.861 | 23.837 | 255004 | 321877 | 22.203 | 22.448 |
| 6 | 240509 | 233208 | 16.772 | 16.263 | 255004 | 255112 | 17.785 | 17.791 |
| 7 | 59046 | 43546 | 04.117 | 03.037 | 52847 | 47481 | 03.684 | 03.311 |
| Total | 1433908 | 1433908 | 99.995 | 99.995 | 1433908 | 1433908 | 99.999 | 100.000 |

**Table 4**

*Band-wise Pixel Value of the Two Images for Selected GCP*

| GCP | Long | Lat | IRS-P6 LISS III | | | | IRS-1D LISS III | | | |
|---|---|---|---|---|---|---|---|---|---|---|
| | | | Band 1 | Band 2 | Band 3 | Band 4 | Band 1 | Band 2 | Band 3 | Band 4 |
| 1. | 84.074524 | 24.10979 | 131 | 89 | 73 | 101 | 106 | 91 | 73 | 228 |
| 2. | 84.077599 | 24.11447 | 137 | 93 | 121 | 131 | 101 | 78 | 95 | 201 |
| 3. | 84.134882 | 24.07490 | 126 | 86 | 77 | 102 | 100 | 85 | 69 | 168 |
| 4. | 84.121045 | 24.10978 | 152 | 106 | 116 | 145 | 120 | 106 | 96 | 227 |
| 5. | 84.043484 | 24.15710 | 145 | 96 | 104 | 144 | 109 | 82 | 77 | 171 |
| 6. | 84.005389 | 24.15918 | 128 | 79 | 104 | 126 | 108 | 83 | 92 | 187 |
| 7. | 84.056108 | 24.17578 | 141 | 102 | 117 | 152 | 114 | 98 | 100 | 179 |
| 8. | 84.042523 | 24.17299 | 165 | 125 | 133 | 168 | 125 | 104 | 110 | 190 |
| 9. | 84.071449 | 24.01889 | 125 | 89 | 65 | 71 | 107 | 91 | 86 | 159 |
| 10. | 84.106875 | 24.04609 | 156 | 109 | 127 | 164 | 115 | 95 | 108 | 208 |
| 11. | 84.123385 | 24.00084 | 139 | 96 | 102 | 142 | 108 | 86 | 89 | 203 |
| 12. | 84.086330 | 24.01654 | 130 | 93 | 89 | 133 | 109 | 90 | 83 | 194 |
| 13. | 84.136620 | 24.14642 | 167 | 127 | 124 | 171 | 122 | 104 | 96 | 206 |
| 14. | 84.116391 | 24.13637 | 169 | 126 | 127 | 176 | 127 | 113 | 102 | 206 |
| 15. | 84.072051 | 24.19501 | 144 | 101 | 119 | 173 | 118 | 99 | 104 | 215 |
| 16. | 84.065659 | 24.21322 | 146 | 110 | 116 | 159 | 101 | 91 | 82 | 191 |

Contd...

Contd...

| | | | | | | | | | |
|---|---|---|---|---|---|---|---|---|---|
| 17. | 84.228341 | 24.10116 | 168 | 138 | 130 | 201 | 124 | 112 | 106 | 247 |
| 18. | 84.178376 | 24.03242 | 136 | 100 | 89 | 129 | 104 | 77 | 79 | 197 |
| 19. | 83.995473 | 24.06623 | 148 | 109 | 120 | 168 | 111 | 96 | 102 | 222 |
| 20. | 84.005235 | 24.00914 | 141 | 99 | 128 | 154 | 110 | 92 | 96 | 202 |
| 21. | 84.053827 | 24.03681 | 167 | 111 | 143 | 173 | 130 | 106 | 119 | 239 |
| 22. | 83.995902 | 24.09738 | 155 | 104 | 124 | 157 | 111 | 87 | 105 | 219 |
| 23. | 84.064414 | 24.23412 | 128 | 83 | 126 | 150 | 103 | 82 | 103 | 200 |
| 24. | 84.021303 | 24.20800 | 165 | 126 | 117 | 169 | 119 | 103 | 97 | 222 |
| 25. | 84.175012 | 24.17550 | 143 | 101 | 118 | 160 | 110 | 92 | 96 | 211 |
| 26. | 84.098598 | 24.26926 | 142 | 104 | 112 | 158 | 111 | 96 | 94 | 215 |
| 27. | 84.239042 | 24.19754 | 149 | 110 | 116 | 168 | 116 | 105 | 105 | 195 |
| 28. | 83.998706 | 24.20124 | 178 | 143 | 148 | 203 | 130 | 121 | 121 | 229 |
| 29. | 84.134064 | 24.20929 | 146 | 105 | 119 | 164 | 119 | 104 | 110 | 214 |
| 30. | 84.116553 | 24.18360 | 178 | 138 | 143 | 183 | 132 | 117 | 112 | 219 |
| 31. | 84.172220 | 24.11746 | 175 | 154 | 142 | 211 | 126 | 132 | 115 | 255 |
| 32. | 84.241975 | 24.15072 | 164 | 117 | 129 | 179 | 122 | 107 | 107 | 232 |
| 33. | 84.161175 | 24.26663 | 116 | 69 | 116 | 131 | 92 | 67 | 96 | 168 |
| 34. | 84.012155 | 24.26019 | 116 | 70 | 117 | 139 | 91 | 64 | 101 | 170 |
| 35. | 84.173501 | 23.99426 | 126 | 83 | 78 | 117 | 101 | 83 | 77 | 160 |
| 36. | 84.209337 | 23.99662 | 137 | 96 | 92 | 134 | 111 | 92 | 83 | 192 |
| 37. | 84.239926 | 24.07705 | 179 | 141 | 139 | 194 | 143 | 137 | 127 | 231 |
| 38. | 84.167842 | 24.06499 | 173 | 133 | 131 | 175 | 136 | 123 | 110 | 209 |

**Figure 1**

*Comparison of DN Value for Band 1*

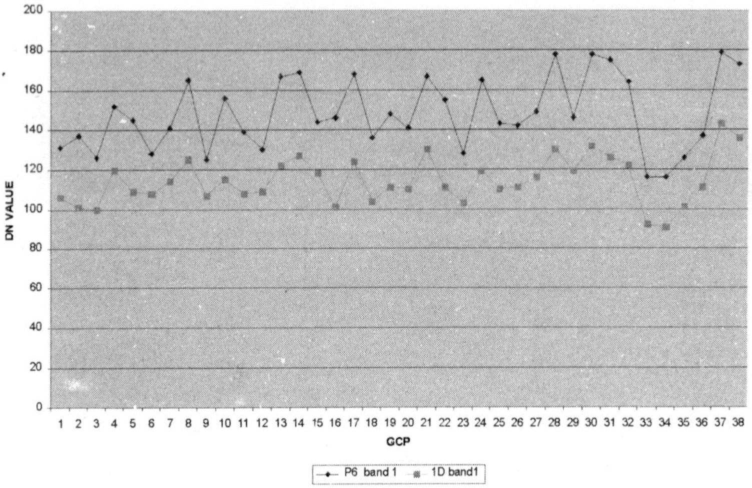


**Figure 2**

*Comparison of DN Value for Band 2*

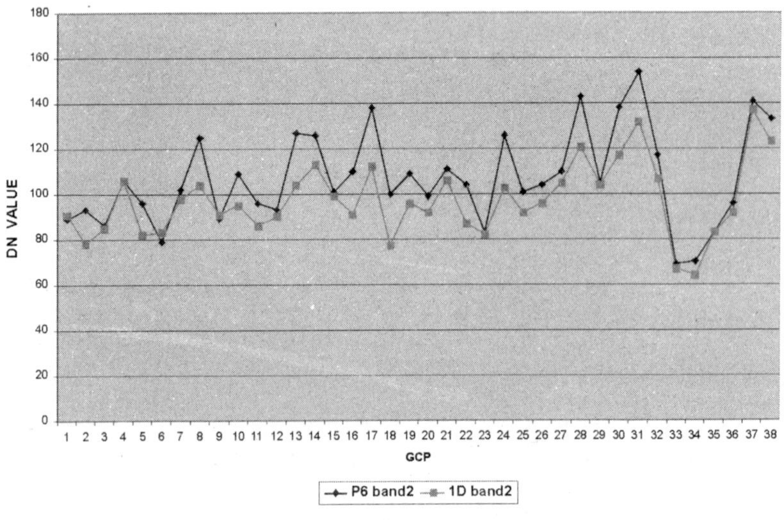

**Figure 3**

*Comparison of DN Value for Band 3*

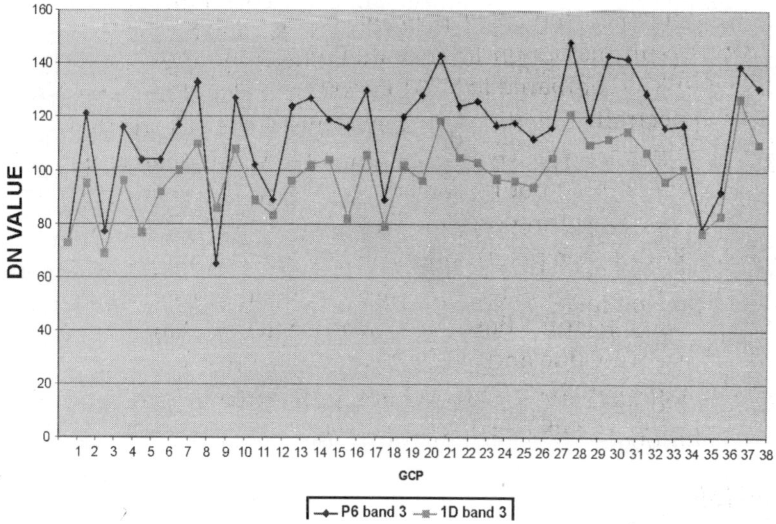


**Figure 4**

*Comparison of DN Value for Band 4*

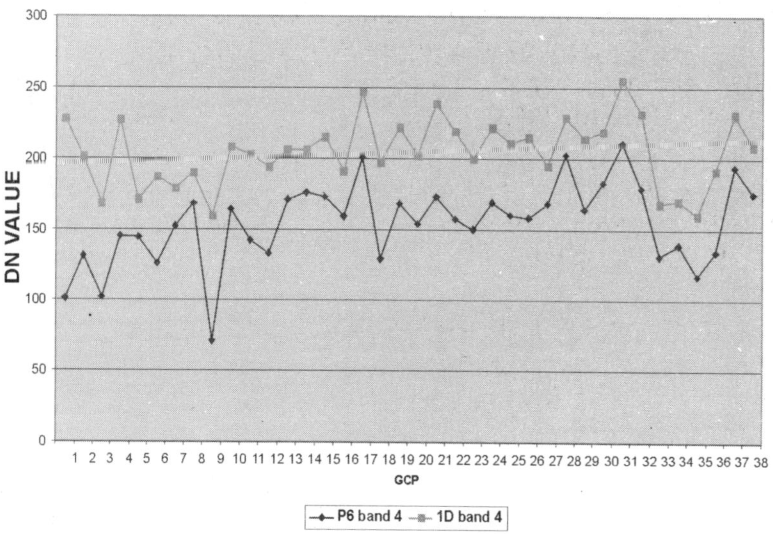

The bandwise comparison of DN values of the two data reveals anomalous behaviour in spectral clustering. The spectral reflectance of GCPs experiences almost similar trend of up and down peaks which is also supported by high correlation coefficient value between the corresponding bands of the two data. The difference in histogram mean value of band 2 and 3 of the two data is less as compared to band 1 and 4. It is clear that band 2 and 3 are spectrally identical as they exhibit high correlation coefficient value and low covariance value. However, high correlation value alone may not be considered a definite criterion for ensuring similar clustering output. The covariance (difference in actual DN value of each pixel) and mean DN value around which spectral clustering takes place, may also be taken into account as determining key factors. Band 4 shows least correlation and highest covariance value and this indicates a distinct behavior of the sensor and may be accounted for the difference in their spatial resolution as well as detector ageing.

## Conclusion

The contrast for IRS-P6 data is better than that of IRS-1D in all the bands except band 4. For comparative visual interpretation FCC should be generated using 321 band of LISS III of both IRS-1D and P6 because some of its detectors, possibly band 4, are showing radiometric error. Furthermore, spatial resolution of band 4 is different for IRS-1D and IRS-P6 which results unlike DN values leading to variation in tone in the FCC. The LISS III data of IRS-1D and IRS-P6 have symmetrical radiometric behaviour in the order of Band 2> Band 3> Band 1> Band 4. It is, thus, inferred that, when data are used in conjunction then a band calibration between corresponding bands of the two data is desirable to obtain a good result.

| | *Correlation Coefficient* | *Covariance* |
|---|---|---|
| Band 1 | 0.9204021 | 523.8292 |
| Band 2 | 0.9161021 | 361.8666 |
| Band 3 | 0.8829111 | 372.7698 |
| Band 4 | 0.8700517 | 1337.4722 |

## *Acknowledgement*

The authors are thankful to the Indian Space Research Organization, Department of Space, Government of India for providing financial assistance under RESPOND project to carry out the work.


## References


Jayabharathi, S., Jayachandraiah, B., Muralikrishnan, S. and Senthil Kumar, A.(2003). Cross-calibration of IRS 1C and IRS 1D: WIFS Using near Synchronous Matching Scene. Proceeding of ISRS

Madhavan Unni, N.V., Roy, P.S., Jadhav, R.N., Tiwari, A.K., Sudhakar, S., Ranganath, B.K. and Dabral, S.I. (1991). IRS 1A Application in Forestry. *Current Science,* Vol. 61 (3 and 4): 189-192.

Roy, P.S., Ranganath, B.K., Vohra T.P.S. and Diwakar, P.G. (1988). Analysis of IRS-1A LISS II multispectral digital data for type discrimination and its comparison with landsat TM and SPOT-1. Proc. IRS symposium, Hyderabad.

Roy, P.S., Ranganath, B.K., Vohra, T.P.S., Diwakar, P.G. and Bhan, S.K. (1990). Forest resource management using Indian Remote Sensing Satellite data. *Asian-Pacific Remote Sensing Journal,* Vol. 3(1): 11-16

Roy, P.S., Ranganath, B.K., Vohra, T.P.S., Diwakar, P.G., Bhan, S.K., Singh, J.J and Pandian, V.C. (1991). Tropical Forest Type Mapping and Monitoring using Remote Sensing. *Int. J. Remote Sensing,* Vol. 12(11): 2205-2225'.

Srikrishna, N., Vindhya, K., Satyanarayana, P. and Sudarshana, R. (2003). Content Based Image Retrieval: A Solution Using ANN Approach. Proceedings of ISRS

Stephen, R. Yool, Jefferey, L Star, John E. Estes, Daniel B. Botkin, David N. Echardt and Frank W.K. Davis (1986). Performance analysis of image processing algorithm for classification of natural vegetation in the mountains of Southern California. *Int. J. Remote Sensing,* Vol. 7 (5): 683-702.

Teillet, P.M., Markham, B.L., Barker, J.L., Storey, J.C., Irish, R.R and Seiferth, J.C. (2000). Landsat Sensor Cross-Calibration Using Nearly-Coincident Matching Scenes, Proceedings of SPIE, Vol. 4049: 155-167.

Thome, K., Markham, B., Barker, J., Slater, P. and Biggar, S. (1997). Radiometric Calibration of Landsat, Photogramm. *Eng. and Rem. Sens.,* 63(7): 853-858.

# 9

# Elephant Movement in Karra C.D. Block: A GIS Analysis

Hari Shankar Gupta and M.S. Nathawat

An increasing number of international large mammal studies are using species distribution maps to convey their message about conservation priorities (e.g., Hatough-Bouran and Disi, 1991; Smith and Mishra, 1992) and elephant is a featured concern of conservationist world wide (e.g., African Elephant Conservation Coordinating Group 1991; DF & C, 1991; FDA 1991; Lindeque 1991; Campbell and Huish, 1992; Environmental Investigation Agency, 1992; National Parks Board of South Africa, 1992; Phanthavong and Santiapillai, 1992).

Critical areas of their habitat are rapidly being fragmented and eroded by competing land uses. Sanderson (1878), has made observation of the long-term perspective on elephant movement patterns.

The use of GIS has been made to systematize, standardize and manage the enormous amounts of spatial data generated by disturbance to the landscape.

## Objective of the Study

The main objective of the study is aimed to answer the following:

(1) How are the elephants of this tract distributed.

(2) The utilization pattern of the different habitats types, in different seasons.

The real time habitat use information along with their spatial distribution of the wild life habitat and temporal changes have been used for habitat evaluation. The limitation of conventional method of physical survey has been avoided by using remote sensing and GIS technology.

### *The Study Area*

The study area is situated between 84°55′E and 85°15′E, 22°55′N and 23°20′N it comprises the forested area of 567.2 10 km$^2$, spread over the Karra C.D. Block of Ranchi district. The cultivated area occupies an additional 31,825 km$^2$ in the study area.

The average elevation of the tract is 600m; the general slope of the tract is running from north-east to south-west, which present a vista of grassland and forest vegetation.

### *Geology*

The major geological formations which are found within the borders of the Ranchi district, are as follows:

Recent
Tertiary to Recent
Upper Cretaceous
Dharwar

### *Hydrology*

The region has one major drainage system: emptying into the Karo river. Given a general northwards aspect in this area, most of the streams ultimately reach the river Karo, which flows in a southward direction.

### *Climate*

The study area has not much of variation of climate, due to relief and topography.

Temperatures begin to drop rapidly from November. December is the coldest month. After February the temperatures rise steadily till May. With the advance of the south-west monsoon by the end of September, temperatures begin to drop.

## Rainfall

The district receives rainfall almost throughout the year, though the concentration is during the monsoon months, from June to September. The average annual rainfall in the district as a whole is 1,481.9 mm.

Three seasons have been defined, based mainly on the rainfall pattern and water availability in the environment. Sukumar (1983):

Dry season (March – May)

This period is characterized by negligible rainfall.

Rainy or Wet season (June – September)

The heavy pre-monsoon showers occur in June and main rain follows in July to September.

Winter season (October – February)

This season is marked by low temperature.

## *Land-use Pattern*

The area being comparatively flat than the other part of the district is honeycombed with cultivation, chiefly of paddy, etc. Besides this the villages have groves.

## *Vegetation Types*

The basic vegetation type as defined by the Champion and Seth (3b/c $B_2a$) Northern Tropical moist deciduous forests containing Sal, etc.

In addition, many plantation forests of Acacia and eucalypts etc. are scattered.

Human impact has converted most of the primary forest into secondary forms, degraded into scrub in the vicinity of settlements.

### *Mammals*

The fauna of the Indian subcontinent are well represented in the study area, including the elephant, the dominant wild mammal in terms of biomass.

Besides elephants, chital (*Axis axis*). Barking deer (*Muntiacus muntjac*) leopard (*Panthera pardus*) wild pig (*Sus scrofa*), sloth bear (*Melursus ursinus*), the monkey (*Macaca mulata*) and the common langur (*Presbytis entellus*) etc., are present.

### *The People*

The total population of 67,329 people live within the study area (District Census Report, 1981).

## Estimation of Seasonal Elephant Movement

In the study area, two twenty-four hour trackers were employed for monitoring the movements of the elephants, starting from April 1998. The tracker used to visit village to another village of the study area and noted down the details of the elephants; basically regarding description and size of the herd and from the village it came; its stay, its departure destination on day-to-day basis. This movement data was compiled for the period April 1998 to December 2000. Later, this movement pattern was overlaid on study area, with the village boundaries, both separately, using ARCVIEW. The length of movement – falling in each village boundary has been considered an important parameter, which has been observed in the field. This length of movement has time attribute also, hence it has also helped in seasonal analysis of movement. This relatively simple approach in the present study has been taken, taking into consideration the constraints faced in the field for such type of study.

## Methodology

Those factors, which most closely correlated with elephant distribution were tested and used to predict their distribution, without a long history of previous survey work, accomplished by using a GIS for integration and analysis, environmental elements associated with the elephant. However, limitations in collecting comprehensive elephant data have made comprehensive trend analysis very difficult,

The Environmental Systems Research Institute's (ESRI) ARCVIEW3.1 is the main GIS software used. Karra C.D. block map along with the village boundaries at a scale 1:50000 were digitiized. Spatial information within the study area was stored in vector format in which features are represented by points, lines and polygons.

## Result and Discussion

For studying the objective of this study; the null hypothesis ($H_0$) that the dependent variable ($Y_1$) i.e., the "elephant's movement in a village" is affected by independent variables, i.e., the "population density", "area available for agriculture in the village", "the forest density", "temperature", "relative humidity", "rainfall" and the alternative hypothesis ($H_1$) that the elephant movement is not affected by the above said independent variable; were set up. The village-wise movement data of the elephants as recorded for the different villages of Karra C.D. Block of Ranchi District are in Table 1.

**Table 1**

*Correlation Coefficient Between Dependant and Independent Variables*

| *Dependant Variable* | *Independent Variables* | *Correlation Coefficient* |
|---|---|---|
| Elephant's movement | Population density | 0.08 |
| Elephant's movement | Area available for agriculture | 0.1 |
| Elephant's movement | Forest density | 0.05 |
| Elephant's movement | Mean monthly temperature | 0.2 |
| Elephant's movement | Relative humidity | 0.06 |
| Elephant's movement | Mean monthly rainfall | 0.1 |

As a first step the correlation coefficient between dependant variable and each of independent variables was calculated separately, as shown in Table 1. The correlation coefficient between "Elephant's movement" and "population density", between "Elephant's movement" and "the area available for agriculture" in the village, between "Elephant's movement" and the "Forest density", between "Elephant's movement" and the "temperature", between "Elephant's movement" and the "relative humidity" and between "Elephant's movement" and the "rainfall" are 0.08, 0.1, -0.05, 0.2, 0.06, 0.1 respectively.

The significance of the above correlation coefficient were tested by the applying the t-test, since the calculated values of $|t|$ for all the correlation coefficients did not exceed their corresponding tabulated values. Hence these correlation coefficients are not significant. So, the null hypothesis ($H_0$) was accepted. Since there were six independent variables for study, so for calculation of the multiple regression, the importance of the independent variables was tested by applying the following F-test.

By this criteria, the independent variables, i.e., the "area available for agriculture" and the "forest density" were found suitable for multiple correlation and other independent variables – "population density", "temperature", "relative humidity" and "mean monthly rainfall" were ignored.

The importance of rainfall on the basis of actual data for considering multiple correlation coefficient was tested. But the calculated value of F = 0.35 where S sum of square of deviation from the mean of Elephant's movement and s=Sum of square of deviation from the "mean of Rainfall" was less than the tabulated value of F, i.e., 3.58 at 5 per cent of level of significance with n=8 and n2=8 d.f. Similarly the importance of temperature for considering multiple correlation coefficient was tested. The calculated value of 'F', i.e., 2.97 is lesser than the tabulated value i.e 3.58 at 5 per cent level of significance. Hence the rainfall and "temperature" were not taken into consideration for multiple correlation.

Thus the multiple regression equation for the dependent variable i.e. elephant's movement and the two independent variables, i.e., "area available for agriculture" and "forest density" was developed.

The regression equations developed for the above variables on the basis of data for summer season 1998, winter season 1998, rainy season 1998 and complete Year 1998 are as following:

(i) $Y = 35.22 - 0.28X_1 - 0.25X_2$ (On the basis of data of summer season 1998)

(ii) $Y = (-)\ 1.72 + 0.23X_1 - 0.25X_2$ (On the basis of data of winter season 1998)

(iii) $Y = (-)\ 58.8 + 2.7X_1 - 0.91X_2$ (On the basis of data of Rainy season 1998)

(iv) $Y = (-)\ 2.7 + 0.28X_1 + 0.25X_2$ (On the basis of data of winter season 1998)

Where $Y =$ Elephant's movement in the Block,
$X_1 =$ Area available for Agriculture,

$$X_2 = \text{Forest density} = \left(\frac{\text{Forest Area}}{\text{Village Area}}\right)$$

By applying the above equation, the predicted values of Elephant's movement for Year 1999 and 2000 were calculated and the difference of averages of actual value and predicted values was tested by applying the t-test.

The difference of mean of actual movement and predicted value for the year "1999" and "2000", "winter season" are not significant. But the average difference in "summer season" and "rainy season" are significant. The data have been shown in Table 1, 2, 3 and 4.

The calculated value of F (1.18) was less than the Critical (tabulated) value of F (3.13, 3.63) for 5 per cent and 10 per cent of level of significance respectively with $n_1 = 2$ and $n_2 = 69$ d.f. Hence the null hypothesis ($H_0$) was accepted. In other words it can be said that the independent variables i.e. the area available for agriculture and the forest density taken together have an effect on the Elephant's movement.

Inference: The linear correlation coefficient between the elephant's movement and the area available for agriculture is positive, elephant's movement and the forest density is negative but too weak in both cases. But the multiple co-relation coefficient between the "elephant's movement", "area available for agriculture" and the "forest density" is positive and significant. Hence it can be concluded that the "elephant's movement" is dependent on "area available for agriculture" and the "forest density" during summer and rainy season only.

**Table 1**

| Sl.No. | Name of village | Elephant's movement in 1998 | Elephant's movement in 1999 | Predicted value for 1999 as per Equation (IV) | Elephant's movement in 2000 | Predicted value for 2000 as per Equation (IV) |
|---|---|---|---|---|---|---|
| 1 | Nagra | 71.1 | 38.0 | 76.2 | — | 76.2 |
| " | " | " | " | " | " | " |
| 72 | Ata | — | 2.4 | 48.3 | — | 48.3 |

## For the year 1999

$Y = -2.7 + 0.28\,X_1 + 1.7\,X_2$

Average of predicted Elephant's movement value in 1999

$(x_1) = 42.01$

Average of actual Elephant's Movement in 1999

$(x_2) = 19.05$

$S = 127.74$

Whereas, the tabulated value of "t" – 1.96
Hence, difference is not significant

## For the year 2000

Average of Elephant's movement in the year 2000 (actual)

$= 15.4$

Average of Elephant's movement in 2000 (predicted)

$= 42$

$S = 135.5$

$|t| = 1.15$

Whereas the tabulated value of "t" = 1.96
Hence, difference is not significant

### *For Calculation of Multiple Correlation*

| *Elephant's movement in 1999* | *Predicted value of Elephant's movement in 1999* | *Difference between Actual value and predicted value* | *Difference between Actual value and mean of actual value* | *Difference between predicted value and mean of actual value* |
|---|---|---|---|---|
| $(Y_i)$ | $(Y_i)$ | $(Y_i - \hat{Y}_i)$ | $(Y_i - \hat{Y})$ | $(\hat{Y}_i - \overline{Y})$ |
| 38.0 | 76.2 | (–)38.2 | (–)18.9 | 57.1 |
| " | " | " | " | " |
| – | 48.3 | (–)48.3 | (–)19.1 | 29.2 |

**Table 2**

| *Sl.No.* | *Name of Village* | *Actual Elephant's Movement in Winter Season (Oct - Feb 98)* |
|---|---|---|
| 1 | Nagra | 12.5 |
| " | " | " |
| 71 | Ata | – |
| | Total | 111.8 |

So, Regression equation between Elephant's movement (Y), area available for agriculture ($X_1$) and the Forest density ($X_2$) is as following.

$$Y = -1.72 + 0.23X_1 - 0.43X_2$$

*For Winter Season*

| *Village* | *Actual Elephant's movement in Winter Season 1999* | *Actual Elephant's movement in Winter Season 2000* | *Predicted value of Elephant's movement in winter (1999)* | *Predicted value in Winter Season (2000)* |
|---|---|---|---|---|
| Nagra | – | – | (–)7.6 | (–)7.6 |
| " | " | " | " | " |
| Ata | 2.4 | – | 8.0 | 8.0 |

Average of actual Elephant's movement per village in Winter 1999 = 4.4

Average of predicted Elephant's movement in winter 1999 = 4.4

$$S = 58534.9$$

$$|\, t\, | = 0$$

Whereas the tabulated value of "t" = 1.96

Similarly, Average of Elephants movement (actual) in winter 2000 = 3.6

Average Elephant's movement (predicted) in winter of 2000 = 4.4

$$S = 22.4$$

$$|\, t\, | = 0.21$$

Where, the tabulated value of | t | = 1.96

**Table 3**

*(For Summer Season)*

| *Sl.No.* | *Name of village* | *Elephant's movement (Y)* | *Population Density ($X_1$)* | *Area available for agriculture ($X_2$)* | *Forest Density ($X_3$)* |
|---|---|---|---|---|---|
| 1 | Nagra | 28.4 | 0.50 | 46.9 | 38.7 |
| " | " | " | " | " | " |
| 30 | Goso | 6.1 | 0.7 | 59.5 | 30.3 |

Regression equation for summer (April-May 1998)

$$Y = a + b_1X_1 + b_2X_2$$
$$= 35.22 - 0.28X_1 - 0.25X_2$$

Applying the above equation the predicated value of Elephant's movement in summer season of 1999 and 2000 was calculated, which is as following:

| *Sl.No.* | *Name of village* | *Actual Elephant's movement in summer season March-May 1999* | *Predicated value of Elephant's movement in summer season March-May 1999* | *Actual Elephant's movement in summer season March-May 2000* | *Predicted value of Elephant's movement in summer season March-May 2000* |
|---|---|---|---|---|---|
| 1 | 2 | 3 | 4 | 5 | 6 |
| 1 | Nagra | 20.0 | 12.5 | – | 12.5 |
| " | " | " | " | " | " |
| 71 | Ata | – | 9.7 | – | 9.7 |

Average of Elephant's movement in Summer season 1999 = 5.5

Average of predicted value of Elephant's movement of summer season 1999

= 14.6

S = 16.5

| t | = 3.2

Whereas the tabulated value of "t" = 1.96

Similarly, Average of Elephants movement (actual) in winter of 2000 = 4.8

Average Elephant's movement (predicted) in winter of 2000 = 14.6

S = 27.8

| t | = 2.07

Where, the tabulated value of | t | = 1.96

Hence, the difference of average of actual and predicted value of elephants movement for summer season 1999 and 2000 is significant.

**Table 4**
*(For Rainy Season)*

| *Sl. No.* | *Name of Village* | *Rainy Season (June-Sept 1998)* | *Actual Elephant's movement* | *Population density* | *Area available for agriculture* | *Forest density* |
|---|---|---|---|---|---|---|
| 1 | Nagra | " | 30.2 | 0.50 | 46.9 | 38.7 |
| " | " | " | " | " | " | " |
| 72 | Ata | " | - | 0.9 | 75.4 | 17.6 |

Regression equation between Elephant's movement; Area available for agriculture, forest density is as following

$$Y = -58.8 + 2.7X_1 - 0.91X_2$$

The predicted value of Elephant's movement in Rainy season (June–September 1999, June–September 2000) was calculated, which are as follows:

| *Sl.No.* | *Name of village* | *Actual Elephant's movement's in Rainy Season of (June-Sept 1999)* | *Predicted value of Elephant's movement in Rainy Season of (June-Sept 1999)* | *Actual Elephant's movement in Rainy Season of (June-Sept 2000)* | *Predicted value of Elephant's movement Rainy Season of (June-Sept 2000)* |
|---|---|---|---|---|---|
| 1 | Nagra | 18.0 | 32.6 | - | 32.6 |
| " | " | " | " | " | " |
| 71 | Ata | - | 128.7 | - | 128.7 |

Average of actual value of Elephant's movement of Rainy season of 1999 = 2.2

Average of predicted value of Elephant's movement = 86.4

$$S = 61.2$$

$$= 8.09$$

Whereas the tabulated value of "t" = 1.96

$$S = 65.2$$

$$|t| = 7.2$$

Where, the tabulated value of $|t| = 1.96$

Hence, the difference of actual value and predicted value of Elephant's movement in rainy season of 2000 is significant.

Similarly, average of actual value of Elephant's movement in 2000 = 5.7

Average of predicted value of Elephant's movement in 2000 = 86.4

## Elephant Movement Intensity and Seasonal Distribution Patterns

Graphical presentation with data attribute on elephant movement in different villages of the study area are depicted in Figure C1 to C8 showing the spatial distribution patterns during Summer season (March-May), Rainy season (June-September) and Winter season (October-February).

(a) Summer season distribution: Figure C5 shows the concentration of elephant's movement in summer season, i.e., for the months of March to May and it is very obvious that particular villages of Karra C.D. Block are preferred by elephants in this season.

(b) Rainy season distribution (June-September): Figure C6 shows the concentration of elephant's movement in rainy season, i.e., for the months of June to September and it is very obvious that particular villages in this season.

(c) Winter season distribution (October-February): Figure C7 shows the concentration of elephant's movement in winter season, i.e., for the months of October to February and it is very obvious that particular villages are preferred in this season.

From GIS analysis, it appears that during summer season, the elephants have to preference for certain habitats, which has been directly correlated by the frequency of visit to these areas in this season. As is obvious in Figure C8 the intensity of movement has been obtained using "Surface-contour" module, Arc view GIS version 3.1. It also enables to identify villages, which are highly frequented by the elephants.

**Figure C1**

*Map of Karra C.D. Block Showing Villages*

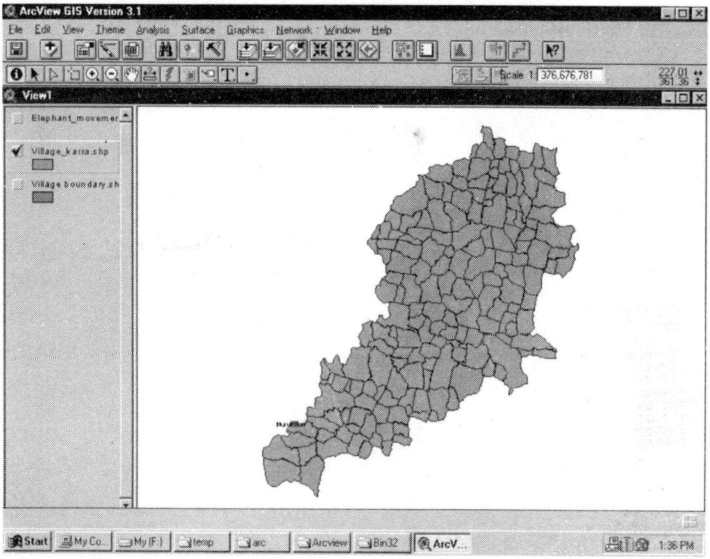


**Figure C2**

*Elephant Movement of Karra Block (intervillage)*

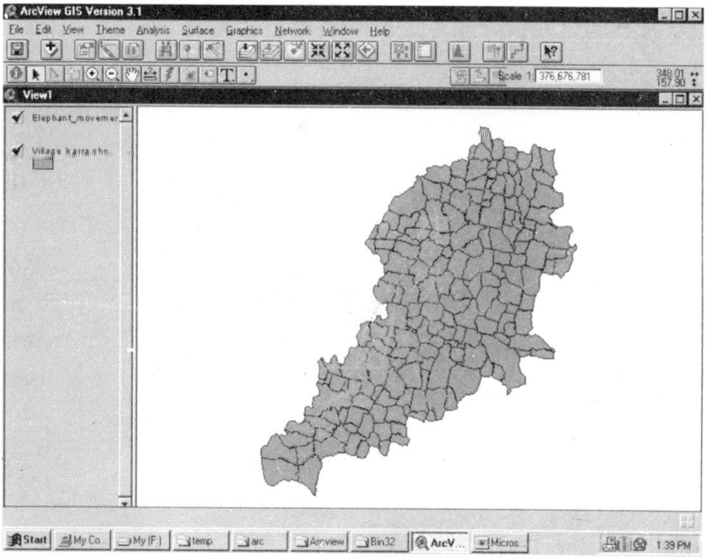

**Figure C3**

*Elephant Movement Showing Datewise Details*

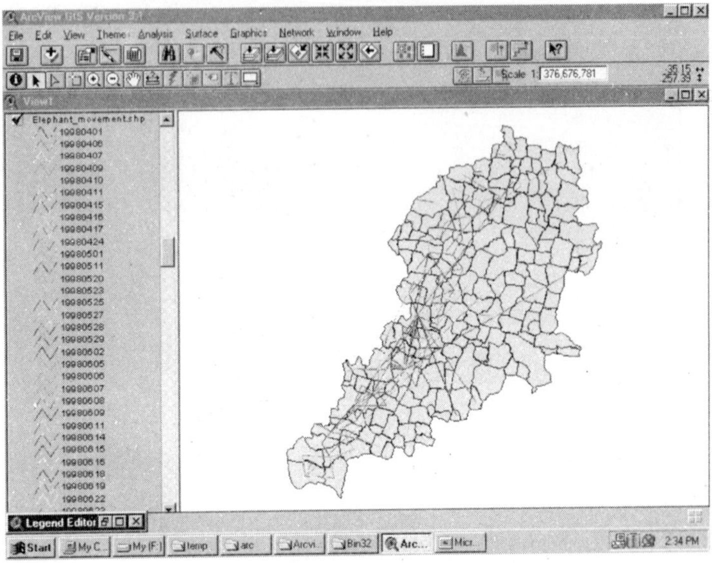


**Figure C4**

*Map of Karra Block and Elephant Movement (showing attribute table, 1-4-98 to 30-7-2000)*

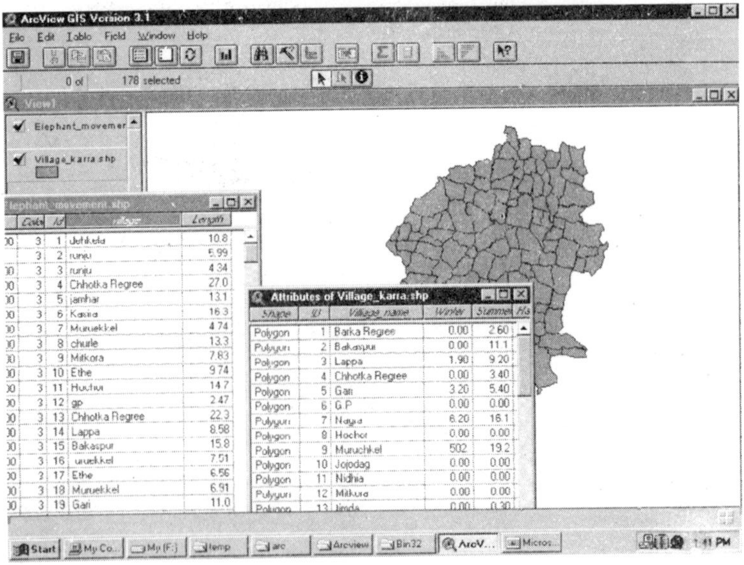

**Figure C5**

*Contours of Elephant Movement in Summer Season*

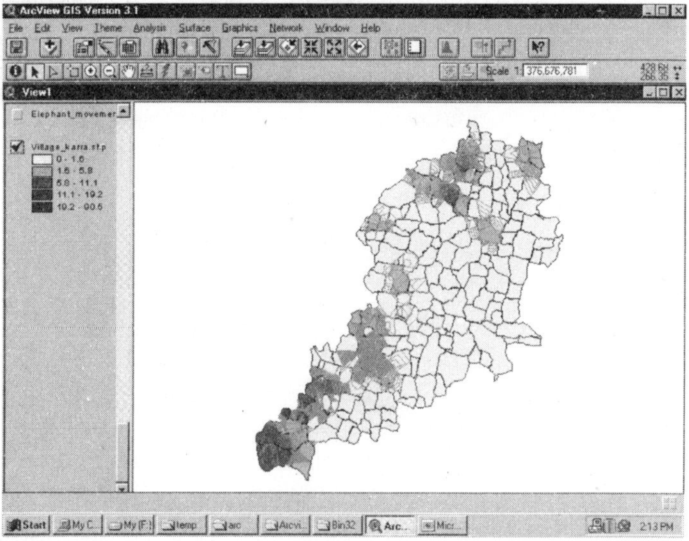


**Figure C6**

*Contours of Elephant Movement in Rainy Season*

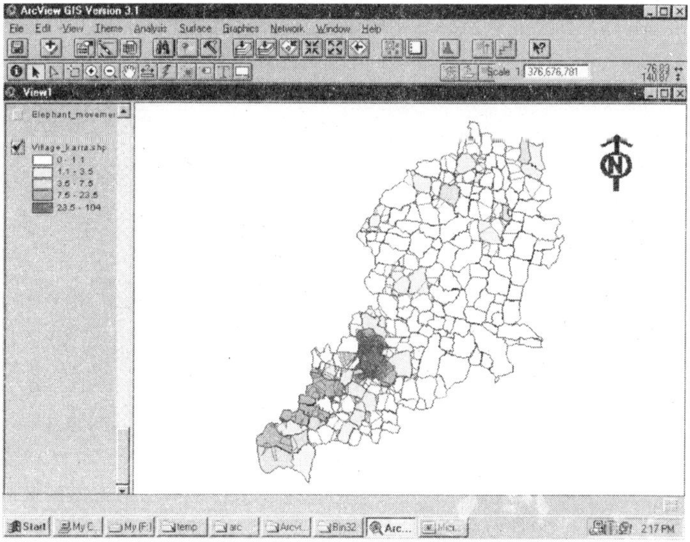

**Figure C7**

*Contours of Elephant Movement in Winter Season*

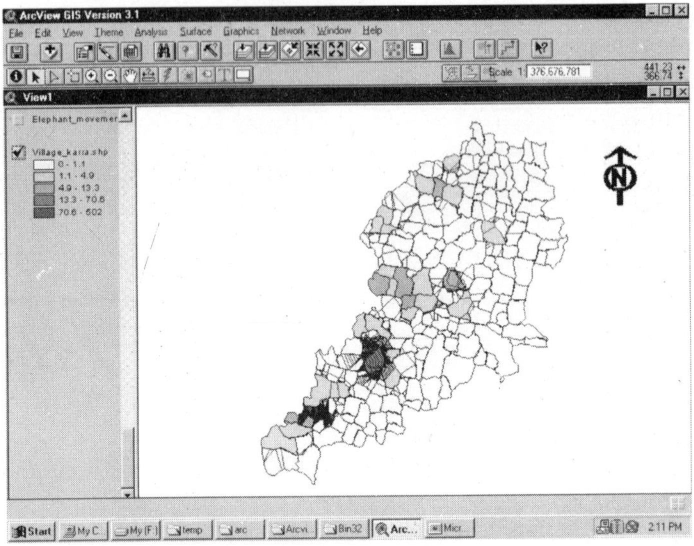


**Figure C8**

*Contours of Elephant Movement in Different Seasons in Karra Block*

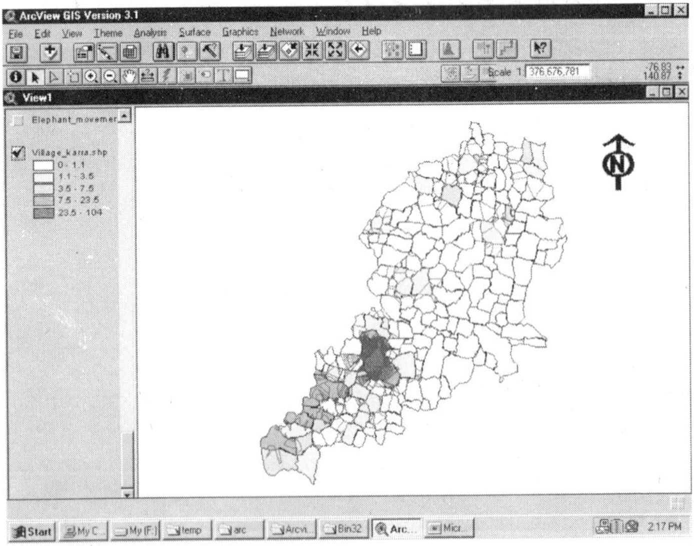

Lapa, Bakaspur, Nagra, Gosa are the some villages which are maximum affected, where some special wildlife management practices should be adopted for elephants and at the same time the people of these villages should be trained and equipped to protect and live with these beasts.

Similarly for winter season, it appears that the elephant have got preference for certain habitats, which has been directly correlated by the frequency of visit to these areas in this season. It also enables to identify villages, which are highly frequented by the elephants. The maximum affected villages in this season are Nagra, Chotka Regre, Barwadag, Limra. In these villages some special wildlife management practices should be adopted for elephants and at the some time the people of these should be trained and equipped to protect and live with the elephants.

Similarly for rainy season, it appears that elephants have got preference for certain habitats, which has been directly correlated by the frequency of visit to these areas in this season. It also enables to identify villages, which are highly frequented by the elephant. The maximum affected villages are Barwadag, Dehkela, Limra.

Also in these villages some special wildlife management practices should be adopted for elephants and at the same time the people of these should be trained and equipped to protect and live with the elephants.

## References

African Elephant Conservation Coordinating Group (1991). *The African Elephant Conservation Review,* AECCG, 21 St. Giles, Oxford, UK, p. 66

Buss, I.O. (1961). Some Observations of Food Habits and Behaviour of the African Elephant. *Journal of Wildlife Management* 25, 131-48.

Campbell, K. and Huish, S. (1992). Recent Trends in Tanzanian Elephant Populations: 1987-1992, *TWCM,* Arusha, Tanzania, p. 7.

Douglas-Hamilton, I. (1972). On the Ecology and Behaviour of the African Elephant. Unpublished D. Phil. Thesis, University of Oxford.

Douglas-Hamilton, I. and Douglas-Hamilton, O. (1975). *Among the Elephants.* London: Collins and Harvill Press.

DF & C (1991). *Elephant Conservation Plan: Gabon,* Direction de La Faune et de La Chasse, Libreville, Gabon, p. 77.

Environmental Investigation Agency (1992). Under Fire: Elephants on the Front Line, *Environmental Investigation Agency,* London, pp. 57.

FDA (1991). *Elephant Conservation Plan: Liberia,* Forestry Development Authority, Monrovia, Liberia, p. 47.

Hatough-Bouran, A. and Disi, A.M. (1991). History, Distribution, and Conservation of Large Mammals and their Habitats in Jordan. *Environmental Conservation,* 18(1), 19-32.

Laws, R.M., Parker, I.S.C. and Johnstone, R.C.B. (1975). *Elephants and Their Habitats*. Oxford: Clarendon Press.

Lindeque, M. (1991). Elephant Conservation and Management Plan: Namibia.

Moss, C.J. (1982). *Portraits in the Wild: Behaviour Studies of East African Mammals.* Chicago: University of Chicago Press.

National Parks Board of South Africa (1992) Draft Elephant Conservation Plan for South Africa, p. 43.

Phanthavong, B. and Santiapillai, C. (1992). Conservation of Elephants in Laos, IUCN/SSC Asian Elephant Specialist Group Newsletter 8, pp. 25-33.

Roy B.K. (1972-73), Cartography of Migration Maps, Bombay. *Geog. Magazine,* Vol.-20-21(1) Roy B.K. (1982), *Indian Census Cartography Geojournal,* Vol. 6 (3)

Sukumar, R. (1983). *The Asian Elephant: Ecology and Management.* Cambridge University Press.

Sanderson, G.P. (1878). *Thirteen Years among the Wild Beasts of India.* London. W.H. Allen.

Smith, J.L.D. and Mishra, H.R. (1992). Status and Distribution of Asian Elephants in Central Nepal. *Oryx,* 26 (1, January), 34-8.

# II

# Geoinformatics for Environmental Issues

# 10

# Geoinformatics for Natural Hazards Mapping with Focus on Seismotectonics: A Case Study from Haryana State

Arvind Chandra Pandey

The state of Haryana is vulnerable to a gamut of natural hazards. Geological and geomorphological set-up of Haryana made it prone to a number of natural hazards, i.e., floods, waterlogging, soil salinity, soil erosion, landslides, drought, etc. These conditions have been further aggravated due to anthropogenic activities. In the recent past, increase in population and unplanned/unscientific development has triggered the intensity and frequency of such natural hazards in Haryana.

Natural hazards and resulting disasters permeate every corner of the society. Natural hazards affect the infrastructure, rendering the utilities haywire. Besides, there is loss of natural resources and human lives. This calls for effective management of disaster. Remote Sensing and Geographic Information System (GIS) technologies provide meaningful solutions to address the problem of natural hazards mapping, mitigation and management.

In the present study author has attempted delineation of natural hazards in Haryana state based on potential areas prone to wind and water-borne erosion, salinity, waterlogging, inundation during peak floods and seismic risk using WiFS satellite images and collateral data. Seismogenic activity in the Haryana state is also discussed with focus on geotectonics of the region. The author has suggested an Earthquake Disaster Management Plan and role of remote sensing and GIS technology in planning and mitigation of pre- and post- earthquake activities.

## Data Used and Methodology

The present work requires delineation of potential areas prone to various natural hazards for the entire state. To delineate such areas, signature of land degredational processes apparent on satellite images are of prime importance. On a regional scale at state level, a broad demarcation of various natural hazard-prone areas can be effectively done by using Wide Field Sensor (WiFS) satellite images. In the present WiFS satellite images of March 1998 and October 1998 were visually interpreted to decipher the extent of land degradation and resultant hazard during pre- and post-monsoon seasons. To delineate the regional extent of geological and tectonic features, IRS-IC data on 1:250,000 and 1:50,000 scale was used. Collateral data on occurrence of earthquakes and their genesis were collected to comment on seismic activity in parts of Haryana state. Survey of India topographical sheets on 1:250, 000 and 1:50, 000 scale were used for the purpose of georeferencing the satellite images.

## Role of Remote Sensing and GIS Technologies

Remotely sensed images provide a wealth of information of large areas of the earth's surface. The high spatial resolution satellite images, viz., LISS-III (23.25m), PAN (5.8m) and IKONOS (1m) provide a plethora of information of terrain features like land use, land cover, groundwater potentiality, geology structure, geomorphology, drainage, slope aspect and infrastructure. The spatial information of these key parameters is essential for building a comprehensive database for hazard mapping and their management. The digital database can be created in scale of 1 : 250,000 for regional study to 1 : 12,500 and 1 : 5000 for detail spatial analysis. Geographical information system (GIS) provides a scientific framework for storage, analysis and retrieval of spatial (thematic) and non-spatial (attribute) data. The analysis of temporal data acquired from remote sensing satellites coupled with ground verification and integration in GIS environment can identify cause-effect relationship of the hazard, which could be utilized to develop predictive models to counter the problem. The spatial distribution of various natural hazard prone areas in the state as inferred from the WiFS satellite images (Plate 1).

**Plate 1**

*WiFS Image of Haryana*

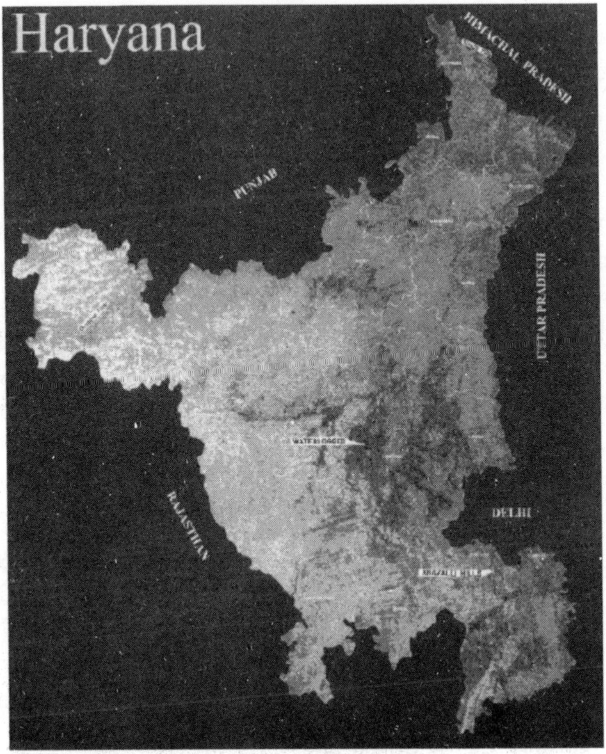

## Waterlogging

Water table rises if the soil is given water more than the vegetation or crop needs. In areas where there is inadequate drainage such as in topographic depressions, the rise of water table is more conspicuous during wetter periods. Excessive irrigation and seepage from canals has aggravated the problem of water logging in the state. To examine the regional scenario of waterlogging in the state, Wide Field Sensor (WiFS) remote sensing data provide ample information. Satellite images revealed that large tracts of Panipat, Hisar, Rohtak were severely waterlogged in addition to parts of Bhiwani, Gurgaon, Rewari, Fatehabad, Jhajjar, Kaithal, Sonipat where waterlogging exist in scattered locations. About 5.95 per cent area of the state is totally waterlogged (0-1.5 m) whereas potential waterlogged (1.5-3m) areas constitute about 15.02 per cent (Arora, B.K. 2001).

## Soil Salinity/Alkalinity

The imprudent irrigation through canals leads to introduction of salts and alkalies in the soil. Due to over-irrigation, the water table rises along with their soluble salts. In Haryana alkaline soils occurs in the districts of Kurukshetra, Karnal and Jind, while saline soils are met towards the end of the surface and subsurface flow directions particularly in areas of topographic depression with insufficient drainage in the districts of Sonipat, Rohtak, Jind, Hissar, Bhiwani and Gurgaon. About 5.78 per cent of the state area is affected by problem of soil-salinity and alkalinity (Sachdev, et al., 1995). The characteristic of high reflection on satellite image together with low sustenance of crops-made salt affected areas mapped accurately by using large scale panchromatic (PAN) and multispectral (LISS-III) images.

## Drought

Droughts result due to deficient rainfall (less than 75% of normal rainfall) causing an acute scarcity of moisture in the soil. The deficiency of moisture leads to loss of soil productivity, deprivation of vegetation and initiation of desertification. Drought is the perennial feature in some parts of the Haryana state. Satellite images provide quick and accurate information of water resources, land resources and fodder resources. The thematic information in conjunction with attribute information on socio-economic and demography of the affected region can help in preparation of drought-relief plans. The areas affected by drought generally encompass a large extent of the state, which could be very well mapped for resource availability using satellite image on 1:250,000 scale for effective management. WiFS images of Kharif season (Oct. 1998) indicated that areas prone to wind-borne soil erosion and sand spread are most vulnerable to occurrence of drought. WiFS images clearly exhibit that western extremities of the state in the districts of Bhiwani, Sirsa, Mahendragarh, Hisar and Rewari were affected by sand spread. Formation of sand dunes and sand spread cause acute moisture stress on agricultural lands producing drought like situation in these areas every year.

## Water-borne Erosion and Land Degradation

Satellite images of LISS-III on 1:50, 000 scale indicated absence of prominent vegetal cover in the steeply sloping Siwalik ranges and undulating piedmont plains of Panchkula and Yamunanagar districts of Haryana. These barren hills and piedmonts are highly susceptible to severe soil erosion due to surface water through process of rainsplash erosion, sheet erosion and gully erosion. Similarly land degradation due to landslides during heavy rains constitutes the most common hazard in the Siwalik ranges. Road network in the higher hilly tracts of Panchkula district sustains heavy damages every year due to the incidence of landslides. A number of villages in the Panchkula district are located on the old landslide zones. This could pose danger to human lives existing in these potential slide prone areas during events of earthquakes or heavy rainfall. Preparation of comprehensive soil erosion and landslide hazard zonation maps is of prime need in these hilly regions. This requires information on various terrain parameters i.e., geology, slope, relief, landuse, drainage, road, settlement, hydrology, rainfall and seiesmicity. The remote sensing based information on said parameters viewed in 3D perspectives under GIS environment could provide evocative measures to combat the ill effects of such natural degradation phenomenon and resultant hazards.

## Floods

Among the major disasters that occur in Haryana state, river floods are most hazardous as it causes heavy damage to agricultural lands. The cause of floods is primarily attributed to peculiarities of rainfall in the state. About 3.31 per cent of the state area suffers from seasonal flooding during monsoon (Sachdev, et al., 1995). The geomorphic setting of the state comprising of Himalayan topographic high in the north and Aravalli mountains in the south resulted in the development of a saucer type of topographic depression in the central part of the state. The floods occur mainly due to heavy run-off in the mountainous terrain and overflow in the major rivers like Yamuna and Ghaggar in the planer areas during monsoon season. WiFS satellite images of Haryana exhibited that active and old flood plain areas around river Yamuna in parts of Karnal, Panipat, Sonipat and Faridabad are dominantly flood prone. Similarly low lying areas in the vicinity of river Ghaggar in the districts of Fatehabad and Sirsa are worst affected by floods during peak rainfall in the catchment areas of Ghaggar and its tributaries. Multi-date radar data together with optical remote sensing data provide real time solution in mapping, monitoring and management of floods and related hazards during flood transgression and regression phase.

## Geological Implication of Seismogenic Activity in Haryana

Earthquakes are considered to be the most dangerous and destructive natural hazards. The impact of this phenomenon is sudden with little or no warning, making it very difficult to predict it or to make preparations to combat against damages it causes. To understand the cause of high seismogenic activity and earthquake impending zone in Haryana state adequate knowledge of geological set-up and seismotectonics is essential.

## Geological Set-up

Haryana state comprising a good assemblage of rocks (Pre-Cambrian to Quaternary) can be divided in three different geological domains:

1. Pre-Cambrian rocks of Aravalli mountains.
2. Tertiary and Pre-Tertiary rocks of Siwaliks and Lesser Himalayas.
3. Quaternary deposits of Indo-Gangetic Plains.

Pre-Cambrian rocks of Aravalli mountains belong to Delhi Super Group (2500 to 1600 million years). The rocks of the Delhi Super Group are confined to the southern part of the state. It consists of Alwar Group of rocks, which are overlain by rocks of Ajebgarh Group. The Alwar Group of rocks represented by thickly bedded quartzites is well developed in the NE-SW hill ranges of Gurgaon district. The Ajebgarh group of rocks are mostly developed in the form of hill ranges as well as parallel isolated hillocks in parts of Rewari and Mahendragarh and Gurgaon districts. The dominant rock types of the group include shale, slate, phyllite and limestone and marble.

Pre-Tertiary rocks of Lesser Himalayas are constituted by Subathu (shale) and Dagshai (clay and sandstone) formations, which are exposed in the northernmost part of Panchkula district. Tertiary rocks of Siwaliks are represented by Siwalik Group (Lower Siwalik, Middle Siwalik and Upper Siwalik). The tectonic contact of Siwaliks with the Indo Gangetic Plains is marked by Main Frontal Thrust whereas the contact of Siwaliks with the Subathu formation is delineated by Main Boundary Thrust.

The rocks of Upper Siwalik occurs in a 5 km wide zone all along the southern margin of the Siwalik belt in the Panchkula district and along the northern margin of Yamunanagar district. It is characterized by coarse boulder conglomerate and associated with alternating sequence of sandstones, grits, clays and conglomerates. Middle Siwaliks exposed in the northern parts of Yamunanagar district consist of alternating sequence of massive sandstone, red shales and clay. Lower Siwaliks constituted by alternating sequence of dark hard sandstone, clays and red-purple shales are exposed in the Siwalik range in Panchkula district. It exhibits tectonic contacts with rocks of Upper Siwlaik and Dagshai formation.

Quaternary deposits of Indo-Gangetic Plain occupy nearly 97 per cent area of the state. It is formed by the deposition of alluvial sediments brought by Himalayan rivers into a deep crustal trough between Himalayas in the north and Deccan plateau in the south. The deposits range in age from Pleistocene to Recent period. The thickness of the alluvium varies from a few metres to about 1,200 metres. The alluvial deposits consist of sand, silt, clays, kankar, gravel and boulders.

## Seismotectonics

It is well known that the largest number of epicentres is located in the Himalayan belt where the strongest seiesmicity is attributed to the movements along the Main Boundary Thrust (MBT) and Main Central Thrust (MCT) (Pandey, 2000). The MBT separates the Siwalik ranges from the Himalayan domain and serves as the intra-crustal boundary along which the Indian plate is presently sliding under the Himalayan plate at a rate of 4-5 cm per year. Simultaneously, the plates have been split up into blocks by transverse tear faults causing seiesmicity in the region (Valdiya, 1987).

Apart from Himalayan belt, Delhi-Hardwar Ridge (DHR) represent a seismically active tectonic feature of the North Indian plain. This is an NNE-SSW trending underground ridge, which is the direct extension of Aravalli ranges. This originates in Gujarat and pass through Mount Abu and Jaipur in Rajasthan and continues via Rewari, Sohna, Faridabad, Gaugaon and Delhi. Further north of Delhi this ridge gradually descends below the surface and passes through Sonipat, Karnal, Saharanpur, Ponta, Hardwar, Dehradun, Mussoorie, Uttarkashi, Harsil. It continues further north via Lanzou in China to meet the Pacific boundary at Okhotsk (Pal, 2000). It is 70-75 km in width and probably lies at a depth of 2-2.5 km below Dehradun city. Its eastern and western flanks are fault controlled which are being followed by Ganga and Yamuna rivers respectively that have cut across the Siwalik ranges to flow down to Indo-Gangetic Plains.

The Delhi-Hardwar ridge also acts as conduit of high seismic energy being generated in the Himalayan basement. The presence of this ridge in the basement of the Himalayan rocks results in its baking, fracturing, rupturing causing melting of the rocks. The development of magma and its upward movement is a continuos process and it produces tremendous amount of energy. This energy is temporarily stored in the open spaces in the Himalayan basement and released towards the earth surface along planes of structural weakness. The Main Boundary Thrust (MBT) and Main Central Thrust and with their subsidiary thrust and faults constitutes the main zones of crustal weakness in the Lesser Himalayan region which facilitates this movement of seismic energy.

## Seismogenic Activity

The state of Haryana falls in the Seismic zone IV/V rendering it vulnerable to earthquakes. The occurrence of earthquakes of moderately high intensity of M-6.5 (Uttarkashi Earthquake, Oct., 1991) and M-6.6 (Chamoli Earthquake, March 1999) in the last decade in the Lower Himalayan regions of Garhwal Himalayas has given warning signature of high seismogenic activities in certain parts of Haryana state. During Uttarkashi earthquake high intensity tremors were felt in alluvial plains surrounding DHR around Delhi and further south up to Alwar in Rajasthan. Similarly in Chamoli earthquake isoseist of intensity VI on Mercalli Scale (MSK) exist in Delhi as well as in Garhwal region (Shankar and Narula, 1998). The higher intensity zone around Delhi-Faridabad-Gurgaon region can be attributed to the existence of ridge on the surface along which the seismic waves traveled from epicenter (NNE, 11-12 km depth) to ground level (SSW).

Prior to Chamoli earthquake several foreshocks of this quake were recorded and felt in the region of Delhi-Hardwar Ridge within a month preceding the main event of Chamoli earthquake on March 29, 1999. A number of these foreshocks were recorded in the alluvial plains of Haryana (Sonepat), Uttar Pradesh (Bagpat) and Delhi (Table 1). It was inferred that the occurrence of foreshocks was mainly confined to the Yamuna flood plain regions in Haryana and Uttar Pradesh. This signifies the high seismogenic activity in the basement structure in the Yamuna flood plain areas in Haryana. This is further corroborated by historical seismic events in the region including 1720 earthquake (Int. VIII) near Delhi and 1803 earthquake (Int. IX) near Mathura. Relatively recent events of magnitude 5 and above are located near Delhi (1960, mag. 5.5), Bulandshehar (1956, mag. 6.0) and Moradabad (1966, mag. 6.0). A cluster of micro events located

**Plate 3**

*Exposures of DHR in Delhi-Faridabad-Gurgaon Regions*

**Plate 4**

*Tear Fault along Yamuna River*

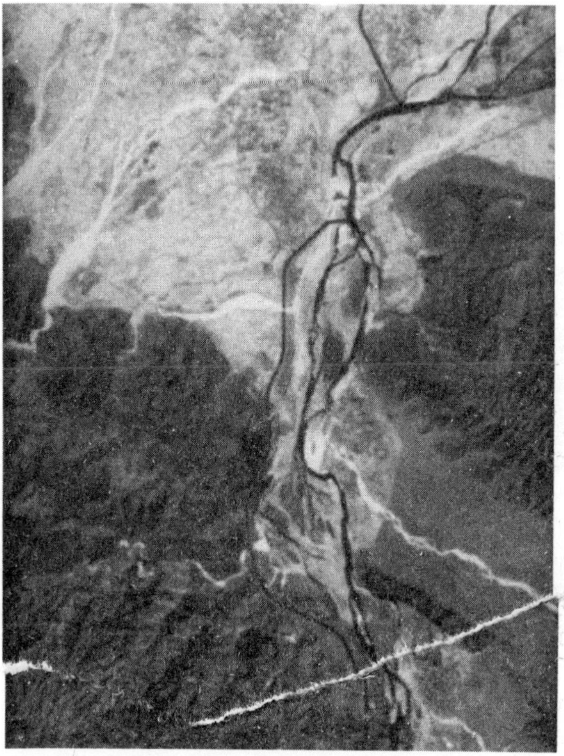

near Rohtak, Sonipat and Gurgaon indicated that several lineaments might be active in the region of DHR (Kamble and Choudhary, 1979). It is to remark that high seismic activity in the Delhi-Faridabad-Gurgaon regions need due consideration as the seismically active Delhi-Hardwar ridge crop out in these areas (Plate 3) making them prone to high earthquake hazards. Seismogenic activity in the northern parts of Haryana state is mainly confined to Main Boundary Thrust (MBT) which passes in the vicinity of Kalka and Morni areas. The tear faults along the rivers Ghaggar and Yamuna (Plate 4), which cut across the Siwalik ranges, represent the youngest faults in Himalayan orogeny.

Neotecnonic activities in the vicinity of these MBT and tear faults need seismotectonic monitoring to evaluate the possible mechanism of neotectonic movements in these areas. In can be presumed that highly populated cities where density of infra-structure is high are at greater risk during earthquake disaster. In Haryana population density is high in the urban townships of Faridabad, Gurgaon and Panchkula, which lie in the high seismic hazard zones. The Faridabad, Gurgaon townships are located in the region of Delhi-Hardwar ridge, which is seismically the most active tectonic feature in the North Indian plains, whereas the Panchkula city is located in the vicinity of active Main Boundary Thrust where neotectonic movements and seiesmicity is commonly manifest.

**Table 1**

*Foreshocks before Chamoli Earthquake (Modified after, Pal 2000)*

| *Date* | *Time (hrs)* | *Mag. on Richter Scale* | *Epicentral Area* | *Felt Area* |
|---|---|---|---|---|
| 18.2.99 | 10.35 | 2.6 | Sonepat | Sonepat-Panipat |
| 20.2.99 | 05.12 | 2.5 | Dharchula-Pithoragarh | Indo-Nepal Border |
| 24.2.99 | 03.17 | 2.2 | Sonepat | Haryana-Rajesthan Border |
| 28.2.99 | 14.28 | 1.7 | Delhi | Not felt |
| 10.3.99 | 07.08 | 4.0 | Kutch | Kutch-Saurastra |
| 16.3.99 | 04.47 | 3.2 | North of Bagpat | Kandhala-Budhana |
| 17.3.99 | 03.57 | 3.1 | East of Meerut | Mawana-Kithore |
| 22.3.99 | 15.26 | 3.4 | Sonepat | Sonepat-Panipat |
| 23.3.99 | 13.00 | 3.0 | East of Meerut | Mawana-Kithore |
| 29.3.99 | 0035 | 6.8 | Mandal -Chamoli | Uttaranchal and H.P. Yamuna plains of Haryana, U.P., Delhi-Faridabad-Gurgaon, Alwar-Jaipur |

## Disaster Mangament Plan

There is no end of disasters. They will continue with same vigour and imparting destruction of similar intensities. To contain them an efficient disaster management plan need to be evolved having a backbone of comprehensive, common and uniform database, real time monitoring and mapping tools, on line modelling system and communication network. Remote Sensing, GIS and Global Positioning System (GPS) are the most efficient tools in making it possible.

The disaster management plan should include three stages, i.e., pre-disaster, emergency and post-disaster. A long-term planning on issues of preparedness and mitigation is essential during pre-disaster stage. Preparing hazard zonation maps, land-use planning and preparedness plan for the area, application and development of new modern forecasting and warning systems, creating awareness through education and communication are the key items, which need due consideration during preparedness stage. To mitigate the disaster event, the area needs construction of disaster resistant houses and re-fortification of the existing weak and disaster prone houses, shifting of the people to safer areas and making people aware of Do's and Don'ts during the disaster. To make the operation of emergency more effective and successful, training of the local people for handling such situation is very important. The rehabilitation constitutes the most important activity in post-disaster stage and it should focus on overall socio-economic development of the region.

## Conclusions

Remote Sensing and GIS technologies provide a suitable platform to map the spatial extent of natural hazards and analyse the processes operative on land leading to such hazards. It provides quick and accurate estimates of the extent of damage caused during an event of disaster. Natural hazards represent the geodynamic phenomenon, their extent and intensity changes with time, hence only satellite data can depict the near real time picture of such phenomenon. The satellite based information coupled with thematic-attribute and socio-economic data available at various departments can be used in deciphering the extent and gravity of problem induced due to various natural hazards. Broadly speaking the northern and southern parts of the state are prone to seismic hazards, whereas the eastern and western extremities are vulnerable to floods and drought respectively. To safeguard the human life from the most disastrous and unpredictable event of earthquake, monitoring of neotectonic and seismic activity in the region of Main Boundary Thrust, Yamuna and Ghaggar tear faults and along Delhi-Hardwar Ridge is the need of the hour. Disaster management plan should be prepared in advance to combat the ill effects of each disaster with respect to socio-economic and agro-climatic conditions in the disaster prone regions.

*Acknowledgements*

The author is grateful to Head, Department of Remote Sensing, Birla Institute of Technology, Mesra for kind permission to present this study during the National Seminar-Geoinormatics and publication in the proceeding volume.

## References

Arora, B.K. (2001), Study Report on Perspective Land-use Plan for Haryana, National Productivity Council, Chandigarh.

Kamble, V.P. and Choudhary, H.M. (1979), Recent seismic activity in Delhi and neighbourhood. *Mausam,* V. 30, pp. 305-312.

Pal, D., (2000), Extension of Aravalli basement below Garhwal Himalaya and its geological control over the occurrence of natural hazards in Uttaranchal state. Proc. National Seminar on Geodynamics and Env. Management of Himalaya, H.N.B. Garhwal University, Srinagar.

Pandey, A.C. (2000), Geology and Morphotectonics of a part of Lesser Garhwal Himalaya, Tehri-Uttarkashi district, Using Remote Sensing and GIS Techniques. Unpublished Ph.D thesis, University of Delhi, Delhi.

Rawat, R.B.S. and Satendra (2000), Role of forestry in disaster management. Proc. National Seminar on Geodynamics and Env. Management of Himalaya., H.N.B. Garhwal University, Srinagar.

Sachdev, C.B., Lal, T., Rana, K.P.S., and Sehgal J. (1995), Soils of Haryana, *Soils of India,* series 3, NBSS Publ. 44.

Shankar, R. and Narula, P.L., (1998), Chamoli Earthquake of 29th March 1999: a prilimnary appraisal. Ind. Min., 52 (1and2), 141-150.

Valdiya, K.S. (1987), *Environmental Geology, Indian Context,* Tata McGraw-Hill Publishing Company Limited, New Delhi.

# 11

# Study on Environmental Degradation Through High Resolution Imagery in Eastern Part of Guwahati City, Assam

A. Bora, A.D. Patgiri and R.N. Sharma

Environment has been defined as "the sum of total of all conditions and influences that affect the development and life of organism". Historically we have been serious minded about environment and the basic philosophy has been, one of harmony with nature (Khoshoo, 1986). But last few decades, our natural capital in air, water, land and biodiversity is continually encroached upon, which has created a serious threat to the entire ecosystem. In cities this problem is more prominent as the concentration of human population is high and thus the usages of natural resources are at the maximum.

Among the natural resources, land is the most important component of life support system. It is a dynamic and complex combination of geology, topography, hydrology, soil, flora and fauna, which influences every sphere of human activities (Ram, 2003). But this resource has been overused and even abused over the centuries. This problem is more acute in the premier city of Guwahati, the gateway of the North-East India. This beautiful city is located on the bank of mighty Brahmaputra and surrounded by picturesque undulating terrain, dotted with small hillocks. At present the main problem has arisen due to the fact that the rapid urban growth is getting concentrated on these hillocks which otherwise should have been left totally free for the nature to nurture independently. However, today the scene is totally different and is effecting the environment to a very great extent causing ecological imbalances – disturbed due to removal of vegetation and increase in concrete cover (Figure 1). Removal of vegetal cover coupled with indiscriminate cutting of hill slopes causing ecological imbalances, accelerates soil erosion which creates havoc to the denizens of Guwahati by resulting flash flood due to blocking drainages by the wash over soil materials, hindering free flow of rainwater (Figure 2). Another resultant out-put of the flash flood is – health hazard during the dry days when the silt and soil materials gets dried and moving vehicles raise dust on the roads – causing pollution (Figure 3). Moreover, this population is unaware or keeping themselves in the dark

**Figure 1**

*Removal of Vegetal Cover for Habitation*

**Figure 2**

*Flash Flood Leading to Waterlogging in a City Road*

**Figure 3**

*Siltation on Roads, Leading to Dust Pollution*

knowingly the extensive damage they are causing leading to frequent landslides in these hill slope areas, which has taken the toll of valuable life and property in recent past.

Remote sensing from space has already established its capabilities of urban mapping, monitoring, targeting and management and to achieve optimization in their utilization and conservation (Joseph, 2003). With the successful launching of IKONOS satellite now it is possible to map urban land use/ land cover details, utilities and infrastructure mapping (better than 1:10, 000 scale) (Raju, et al., 2001). In this study an attempt has been made to demarcate the areas affected by one of the important environmental degradation factor, i.e., soil erosion and its relation to the unplanned urbanization in the hilly slopes from a small area of Guwahati city, as a pilot study to demonstrate the applicability of high resolution IKONOS image for such studies.

## Study Area

The study area covers eastern part of Guwahati city in Kamrup district, namely east Jyotinagar and its adjacent areas. Geographically it covers an area of about 1 sq. km and lies between north latitudes 26°11′26″ and 26°11′57″ and east longitudes 91°47′16″ and 91°47′51″. The location map of the study area is given in the Figure 4.

## Data and Methodology

IKONOS 4m Multispectral and 1m PAN merged orthogeo bundled data (January, 2003) was used for this study. Survey of India topographic sheet no 78 N/16 was used for digitization of contours required for generation of DEM of the study area.

At first single frame orthocorrection (Beaudoin, et al., 2001) of all the data was performed with Rational Polynomial Coefficient (RPC) files and the merged product of the same was genetrated. Then signatures were created by visual interpretation. Supervised classification is then performed by using these signatures and land use/land cover map was generated (Figure 5). During the field investigation, the shadow areas were

**Figure 4**
*Location Map of the Study Area*

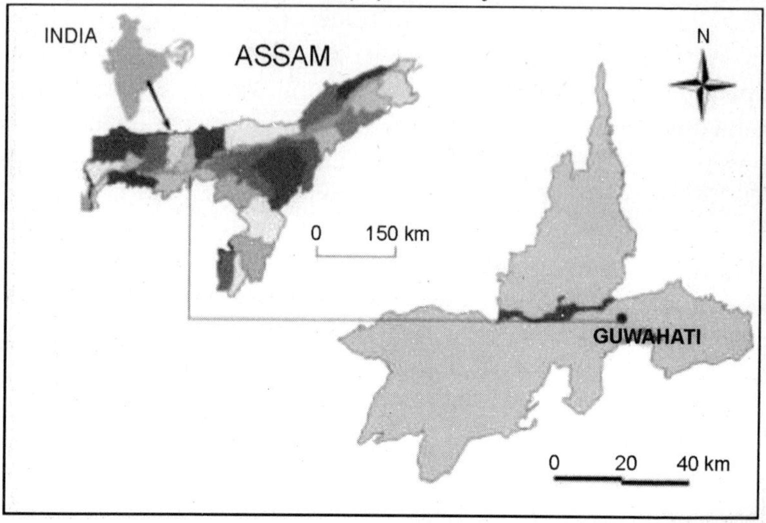


marked carefully and necessary correction of the data was done after supervised classification. The house footings were extracted by on-screen digitization of entire study area (Figure 6). Then digitization of contours was done, where height values were incorporated for DEM generation (Figure 7). After the above steps the satellite image was draped on the 3D surface to have an idea on the undulations of the different localities.

Slope gradient and length of slope significantly influence soil erosion by surface water movement. The slope of the study area was derived from the digital elevation model (DEM) generated from the contours (Figure 8). The specification suggested by AISLUS (All India Soil and Landuse Survey) (Singh, et al., 1990) has been followed in categorizing the slope classes. The slope map was generated in Arc Info 3.5.2 and Arc View GIS 3.2a.

**Figure 5**

*Land Use/Land Cover Map of the Study Area*

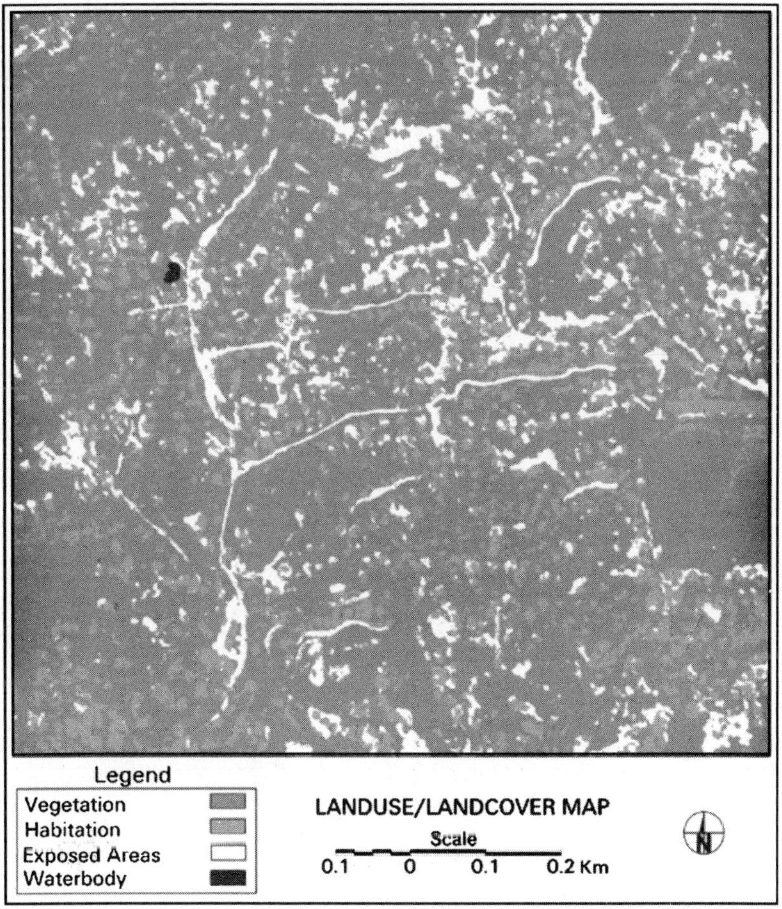


## Results and Discussion

The land use/land cover (Figure 5) classes are divided into four classes, namely, vegetation, habitation, exposed area and water bodies. The statistical results of this classification are given in Table 1.

**Figure 6**

*Settlement Map of the Study Area*

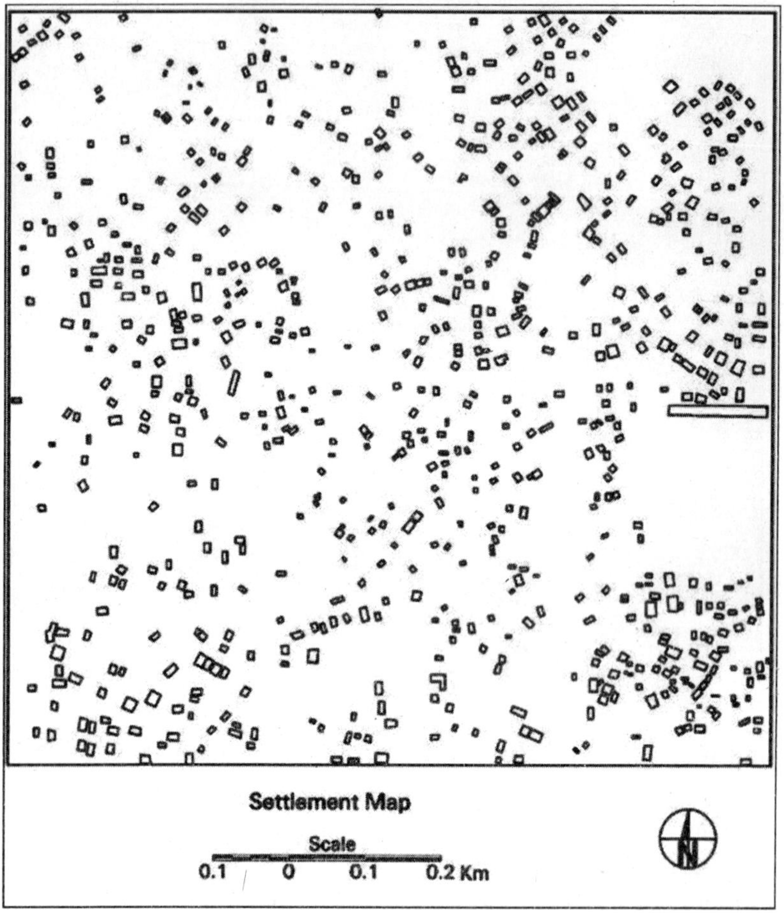


**Table 1**

*Land Use/Land Cover Classification of the Study Area*

| *Class Name* | *Area in sq. km* | *Area in percentage* |
|---|---|---|
| Vegetation | 0.662 | 66.2% |
| Habitation | 0.161 | 16.1% |
| Exposed area | 0.193 | 19.3% |
| Water body | 0.004 | 0.4% |

**Figure 7**

*Digital Elevation Model (DEM) of the Study Area*

The exposed area of 0.193 sq. km of the area under study is primarily responsible for the soil erosion. However, 66 per cent of the area is still covered with green vegetation cover. But the field study shows that the vegetative area is less wooded due to climatic factors, non-existence of good soil pockets and human interference. Haphazard growth of the urban areas can be clearly understood from the Figure 3. People used to build houses mainly in the lower reaches of the hill slopes; moreover, from field visit it was observed that most of the exposed areas (except earthen roads) are confined in and around the existing houses. In most cases, inhabitants are found to be completely unaware of the soil erosion problem and also ignorant to the danger of landslide and are therefore, not taking any measures for prevention of the soil erosion. So these areas are the main contributors to the soil erosion. Closer observation of the DEM will indicate that the locations of the exposed areas are situated in the

**Figure 8**
*Slope Map of the Study Area*

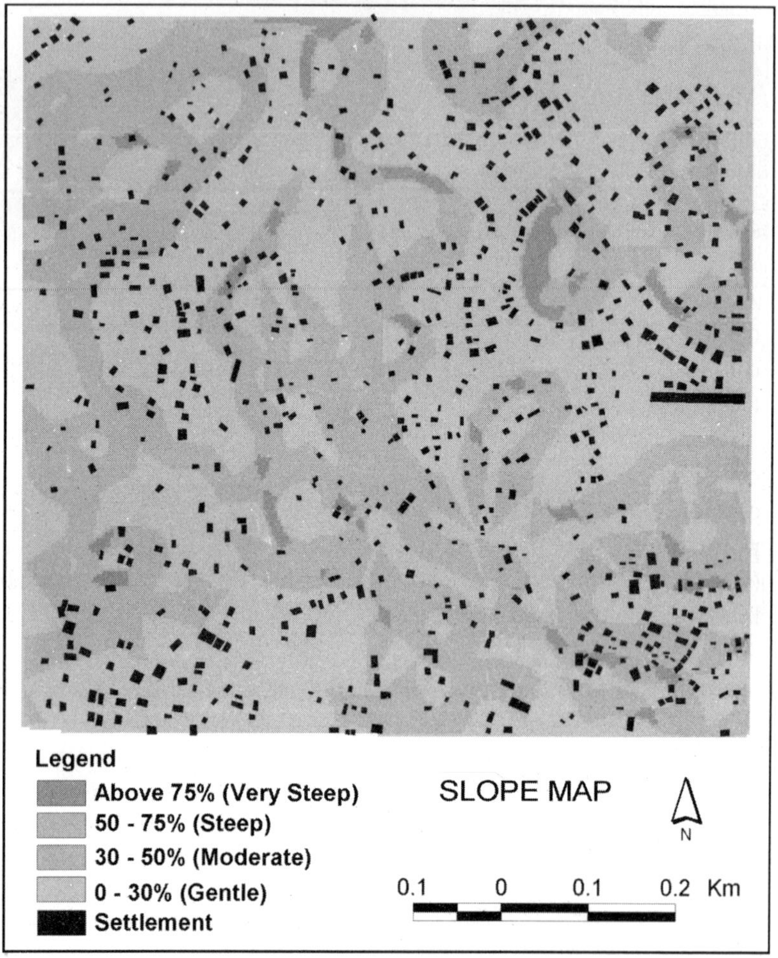


lower reaches of the hill slopes, only few are seen to have occupied the upper reaches of the hills.

The slope map (Figure 8) indicates various slope classes and relative position of the houses in these slope classes. Table 2 gives the slope classes with settlement percentage:

**Table 2**
*Slope Classification of the Study Area*

| *Slope Class* | *Area in sq. km* | *Area in percentage* | *Settlement percentage* |
|---|---|---|---|
| Very steep | 0.03 | 2.97% | 0.89% |
| Steep | 0.15 | 14.85% | 9.27% |
| Moderate | 0.29 | 28.72% | 26.56% |
| Gentle | 0.54 | 53.46% | 63.28% |

Majority of the houses are seen to be located in gentle slope class and these areas are generally found to be stable. Moreover, due to the gentle slope the amount of soil erosion are comparatively less than the higher classes. But during the field visit it was observed that majority of the exposed areas falls within 0-25 per cent and 25-30 per cent slope class and their contribution to soil erosion is also significant due to human interference, which is a point of major concern. Next to this, the moderate slope class (30-50%) with 28 per cent settlement is comparatively in landslide hazard prone, besides being prone to soil erosion. About 10 per cent of buildings are situated in steep and very steep slope classes which are very prone to landslide, as stability of the slope has already been disturbed by human interference.

## Conclusion

The prime objective of the study is to show the application of high resolution satellite images for mapping of areas affected by soil erosion which are being degraded due to haphazard settlements and also to suggest some remedial measures to curtail this environmental hazard. However, to manage the activities of the people to conserve habitats, resources and at the same time to promote development for urbanization are two contrasting aspects of modernization, which are required to be very judiciously managed to achieve the goal. But to ensure sustainability, there is also a need to integrate environmental concerns with economic, social and developmental activities. So, proper planning is the utmost requirement to make sure of proper development as well as to conserve nature with less degradation as far as possible. The planning process shall have to be in an integrated fashion taking note of the requirements in relation to escalating population. At the same time the law enforcement agencies should strictly amend the present laws to conserve the existing resources.

Based on the present study following suggestions are recommended:

(1) Aforestation is the primary need, as trees will restrain soil erosion.

(2) Examine the soil erosion prone areas; the unauthorized inhabitants should be removed/resettled suitably by the District Authority and see that the settlement in the slopes/hills need to be stopped totally.

(3) Educate people who are staying in this area, about danger of landslide and give necessary suggestions in hazard prone areas on steps for stabilizing the slope.

(4) In steep slope areas impose ban on building of houses and if necessary evacuate people from such areas.

(5) Awareness drive is very important for the general public—that the hills natural slopes should never be disturbed. In case if they are at all, planning to reside in these slopes the cutting of the hill slopes must be avoided and

**Figure 9**

*Cantilevers' Type of Projection on Hill Slopes*

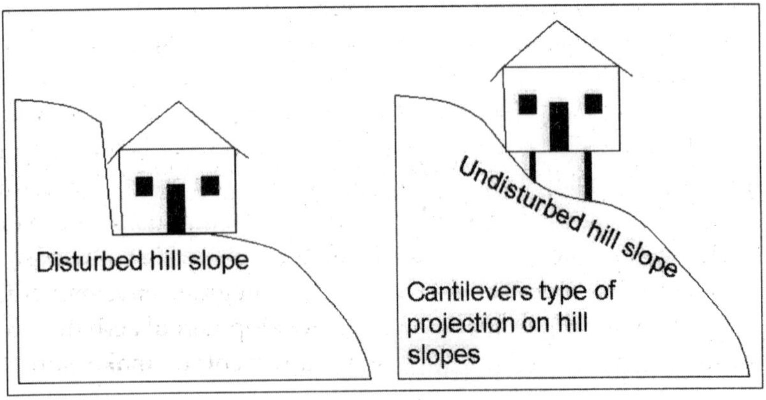

cantilevers' type of projections need to be encouraged (Figure 9).

## References

Beaudoin, C., Booth-Lamirand, J., Curry, K., Pongonis, J. and Pouncey, R. (2001) IKONOS Sensor Model Support, ERDAS Imagine Tour Guides, ERDAS Inc., Gorgia, Atlanta: 7-14.

Gupta, K., Pathan, S.K. and Kumar, G.A. (2003). Use of High-Resolution Satellite Data for Micro Level Human Resource Enumeration. ISRS Annual Convention 2003, 9-12, December, Thiruvananthapuram.

Joseph, G. (2003) *Fundamentals of Remote Sensing*, Universities Press (India) Private Limited, Hyderabad: 433.

Khoshoo, T.N. (1986) Environmental Priorities in India and Sustainable Development, Indian Science Congress Association, New Delhi: 224.

Ram, G. (2003) State of the Environment Report and Action Plan-2003. Department of Forest, Ecology and Environment. Government of Karnataka: 280.

Raju, P.V.S.P.P., Ghosh, S., Chandrakanth, R. and Ramachandran. (2001) Some Perspectives of Information Extraction Techniques Using High Resolution Satellite Imagery. National Symposium on Advances in Remote Sensing Technology with Special Emphasis on High Resolution Imagery, 11-13, December, ISRS, Ahmedabad.

Singh, G., Venkataramanan, C., Sastry, G. and Joshi, B.P. (1990) *Manual of Soil and Water Conservation Practices*. Oxford and IBH Publishing Co. Pvt. Ltd.: 325.

# 12

# Role of Geoinformatics in Hazard Management for Better Governance

Kiran Kumar Vemula, Deva Pratap, K. Vindhya and Kiran Kumar P.

In the present day environment, the advanced techniques like GPS and Remote Sensing, if integrated with GIS, can provide valuable information on earth surface features, for better governance.

A hazard is a condition of the environment, which can exert an adverse influence on human life, property or activity. A disaster is a serious, damaging effect on human life, property or activity, which results from the impact of a hazard that has exceeded its critical levels. While the hazards cannot be eliminated totally, the limit of tolerance of every society to their effects can be increased and hence the potential for the disaster reduced.

The impact of disasters on modern day society is rapidly increasing. There is an inverse relationship between the level of development and the loss of human lives in the case of a disaster. However, natural disasters are inevitable, but their impact can be reduced through a proper disaster management, which includes disaster prevention (hazard and risk assessment, land use planning and legislation, building codes), disaster preparedness (forecasts/warning/prediction) and rapid and adequate disaster relief.

Various suggestions have been offered to explain this apparent anomaly. Some of them are:

— Poor decision criteria at the outset
— Traditional dependence upon engineering solutions
— Limited attention devoted to non-structural measures

The advanced techniques such as remote sensing, GPS, when integrated with GIS, had proved their usefulness in disaster management. The techniques such as satellite remote sensing, GIS, GPS, electronic total stations, digital photogrammetry, laser rangers, etc., combine to form a new discipline known as Geoinformatics. Geoinformatics is a term that appears to have been independently coined by several groups around the world to describe a variety of efforts to promote collaboration between computer science and the geosciences to solve complex scientific questions.

## The Role of Geoinformatics in Natural Disaster Management

Nowadays we have access to information gathering and organizing technologies like Remote Sensing and GIS, which have proven their usefulness in disaster management. So, in more ways than one, Geoinformatics is ideally suitable for disaster management.

— First of all it provides a data base from which the evidence left behind by disasters that have occurred before can be interpreted and combined with other information to derive hazard maps, indicating which areas are potentially dangerous. Satellite images give a synoptic overview and provide very useful geomorphological information, for a wide range of scale, from entire continents to details of a few metres.

— Secondly, many types of disasters, such as floods, drought, cyclones, volcanic eruptions, etc., will have certain precursors. Satellites can be used to detect the early stages of these events as anomalies in a time series. Images are available at regular intervals, and can be regular short-time intervals, and can be used for the prediction of both rapid and slow disasters.

— Then, when a disaster occurs, the speed of information collection from satellites and the possibility of information dissemination with a matching swiftness makes it possible to monitor the occurrence of the disaster. Many disasters may affect large areas and no other tool than Remote Sensing would provide a matching spatial coverage. Remote Sensing also allows us to monitor the event during the time of occurrence while the forces are in full swing.
— The impact and departure of the disaster event leaves behind an area of immense devastation. Remote Sensing can assist in damage assessment and aftermath monitoring, providing a quantitative base of relief operations.
— Satellite data can be used to map the new situation and update the databases used for the reconstruction of an area, and to prevent that such disaster occurs again.

The use of Remote Sensing has become an integrated, well-developed and successful tool in disaster management. Also India has its own earth observation programme, and expertise is built-up in a wide variety of Remote Sensing applications and one of these fields is natural hazard management. In order to effectively utilize Remote Sensing for natural disaster reduction in India it is equally important to set up a network of organizations dealing with this problem, to develop techniques specifically adapted to the regional situation, and to train many persons to utilize these techniques at various institutional levels.

## Remote Sensing Data Required

The Remote Sensing systems that can be used for disaster management are – optical, thermal and radar. Their use for hazard monitoring, hazard assessment and damage assessment will be the prime focus. Some of the Remote Sensing data that can be used for analysing hazards are as follows:

- Use of meteorological satellites for cyclone disasters
- Use of AVHRR for drought monitoring
- Use of optical multi-spectral images for the multi-temporal analysis of land use changes
- Use of radar images in flood hazard assessment
- Use of high-resolution imagery for landslide mapping
- Use of thermal imagery for volcanic hazard, coal fires and forest fire assessment
- Use of different types of aerial photographs for mapping hazard features and for damage assessment
- Combination of different images for image fusion

## Methodology

Landslides are one of the most prominent natural hazards affecting human life and property in many parts of the world, particularly in these mountainous regions. In the recent past various methods and techniques have been proposed to analyse these factors and produce maps portraying the probability of occurrences of similar phenomena in future. Broadly there are various methods, which are as follows:

1. Landslide Inventory
2. Heuristic Approach
    - Geomorphic Analysis
    - Qualitative Map combination
3. Deterministic Approach
4. Statistical Approach
    - Bi-variate Statistical Analysis
    - Multi-variate Statistical Analysis

Several statistical methods have been termed, such as the Landslide Susceptibility Method, Information Value Method, and Weight of Evidence Modelling Method. Among the above statistical method was chosen for the present study. The statistical method is a data driven method which is an indirect method calculated from trained data. Among different statistical methods information value method was chosen for the present study to carry out landslide hazard zonation. The data integration has been carried out in raster GIS environment in Arc/Info.

## Information Value Method

The conditional analysis based on the information value method given here under with seven input variables (Aruna, 2002). Information values were calculated by considering one landslide map and one parameter map. Some calculation steps are given below.

One landslide map and one parameter map were crossed and the area of landslide per class (Nslclass) and total area of each class (Nclass) were calculated using various steps suing Arc/Info 7.2.1.

Information value can be calculated using the following expression as

Information value = In Wi

In Wi = ln (Densclass/Densmap)

$$= \ln \left( \frac{\text{Nsclass} / \text{Nclass}}{\text{Nslide} / \text{Nmap}} \right)$$

Where,

Wi = the weight given to a certain parameter class (e.g., a rock type)

Densclass = the landslide density within the parameter class

Densmap = the landslide density within the entire map

Nslclass = Area, which contain landslides, in a certain parameter class

Nclass = total area of a certain parameter class

Nslide = total slided area (active) in the landslide map

Nmap = total area of the map

## Data Integration Using Information Value

The simple data integıation technique is applied, as it requires less statistical assumptions and is easier interpreting the results. Thematic values were replaced by the respective information values and then the maps are summed up to give a combined information value map. The whole range of values, in the resultant information value were classified into six classes as per the histogram and the histogram limits are given as:

| *Hazard classes* | *Lower limit* | *Higher limit* |
|---|---|---|
| Very low | –11.0 | –5.5 |
| Low | –5.5 | 0.0 |
| Low/moderate | 0.0 | 2.0 |
| Moderate/high | 2.0 | 4.0 |
| High | 4.0 | 6.0 |
| Very high | 6.0 | 10.0 |

*Methodology of Information Value Method*

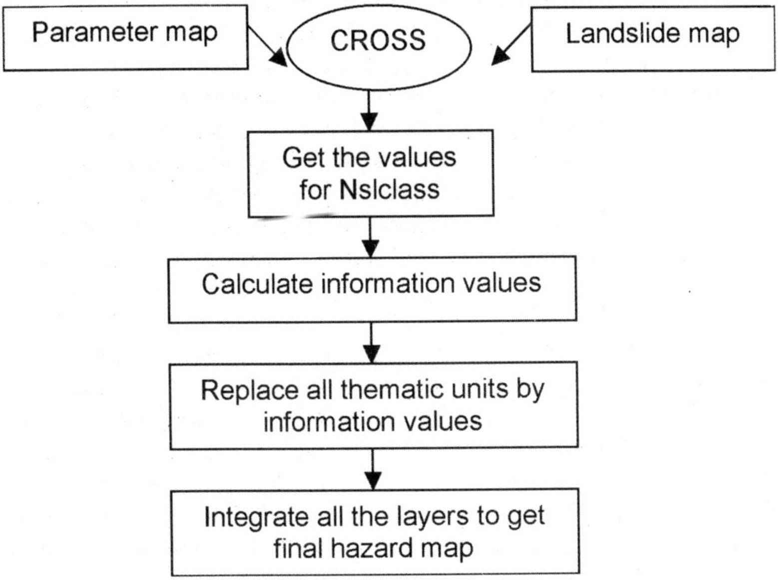

To test the reliability and predictive power of the model, LHZ map is again crossed with the landslide map and various statistical measures were obtained, which are correlated with the active landslides. Flow chart of the methodology is given above.

## Suggestions

Although information value method produces a LHZ map, that can predict landslides in high hazard zone, in some situations it leads to incorrect prediction of hazard areas. However, the main advantage of the model is that it is computationally simple and can be implemented very easily in raster-based GIS like ILWIS. It provides values that can be easily interpreted in terms of average density of landslides in the region.

The reliability of statistical models and resultant hazard zonation map depends on the accuracy of the input data layers that are always affected by uncertainty and errors which cannot generally be evaluated and controlled.

The method can be improved by giving weightages to the individual parameters as per the order of their importance and integrating together to get final hazard map.

## Conclusions

Remote Sensing and GIS can be considered as powerful tools for representation and analysis of various spatial and non-spatial features associated with natural disasters. These not only facilitate timely data acquisition and co-registering, but also offer powerful computational tools that facilitate multi-map integration using statistical and deterministic modelling techniques. Many advance tools even exist to include a fourth dimension, i.e., the concept of time in the 3-D spatial database. This has tremendous potential in real time mapping and monitoring of pollutant movement. Analytical capability to simulate and emulating different natural phenomena and man-made situation much before these really occur is one of the most fascinating features that GIS offers to planners and scientists. Future computing environment will see real time integration of Remote Sensing and GIS, thus allowing effective monitoring of geo-environmental changes, which will facilitate the scientific and management decision-making. In this connection, the most crucial concept is data sharing and system integration. A GIS developed for one hazard mitigation programme should be available to other users, and it should not face the constraints of data convertibility and portability. Standardization should be maintained during database creation and methodology development, and every attempt should be made towards creating an open GIS for easy access and experimentation using internet. Then only all concerned for geological hazard mitigation can utilize the true benefit of GIS.

Remote Sensing and GIS analysis techniques have opened up new vistas of data collection. Various forms of Remote Sensing have been used for these activities. There is a tremendous potential for Remote Sensing for more efficient and economic monitoring of environmental hazards, and therefore, for the increased avoidance or mitigation of natural disaster. Use of Remote Sensing will undoubtedly increase as we improve our ability to interpret and understand the imagery. Once the database has been created in GIS using Remote Sensing and secondary information, it becomes easy to analyse various situations of hazard by generating scenarios.

Thus, the role of geoinformatics in disaster management can be summarized as below:

— In general, Remote Sensing, GIS and GPS provide database that can be interpreted and analysed to produce hazard maps, which have immense value in any kind of activities related to disasters.

— Secondly, many types of disasters have precursors, which can be detected at an early stage by orbiting satellites and can be used for the prediction of both rapid and slow disasters.

— When a disaster occurs, the aerospace media is the most authentic and quickest mode of information collection and can be utilized for monitoring during the event.

— Finally, Remote Sensing can also assist in damage assessment and aftermath monitoring, providing a quantitative base for relief operations.

## Select References

Akhouri, Pramod Krishna (2000) *Landslide Management in the Himalayas.*

Aruna, B. (2002) Multi-Criteria Decision Analysis for Landslide Hazard Zonation Using Information Value Method in Parts of H.P. Himalaya Using Remote Sensing and GIS (Unpublished Thesis).

Carrara, A. (1983) *Multivariate Models for Landslide Hazard Evaluation,* Vol. 15, No.3, pp. 403-427.

Carrara, A. (1988) *Landslide Hazard Mapping by Statistical Methods.*

Foster, H.D. (1980) *Disaster Planning: The Preservation of Life and Property.*

Lillesand, T.M., and Kiefer, R.W. (1994) *Remote Sensing and Image Interpretation,* John Wiley and Sons, Inc. New York, pp. 151-152.

Vitek, J.D. and Marsh, W.M. (1978) *Landslides Hazards Mapping for Local Land Use Planning.*

# 13

# Role of Geoinformatics in Micro-level Planning for Natural Resources and Environmental Management

G.P. Obi Reddy, A.K. Maji and K.S. Gajbhiye

Information on the nature, extent, and spatial distribution of various natural resources is a prerequisite for achieving the eco-friendly management of natural resources and environment. Lack of reliable information on natural resources has also contributed to the aggravation of geo-ecological problems. Therefore, for better management of natural resources and environment, there is a need to generate reliable geo-database to prepare action plans for sustainable development at micro-level. 'Geoinformatics' has emanated in the form of satellite remote sensing, Global Positioning System (GPS) and Geographical Information System (GIS). These technologies revolutionized the natural resources inventory, mapping and generation of databases on a regular basis for better planning, management, monitoring and implementing the land use plans more efficiently at different levels (Kasturirangan, et al., 1996). Recent advances in remote sensing technology provide data for detailed inventory and mapping of natural resources in large scale. High resolution satellite data are extensively used in landform and soil mapping, land degradation assessment, studies on land use/land cover changes, evaluation of water resources, mineral resources, disaster management and environmental impact assessment. Today, GPS has edge over the traditional surveying and mapping through triangulation, traverse and controls with high accuracy. These techniques are being employed increasingly for detailed topographic, soil, geological, engineering and environmental related surveys at different scales. The information generated both by conventional methods and modern techniques are being used to create database in GIS environment for performing integrated analysis to generate environment friendly and sustainable action plans at watershed level.

Watershed is a geo-hydrological unit area that drains to a common point and is considered as an appropriate physical unit for natural resources evaluation, planning and management at micro-level. Watershed management implies the rational utilization of land and water resources for optimum production with minimum hazard to natural environment (Bhat, 1989). The concept of watershed management involves accurate and timely monitoring of natural resources. This approach aims at proper utilization of land and water resources, protecting land against all forms of deterioration, and monitoring and maintaining soil health, on site conservation of water, proper management of local drainage and sediment reduction and increasing productivity from all land uses through suitable land use planning at micro-level. Since watershed management involves decision-making for efficient utilization of natural resources, a multidisciplinary approach is essential. This includes optimum production with minimum hazard to natural resources and assured economic benefits. It also helps in restoring the fragile geo-ecological balance at micro-level through adoption of various conservation measures. In the present paper an attempt has been made to show the utility of geoinformatics in some of the potential areas for micro-level planning.

## Objectives and Principles in Micro-Level Planning at Watershed Level

The main objectives in watershed development in micro-level planning are to promote the economic development of the local community, which directly or indirectly depend on the land and water resources of the watershed. These objectives can be achieved through:

- Optimum utilization of available resources that will mitigate adverse effects of drought and prevent ecological degradation.
- Bring more marginal and other wastelands under use to enhance the productivity of lands, thus enhancing employment generation and other economic development at grass-root level.
- Restoration of degraded ecosystems at the micro-level through sustained community participation.
- Utilizing the land and water resources according to their capabilities to support agriculture and allied activities.
- Maintaining adequate vegetative cover on the soil for controlling soil erosion.
- Conserving maximum possible rainwater through rainwater harvesting by contour bunding and water harvesting structures for future use.

## Digital Elevation Model (DEM)

Delineation of terrain parameters, such as drainage network and watershed boundaries, from space-borne remotely sensed data crucially depend on generation of an accurate DEM (Mahtab, et al., 2003). Capturing surface elevation information in a digital form suitable to input into a computer involves sampling points from a model. These samples of the terrain surface are usually called a digital elevation model (DEM). Other surface attributes such as gradient and aspect may be derived from the DEM and stored in a computer as a digital terrain model (DTM). DTMs are means of terrain representation in which elevations are recorded at specific horizontal positions. As an important component in the database of a GIS, DTMs have been applied in a wide variety of fields such as morphometric analysis, geomorphic analysis, terrain analysis, hazards studies, etc. The land types can be mapped from DTM by applying decision tree classification procedures based on pertinent geomorphometric parameters. The classification processes are both interactive and interpretative, requiring repeated visualization of the resulting classified maps and adjustment of classification parameters. Moreover, it is specific to physiographic provinces having similar topographic and geologic characteristics affecting geomorphic and hydrologic processes. However, DTMs may be represented in contours and profiles traingulated irregular networks and regular grids. Grid DTMs are the most common form because they allow other geomorphic variables to be derived from them more rapidly. The slope map of the area can be generated from the digital elevation model (DEM) using the standard inbuilt algorithms in GIS and it is very useful in watershed management.

## Drainage Morphometric Analysis

Quantitative analysis of drainage parameters is an important aspect in characterization of any hydrological basin or watershed (Strahler, 1964). The various morphometric parameters of a landscape like stream length, bifurcation ratio, drainage density, stream frequency, texture ratio and basin shapes can be calculated using various options in GIS (Reddy, et al., 2002b). These morphometric analyses of any given landscape will be useful in assessing and evaluation of soil parameters, erosion risk, soil and water conservation strategies groundwater potential, watershed characteristics and other environmental parameters. Evaluation of the morphometric parameters necessitates preparation of a drainage map, ordering of the various streams, measurement of the catchment area, length and perimeter of drainage channels, drainage density, drainage frequency, bifurcation ratio, texture ratio and circulatory ratio (Kumar, et al., 2000). Quantitative description of the basin morphometry requires measurement of linear and areal features, gradient of channel network and contributing ground slopes of the drainage basin. GIS techniques are being effectively used in recent times in determining the quantitative description of the basin geometry. It also provides a powerful mechanism not only to upgrade and monitor morphometric parameters but also permits the analysis of drainage information in association with other resources and environmental parameters.

## Landforms Mapping and Analysis

Reliable information on landform units of the area with respect to their nature, extent, spatial distribution, potential and limitations is very useful for evaluation and optimal utilization of natural resources on sustained basis. Detailed landforms mapping also helps in soil resources mapping, groundwater potential zones identification, landscape ecological planning, hazard mapping and other environmental applications and best done by applying remote sensing and GIS technologies. High-resolution satellite data provides economical and reliable source of information to prepare comprehensive and detail inventory of geomorphology in an area. Visual interpretation techniques can be followed in delineation of landform units based on the image characteristics of the satellite data in conjunction with drainage and contour information. The boundary delineation of landform units can be carried out based on the changes in topographic slopes, relief patterns, crest type, drainage texture and image characteristics of the area. Subsequently, detailed landform analysis can be carried out with the help of relief, morphometry and image characteristics.

## Soil-Landscape Analysis

The relationship between landform position and pedological processes determine the soil attributes in any given landscape (Reddy, et al., 1999). The development of soils in a toposequence often follows the way water moves through and over the landscape. Water movement is, in turn, controlled by the geometry of the land surface and underlying lithology. The geometry of the land surface, therefore, can be used as a first approximation for analysis and predicting the movement of water and related material over the landscape (Moore, et al., 1991). Many of the attributes of landscape have potential use for spatial analysis of the soil properties. The GIS-based multi-criteria overlay analysis of various landscape parameters like elevation, slope, aspect, morphometry, hydrological parameters, land use, land cover allows us to establish broad interrelationship between landscape and soil attributes in any given landscape. The analysis of landform-soil relationship in association with drainage and elevation properties in GIS environment can also be effectively used to understand spatial patterns of soil attributes in similar geological and climatic terrain conditions (Gessler, et al., 1995).

## Land Use/Land Cover Analysis

The remote sensing, GPS and GIS are the promising technologies in analysing land use/land cover changes and related environmental studies. The information on land use/land cover dynamics at watershed level is very important for monitoring and improving ecosystems, landscape, hydrology, human activities, habitats and ecological studies for sustainable planning (Rao, et al., 1996). To define conservation of forest ecosystem, understanding of the biophysical processes, their spatial inter-linkages and continuous monitoring of the human interventions are necessary. The integrated information derived from the remotely sensed data, inventories with the help of GPS and generation of spatial database in GIS environment assist in evolving appropriate action plans for initiating sustainable forest management monitoring and environmental protection.

## Hydro-Geomorphological Analysis

Hydro-geomorphological analysis is very important for watershed development as it offers scope to identify groundwater potential zones. These can be carried out through overlay analysis of drainage, landforms, lithology and lineament thematic parameters generated from analysis of remotely sensed data in conjunction with field survey data (Das, et al., 1997). The methodology involves generation of thematic maps on drainage pattern, geomorphology (landforms), lithology (rock types), lineaments and hydro-geomorphology from visual interpretation of satellite data. The drainage pattern maps can be prepared with the help of SoI topographical map on 1:50, 000 or large scale maps. Interpretation of the satellite imagery based on the photo elements like tone, texture, shape, drainage pattern, colour, differential erosional characteristics can be followed for preparation of geology, geomorphology and lineament maps. By integrating the drainage, geomorphology, geology and lineament thematic layers in GIS environment, the composite hydro-geomorphological map at watershed level can be prepared showing the qualitative groundwater potential zones (Reddy, et al., 2001b). However, the necessary field verification and validation of the results with available collateral data are essential for accurate assessment of the groundwater prospects at watershed level.

## Hydrological Models

The representation of hydrologic information is important not only for the reliability of the representation in the map but also for terrain and resources analysis, environmental management and hydrological studies. With the advancement of the computer technology, remote sensing and GIS have become efficient tools for the collection, processing and to integrate the spatial and non-spatial databases for the surface hydrological modelling. It is now possible to obtain data generated routinely and temporally. With these recent innovations, it is now feasible to deal with dynamic surface hydrological processes. In order to perform this modelling, a great deal of information must be known about the area, and thus can be incorporated into the simulation. These data include soil information (infiltration rates, hydraulic conductivity and storage capacities), surface characteristics (pervious, impervious, slope, roughness), geometry and dimensions of flow planes, routing lengths (overland, gutter and sewer) and the geometry and characteristics of routing segments.

## Quantitative Soil Loss Estimation

Checking soil erosion necessitates accurate and timely mapping, monitoring and prioritization of areas based on the their susceptibility to erosion. Soil erosion can be assessed at different levels using Universal Soil Loss Equation (USLE) (Wischmeier and Smith, 1978). The GIS-based techniques have been used extensively in quantification of USLE parameters to estimate soil erosion (Jain and Kothari, 2000). The factors, viz., rainfall erosivity (R), soil erodibility (K), slope length (LS), cover (C) and management (P) are the components of USLE and can be computed using rainfall data, series-wise soil information, slope generated from contours, land use/land cover interpreted from remotely sensed data and field data, respectively (Reddy, et al., 2004). This remotely sensed data and GIS-based integrated approach can also followed in identification of feasibility sites for insitu soil and water conservation to optimum utilization of available land and water resources with minimum hazard to geo-ecological balance at watershed level.

## Prioritization of Watersheds

Soil and water resources conservation is gaining immense importance in rain fed ecosystem. Soil loss is posing great threat in crop production. Supplementary irrigation through conserved water as well as increase of groundwater table plays a very dominant role in crop growth. Prioritization and evaluation of conservation strategies based on integrated analysis in GIS environment proved to be an effective tool (Reddy, et al., 2003). Geo-spatial database on drainage, slope, landforms, soils, land use/land cover and management practices is a prerequisite to prioritize and evaluate the conservation strategies for landscape management. Information on landforms and land use/land cover can be obtained through analysis of satellite data. Soil information on soil depth, texture maps can be generated through

**Figure 1**

*Priority Watersheds in Vena River Basin of Nagpur District*

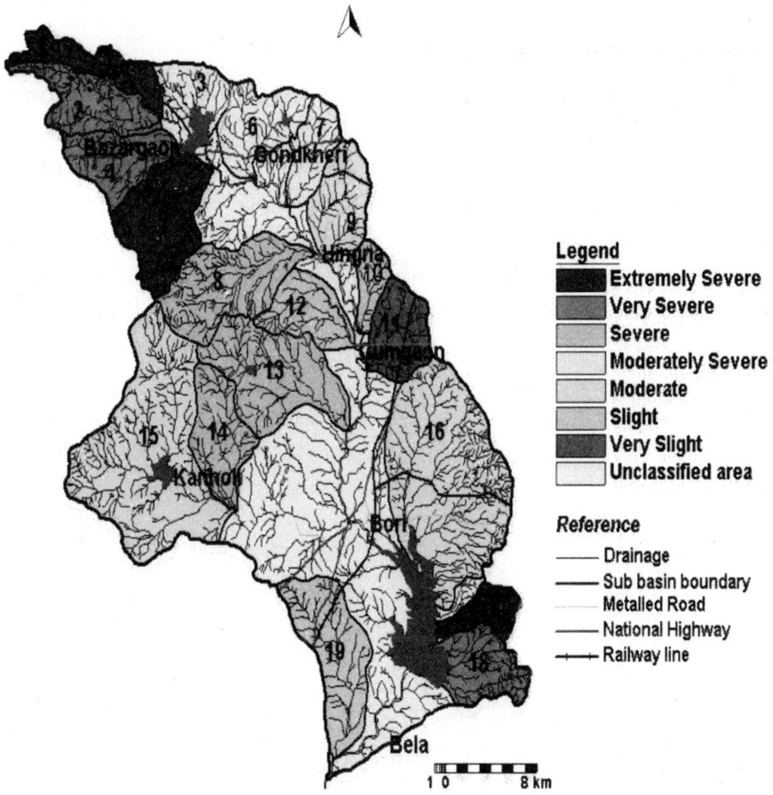


field surveys. Soil Available Waterholding Capacity (AWC) can be estimated using soil analysis data. The USLE parameters can be computed using the above generated information to estimate soil erosion through integrated analysis of USLE parameters in GIS. Potential soil erosion at watershed level can be estimated based on the integration of the R, LS and K-factors in GIS. The actual soil erosion can be estimated by integrating potential erosion with cover (C) and management practices (P) factors in GIS. The watershed can be classified into different erosion susceptibility classes, i.e., slight, moderate, moderately severe,

severe, very severe and extremely severe for prioritization. The depicted Figure 1 shows the priority watersheds.

## Landscape Ecological Zoning

Landscape ecological planning at watershed level can be carried out through terrain analysis in conjunction with high resolution remotely sensed and detailed field surveys. Geomorphological units can be considered as basic units for evaluation of landscape ecology of the watershed. The geo-spatial database on slope, geomorphology, soil depth, soil texture, soil AWC, land use/land cover and groundwater prospects zone layers can be generate to

**Figure 2**

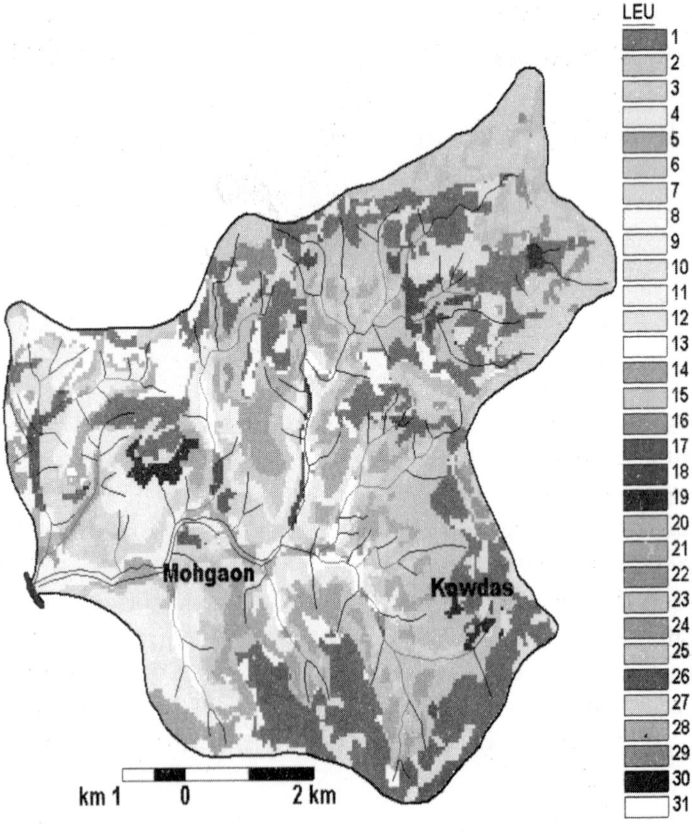

integrate in GIS to arrive distinct landscape ecological units at watershed level (Figure 2) (Reddy, et al., 2001a). Each landscape ecological unit refers to a natural geographic entity having distinctive properties of slope, geomorphology, soil depth, soil texture, soil AWC, land use/land cover and groundwater prospects. The landscape ecological stress zone map of the watershed can also be generated based on the analysis and reclassification of landscape ecological units. The units having minimum ecological impact in terms of slope, geomorphology, soil depth, soil texture, soil AWC and land use/land cover can be delineated under very low stress landscape ecological zone. The units which are having maximum ecological stress in the form of very high slopes, extremely shallow soils, coarse texture, low soil AWC, waste lands and very poor groundwater prospects can be delineated under very high stress landscape ecological zones. The delineated landscape ecological stress zones in the watershed can be considered for landscape ecological planning for better utilization of natural resources without harming the natural geo-ecosystem of the area.

## Geo-Spatial Analysis

GIS provides a very large range of analytical tools to analyse topology or spatial aspects and attributes of the geographical data (Burrough, 1986). GIS includes storing, restructuring, analysing and displaying resource data for providing management information or for the development of better understanding of environmental relationship. The geographical data is represented as spatial and non-spatial, or attribute data. The spatial data features geometrical orientation, shape, size and relative position on the earth surface; and the attribute data represents features like, soil, its depth, erosion, slope, nutrient content, etc., and can be integrated through polygon processing. The capabilities of GIS in natural resource management has been demonstrated by Nair, et al., (1996) and Maji, et al., (2001 and 2002) at different levels. Watershed based research is frequently concerned with the measurement of variables whose magnitude and spatial distribution vary over time. Two distinct characteristics of GIS facilities that analyse the spatial data are topographical structure and the incorporation of a database management system (DBMS) linked to unique spatial entities. Topological structure allows the examination of the spatial relationship between variables and facilitates processing of data using spatial criteria. The linking of a DBMS and spatial entities allows assignment and processing of attribute information. This allows us to display and interpret the spatial natural resources data and the results of modelling.

## Spatial Decision Support System (SDSS)

Decision Support Systems (DSS) are defined as computer based information systems designed to support decision makers interactively in thinking and taking decisions about relatively unstructured problems. Traditionally, DSS have three major components, a database, a model base and a user interface. An extension of the DSS concepts Spatial Decision Support Systems (SDSS), which are the integration of DSS and GIS was initiated by Densham and Goodchild (1988). A significant capability of the SDSS is the ability to use spatial analysis and the display tools with the sectorial models that would form the model base of SDSSs. The modelling capability of GIS allows the user of the SDSS to stimulate changes in objects and attributes. The database components of the SDSS can supply input data for the models. After the models are run, the resulting output can be returned to the database for later display through user interface, in tabular, chart or map form. For planning purposes, this ability to dynamically change information, forecast on perform sensitivity analysis is essential. Both GIS and DSS have been widely used in natural resource management. Goodchild, et al., (1996) described a comprehensive study of GIS in water resources and environmental engineering. This GIS-based SDSS helps in modelling capability, which can integrate with simulation techniques to solve complex natural resource management problems.

## Conclusions

The use of modern technologies such as high resolution satellite data, GPS and GIS can be effectively used in natural resources survey, planning, monitoring and management. The satellite data provides integrated information on landforms, geological structures, soil types, erosion, land use/land cover, surface water bodies and qualitative assessment of groundwater potentials at watershed level. This approach is highly useful in groundwater resources estimation, identification and mapping of perspective groundwater zones, systematic planning, identification of over exploitation zones, prioritization of areas for conservation and optimum use of available water resources. The current and new generation of high resolution satellites imageries, GPS and GIS ease the task of natural resources planners and managers in planning and management of them on sustainable basis. The generated accurate database in GIS on natural/physical resources, socio-economic and contemporary technologies and integrated analysis provide a sophisticated mechanism to generate alternative action plans for watershed development. The evaluation and management of natural resources at watershed level based on remotely sensed data, GPS and GIS can be effectively utilized to arrest various types of degradation, viz., waterlogging, soil erosion, salinity/alkalinity.

## References

Bhat, L.S. (1989). Status Report on Land System Analysis for Evaluation of Resources at Micro-level. Discussion paper for the Fifth Review Meeting for NRDMS Project, DST, Government of India.

Burrough, P.A. (1986). *Opportunities and limitations of GIS based modeling of Solute transport are the regional scale*. In Applications of GIS to the modelling of non-point source pollutants in the Vadose zone, ASA-CSSA-SSSA, Bouyoucos conference, mission inn, rivers side C.A, May 1995.

Das, S., Behera, S.C., Kar, A., Narendra, P. and Guha, S. (1997). Hydrogeomorphological mapping in groundwater exploration using remotely sensed data – A case study in Keonjhar District, Orissa. *J. Indian Soc. Remote Sensing*, 25(4): 247-259.

Densham, P.J., Goodchild, M.F. (1988), Spatial Decision Support Systems. A research agency. GIS/LIS' 88 Proceedings Vol. 2, Bethesda, Maryland: American Congress on surveying and mapping, November-December 1988.

Gessler, P.E., Morre, I.D., McKenzie, N.J. and Ryan, P.J. (1995). Soil-landscape modelling and spatial prediction of soil attributes, *International Journal of Geographical Information Systems,* Vol. 9, pp. 421-432.

Goodchild M.F., Steyaert, L.T. in Parksbo, Johnston, and Midmend, Crane M. and Glendening S. (1996). *GIS and Environmental Model.* Progress and research issues, (eds.) Goodchild M.F., Steyaert, L.T. in Parksbo, Johnston, and Midmend, Crane M. and Glendening S. (1996).Fort Collins, Collins, Colorado. GIS world publications.

Jain, M.K., and Kothari, U.C., (2000). Estimation of soil erosion and sediment yield using GIS, *Hydrological Sciences Journal,* 45 (5): 771-786.

Kasturirangan, K., Aravamudan, R., Deekshatulu, B.L., Joseph George and Chandrasekhar, M.G. (1996). Indian remote sensing satellite IRS-1C – The beginning of New Era, *Current Science,* 70(7): 495-500.

Kumar, R., Kumar, S., Lohani, A.K., Nema, R.K. and Singh, R.D. (2000). Evaluation of Geomorphological characteristics of a Catchment using GIS, *GIS India,* Vol.9 (3) pp. 13-17.

Mahtab, A., Narender, B. and Ajai. (2003). Satellite derived Digital Elevation Model (DEM) and Terrain parameters – Generation, Accuracy, Assessment and Validation, *J. Indian Soc. of Remote Sensing,* Vol. 31 (1): 19-24.

Maji, A.K., Nayak, D.C., Krishna, N.D.R., Srinivas, C.V., Kamble, K., Obi Reddy, G.P and Velayutham, M. (2001). Soil Informtion System of Arunachal Pradesh in a GIS Environment for Land Use Planning. *International Journal of Applied Earth Observation and Geoinformatics (JAG, ITC Journal*), Vol. 3(1), pp. 69-77.

Maji, A.K., Srinivas, C.V., Dubey, P.N., Obi Reddy, G.P., Kamble, K.H and Velayutham, M. (2002). Soil Resources Information of Nagaland in GIS for Land use planning in Mountainous Region, *GIS India,* Vol. 11, No. 4, pp. 13-16.

Moore, I.D, Grayson, R.B and Ladson, A.R.(1991). Digital Terrain Modelling: Review of hydrological, geomorphological and biological applications, *Hydrological Processes,* 5, 3-30.

Nair, K.M., Shivaprasad, C.R., Nagabhushana, S.R., Srinivas, S., Ramesh, M., Raghumohan, N.G. and Sehgal, J. (1996). Soil information for district planning. *Fertil. News*, 41, 21.

Rao, D.P., Gautam, N.C., Nagaraja, R. and Ram Mohan, P. (1996). IRS-1C application in land use mapping and planning, *Current Science*, 70(7):575-581.

Reddy, G.P.O., Esther Shekinah, D. Maurya, U.K., Thayalan, S., Jagdish Prasad, Ray, S.K. and Bhasker, B.P. (1999). Landscape-Soil Relationship in part of Bazargaon Plateau, Maharashtra, *Geographical Review of India*, Vol. 61, No. 3, pp. 280-291.

Reddy, G.P.O., Maji, A.K., Srinivas, C.V., Thayalan, S. and Velayutham, M. (2001a). Landscape Ecological Planning in a Basaltic Terrain, Central India, Using Remote sensing and GIS Techniques, *Jour. of Indian Society of Remote Sensing*, Vol. 29, No. 1 and 2, pp. 3-16.

Reddy, G.P.O., Maji, A.K., Krishna, N.D.R., Srinivas, C.V. and Velayutham, M. (2001b). Integrated remote and GIS approach for delineation of groundwater potential zones and identification of sites for artificial recharge – A case study, *Proceeding on International Conference On Remote Sensing and GIS/GPS, (ICORG)*, BSP Publications Vol. I, pp. 649-658.

Reddy, G.P.O., Maji, A.K. and Gajbhiye, K.S. (2002b). GIS for Morhophometric Analysis of River basins. *GIS India Journal*, Vol. 11, No. 9, pp. 9-14.

Reddy, G.P.O., Maji, A.K., Chary, G.R., Srinivas, C.V., Tiwary, P. and Gajbhiye, K.S. (2003). GIS and Remote sensing Applications in Prioritization of River sub basins using Morphometric and USLE Parameters A Case study. *Asian Journal of Geoinformatics*, Vol. 4, No. 4, pp. 35-49.

Strahler, A.N. (1964). Quantitative geomorphology of drainage and channel network, *Handbook of Applied Hydrology*, (ed. by Ven Te Chow) Mc Graw Hill Book Company, New York, pp. 4-39-4-76.

Wischmeier, W.H., and Smith, D.D. (1978). Predicting rainfall erosion losses – A guide to conservation planning, *USDA Agricultural Handbook*, No. 587.

# 14

# Agroclimatic Regions of Jharkhand State

Anuja Tigga and Hema Malini B.

Rapid growth of population has exerted tremendous pressure especially on land, soil, and water resources, to attain self-sufficiency, in terms of food. In recent years the concept of agroclimatic regionalization is gaining popularity, because the climate of a given region plays a vital role in determining agricultural capability and suitability (Hema Malini, 1991). Scientific assessment of existing land, moisture, and soil resources can be achieved through regionalization which in turn helps in understanding the agricultural potentialities of the region (Quereshi, 1990). As soil moisture is a decisive factor for crop growth, estimation of the same is very essential. In the absence of irrigation and long spells of dry conditions, moisture in the soil gets depleted. The resultant water deficit inhibits the healthy and prosperous growth of plants, thereby, reduction in yields.

Wilsie (1962) in his studies indicated that the crop suitability of an area is entirely based on the availability of water in that particular area. With high water potential a variety of crops could be grown whereas in water stress conditions there will be limited scope and there would be more risk of climatic hazards for efficient productivity of the crops. Mather (1974) also found that the variability in moisture conditions such as water storage, water surplus, and water deficit allows different crops to be grown in a given region as all crops and plants has specific water requirements in various seasons. Hussain (1996) while analysing performance of high yielding variety seeds found that in sub-tropical India, rice and wheat were successfully grown in areas where timely irrigation is maintained to replenish soil moisture whenever required. It is also found that during severe water deficit conditions there was negative deviation in the crop yields from the normal (Hema Malini, 1991). Singh and Dhillon (1998) indicated that moisture is the decisive and fundamental factor for the agricultural prosperity of a region. In a particular area, if the selection of crops is based on the available moisture conditions, there will be scope to get better yields as well as to avoid drought impacts.

For the present study, hydrological conditions of Jharkhand state were scientifically analysed and agroclimatic regions were delineated.

## Study Area

Jharkhand state (Figure 1) occupies a total geographical area of 79.7 km$^2$ in India. It is bounded between 21°58′ N to 25°18′ N latitudes and 83°22′ E and 87°57′ E longitudes. Physiographically the state is a plateau with undulating topography. The Tropic of Cancer passes through the state; hence the climate is tropical with hot summers and cool winters. The average mean temperature of the region is 25°C with the mean maximum and minimum temperatures being 31°C and 19°C respectively. However, the higher elevations experience much cooler climate. January is the coolest month and May is the hottest month. Annual rainfall is 1328 mm (Anuja and Hema Malini, 2003). Jharkhand receives more than 80 to 90 per cent of the total rainfall during south-west monsoon season (June to September). Eastern parts of the state receive comparatively more rainfall (1400 mm).

**Figure 1**
*Location Map of Jharkhand State*

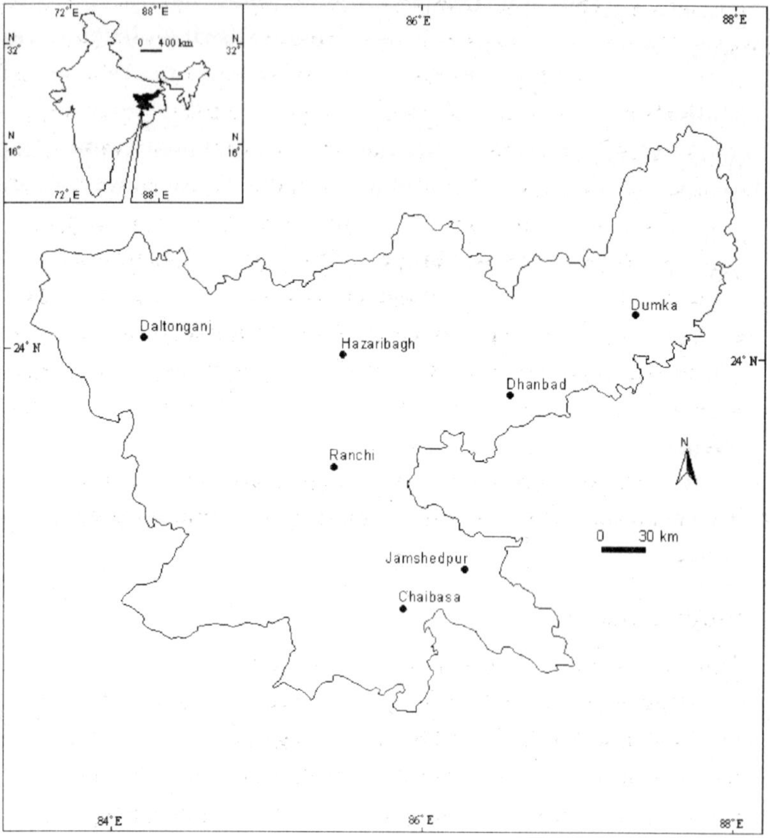


The total population of the state is 2.69 million (Census, 2001) for which about 46 lakh metric tonnes of food grains is required. But the normal production of the state is only 26.50 lakh metric tonnes (Prasad, 2003). Paddy is the chief crop and occupies as much as 77 per cent of the total cropped area during kharif season (June-October). Other kharif crops are maize, pulses, oilseeds, etc. During rabi season (October-March) a few crops are grown, hence the total extent of cropped area is 187.3 thousand hectares against 1825.2 thousand hectares during kharif season. The chief crops of rabi season are pulses, oilseeds, maize and wheat. The area under irrigation is 157 thousand hectares, which is only 8 per cent of the net sown area. Hence, the agriculture in Jharkhand is almost rainfed.

## Methodology and Data Collection

Water balance technique as proposed by Thornthwaite and Mather (1955), has been adopted to evaluate agricultural potentialities of Jharkhand state. It is a useful technique to assess the hydrological conditions of any region. Estimation of water balance elements provides an accurate and meaningful decision with regard to numerous climato-agro-hydrological problems for a purposeful agricultural planning (Mather, 1974). Water balance technique provides a quantitative basis to comprehend the wetness or dryness of a climate in a realistic way. Moreover, several indices derived from various water balance components, help in delineation of agroclimatic regions (Hema Malini, 1991).

To carry out water balance, mean monthly temperature and rainfall data were collected from the records of India Meteorological Department (IMD) of seven stations, for the period ranging between 50 to 100 years (Table 1). Figure 1 shows the location of existing IMD stations in Jharkhand, from which data has been collected.

**Table 1**

*IMD Stations of Jharkhand with Geographical Location*

| *S.No.* | *Name of the Stations* | *Location* | *Period of study* |
|---|---|---|---|
| 1. | Chaibasa | 22°33' N and 85°51' E | 1901-1996 |
| 2. | Jamshedpur | 22°50' N and 86°10' E | 1924-2000 |
| 3. | Ranchi | 23°23' N and 85°23' E | 1901-2000 |
| 4. | Hazaribagh | 23°59' N and 85°25' E | 1933-2000 |
| 5. | Dhanbad | 23°47' N and 86°30' E | 1941-1990 |
| 6. | Dumka | 24°17' N and 87°15' E | 1901-1991 |
| 7. | Daltonganj | 24°02' N and 84°04' E | 1901-2000 |

In the bookkeeping procedure of Thornthwaite, monthly water balance elements can be derived by comparing precipitation (P) with potential evapotranspiration (PE). Initially, monthly values of potential evapotranspiration have been computed by using a series of tables prepared by Thornthwaite and his associates (1957). Further, by comparing PE and P other elements namely soil moisture storage (ST), actual evapotranspiration (AE), water deficit (WD), water surplus (WS) can be derived considering the field capacity of the particular soil. Field capacities of the representative stations were determined based on the table prepared by Thornthwaite and Mather (1957) using various combinations of soil and vegetation types. From the derived water balance elements, indices such as index of aridity (Ia), index of humidity (Ih), index of moisture (Im), and Index of moisture adequacy (Ima) were computed. The formulas used for the calculation were mentioned in the earlier studies (Thornthwaite and Mather, 1955; Subrahmanyam, 1956; Hema Malini, 1991).

## Water Balance Elements and Climatic Regimes

Based on thermal and moisture regimes, climates of Jharkhand were demarcated. In Jharkhand state, the average thermal efficiencies (TE) ranges between 1271 mm and 1548 mm and thus the state enjoys mega thermal climate (Anuja, 2003). Analysis of moisture regime indicated that the state experiences moist sub-humid ($C_2$) type of climate in the upland areas namely central Ranchi and Hazaribagh plateaus. In these areas Im percentage ranges between 0 and + 20. The thermal efficiencies are ranging between 1271 mm and 1289 mm. The average annual rainfall is about 1380 mm. The water deficit is between 240 mm and 246 mm, with more intensity during April and May.

In the remaining parts of the state the climate is dry sub-humid ($C_1$), where the Im percentage ranges between –33.0 and 0. Thermal efficiency ranges between 1457 mm to 1548 mm. Rainfall varies between 1169 mm in the north-western parts to 1438 mm in the eastern parts indicates decrease towards west. Water deficit also varies regionally from 341 mm in the north-eastern parts to 427 mm in the north-western parts. The seasonal distribution of water deficit indicates that water deficit is more during summer months (March to May). In the extreme north-western parts deficit conditions begins from March and extends up to June month.

## Agro-climatic Regions

Agricultural potentiality of the region in terms of water balance can be achieved by assessing the soil moisture adequacy of the region, which is percentage ratio of actual evapotranspiration (AE) to potential evapotranspiration (PE). Where AE is the amount actually lost from the soil storage to meet the atmospheric demands, while, PE is the total water need of the atmosphere.

In Jharkhand state PE ranges between 1271 mm to 1548 mm; while AE ranges between 1031 mm to 1180 mm. Thus, it shows that PE is comparatively more than that of AE. The spatial distribution of PE indicates that it is relatively lower in the central parts, where as it is higher in the south-eastern parts of the state. The spatial pattern of AE shows more or less similar trend.

Since AE denotes the actual availability of moisture at any given time, the index of moisture adequacy (Ima) thus obtained may be considered as an efficient parameter to assess the crop potentialities in a climatic zone (Subrahmanyam, et al., 1963). The high values of moisture adequacy indicate the maximum efficiency to grow high water requirement crops. On the contrary, low Ima values indicate low moisture efficiency of the region and thereby supports only low water requirement crops, which can withstand water stress conditions.

Keeping that in view, agroclimatic regions were delineated to understand the agricultural potentiality of Jharkhand. For that purpose, index of moisture adequacies (Ima) have been derived for all the stations. The study reveals that the moisture adequacy is more in the central parts (81.1%) and it decreases towards the north-western parts (71.8%) of the state. Further, the agroclimatic regions of Jharkhand state have been demarcated based on the Subrahmanyam, et al., (1963) scheme of suitability of crops. Table 2 indicates crop suitability of Jharkhand state for individual stations. Thus, Jharkhand state can be identified into 60-80 per cent Ima zone and above 80 per cent zone.

**Table 2**
*Crop suitability of Jharkhand State*

| *Stations* | *Index of Moisture adequacy (%)* | *Crop Suitability* |
|---|---|---|
| Chaibasa | 77.0 | Suitable to paddy but yields are low: Efficiently suitable to millets |
| Jamshedpur | 74.5 | Suitable to paddy but yields are low: Efficiently suitable to millets |
| Ranchi | 81.1 | Efficiently suitable to paddy. |
| Hazaribagh | 80.9 | Efficiently suitable to paddy. |
| Dhanbad | 74.9 | Suitable to paddy but yields are low: Efficiently suitable to millets |
| Dumka | 77.5 | Suitable to paddy but yields are low: Efficiently suitable to millets |
| Daltonganj | 71.8 | Suitable to paddy but yields are low: Efficiently suitable to millets |

## Spatial Analysis

In order to obtain spatial patterns of agroclimatic zones of Jharkhand, GIS technique has been adopted for the present study. For this purpose Jharkhand state boundary along with its IMD Stations were digitized, and geo-referenced in Auto CAD and Arc Info environments respectively. Then, unique values pertaining to Ima, were assigned to the respective points in Arc View environment and contours were interpolated to get the spatial pattern of agroclimatic regimes in Jharkhand.

The analysis of the spatial pattern as presented in Figure 2 reveals that moisture adequacy efficiency decreases towards west and south-eastern parts, while, it is comparatively higher in the south-central and north-eastern parts of the state. The higher

**Figure 2**
*Agroclimatic Regions of Jharkhand*

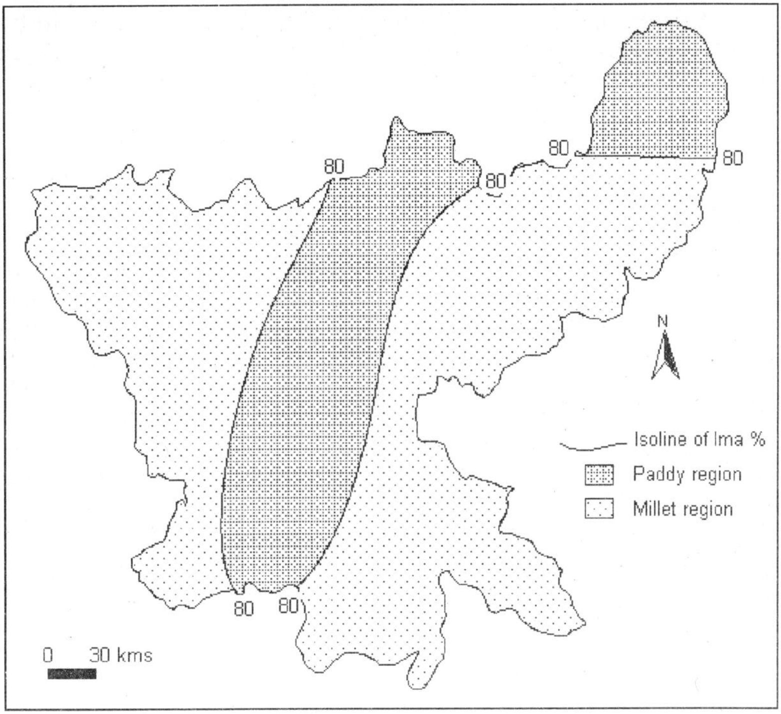


moisture adequacy zone with values above 80 per cent exists in the central Ranchi and Hazaribagh plateau and towards the extreme northeastern parts of the state. The remaining parts of the state come under 60 to 80 per cent of moisture adequacy zone. Climatically, the former zone is efficiently suitable for paddy cultivation and the later to millets. Even though paddy can be cultivated in the 60-80 per cent zone, comparatively millet cultivation is more prosperous.

At present the crops that are being cultivated in the state are paddy, wheat, maize, pulses and oilseeds during kharif and rabi season. When existing cropping pattern is compared with the climatic suitability patterns, it is revealed that the existing cropping is more or less agreeable with the climatic conditions. However, in the extreme eastern and western parts the crops with low water requirement can be grown to get better yields rather than paddy.

## Conclusion

Based on the study it may be concluded that in Jharkhand state, which at present producing only about half of the food requirement has much better agricultural potentialities in general and in the south-central and north-eastern parts in particular. In the highly efficient paddy zone, potentialities must be exploited fully by bringing areas devoted to other crops also under paddy, wherever possible. In the millet zone, paddy cultivation may be suitable wherever irrigation facilities are available so as to protect the crops during the periods of water deficiency and to get ensured yield. In the remaining areas, millets such as maize, jowar, bajra and ragi, etc., may be grown successfully without irrigation, as these crops are efficiently suitable to the region.

## References

Anuja, T., 2003: Agroclimatic regions in the tribal belt of Jharkhand (UGC-minor research project report submitted to Andhra University, Waltair), pp. 13-38.

Anuja, T., and Hema Malini, B., 2003: Climatic variability in Jharkhand State, India, *The Deccan Geographer,* Vol. 41, No. 2, pp. 13-19.

Hema Malini, B., 1991: Studies in ecoclimatology of Andhra Pradesh (unpublished Ph.D. thesis) Andhra University, Waltair.

Hussain, M., 1996: *Systematic Agricultural Geography,* Rawat Publications, New Delhi, pp. 90-103.

Mather, J.R.,. 1974: *Climatology Fundamentals and Applications,* Mc Graw-Hill series in Geography, pp. 87-119.

Mather, J.R., 1977: *Application of Water Budget to a Farming Programme:* Crop Yields, Publ. In Cli. Drexel Inst. of Tech. Elmer, N.J. Vol. 30, No. 1, pp. 76-81.

Prasad, Kushwaha, 2003: *Kharif 2003-2004,* In the report of state level Kharif workshop, Department of Agriculture, Government of Jharkhand (Hindi version), pp. 1-4.

Singh, J. and Dhillon, S.S., 1998: *Agricultural Geography,* Tata McGraw-Hill Pub. Co. Ltd., New Delhi, pp. 72-73.

Subrahmanyam, V.P., Subba Rao, B. and Subramaniam, A.R., 1963: Moisture adequacy in relation to the distribution of a few crops in India, Proc. Berkley (California) Symp. of the Internat. Assn. Sci. Hydrol. (IUGG), pp. 462-467.

Subrahmanyam, V.P., 1956: Climatic types of India according to the rational classification of Thornthwaite, *Ind. Juor. Met. and Geophys.* Vol. 7, No. 3.

Thornthwaite, C.W. and Mather, J.R., 1955: The water balance, Published in *Climatology, Lab of Climat.,* Vol. 8, No. 1, pp. 104.

Thornthwaite, C.W. and Mather, J.R., 1957: Instructions and tables for computing Potential evapotranspiration and the water balance, Published in *Climatology, Lab. of Climat.,* Vol. 10, pp. 185-311.

Qureshi, M.H., 1990: *India-Resources and Regional Development,* NCERT, New Delhi, pp. 79-81.

Wilsie, C.P., 1962: *Crop Adaptation and Distribution,* Eurasia Publishing House, New Delhi, pp. 133-150.

# 15

# Ecotopic Changes Due to Human Pressure on An Ecosystem: A Geoinformatics Approach

Kirti Avishek and M.S. Nathawat

There has been a growing realization during the last decade that landscape will hold a vital position due to its role ineffective integration of different fields of human activity. Due to rapid industrialization, urban population growth, scientific progress but thoughtless utilization of resources has disturbed the equilibrium between nature and man, thus leading to several ecological problems. Forging an ethical relationship with landscape and its people is the challenge of our time. A suite of complex factors including policy, population change, culture, economics and environmental characteristics drive landscape change. Our classification system is based on the ecotopes, defined as "the smallest homogeneous ecosystem units within landscapes". The ecotope classification system has four levels:

- *Ecotope:* The smallest holistic land unit characterized by homogeneity of at least one land attribute of geosphere.
- *Land Facet:* A combination of ecotypes forming a pattern of spatial relationship, being related to at least one attribute.
- *Land System*: A combination of land facet to form one mapping unit.

- *Main Landscape*: A combination of land facets to form one mapping unit.

The Landscape System comprises of natural and anthropogenic system. These are further categorized at land facet level and followed by ecotope level. The entire work is based on this ecotope level classification.

## Study Area

The study area is located as a part of Chotanagpur plateau of Jharkhand.The area is bounded by latitude 23°14′422″N to 23°30′05″N and 85°10′47″E to 85°31′02″E and is covered in SoI toposheets No.73E/2, 3, 4, 6, 7, 8, 11, 13. Area chosen for study includes the three major blocks of Ranchi District, i.e., Kanke, Ratu and Namkum Block. Total area was found out to be 1162.606km$^2$. Plate 1 shows the location map of the study area and Plate 2 shows the overlay of block boundary on the satellite imagery.

## Database Used

### *Primary Data/Spatial Data*

1. Satellite imagery
   Satellite: IRS-Resource SAT
   Sensor: LISS III (Multispectral)
   Path/ROW: 105/56
   Date of Acquisition: March 02, 2004
   Bands: 2(Green), 3(Red), 4(NIR), 5(MIR)
   Scale: 1:50, 000
2. Survey of India Toposheet
   No. 73E/2, 3, 4, 6, 7, 8, 11, 12
   Scale: 1:50, 000

### *Secondary/Collateral Data*

1. Census Data: 1991 and 2001 Data
2. Block Maps:

**Plate 1**
*Location Map of Study Area*

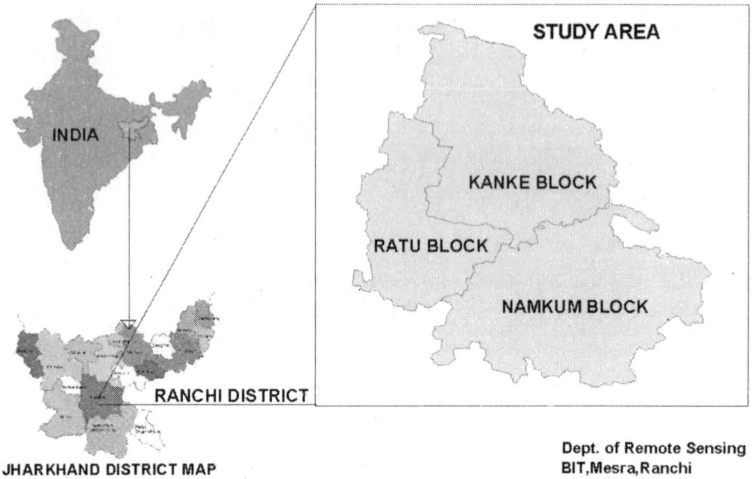


**Plate 2**
*Imagery of Study Area*

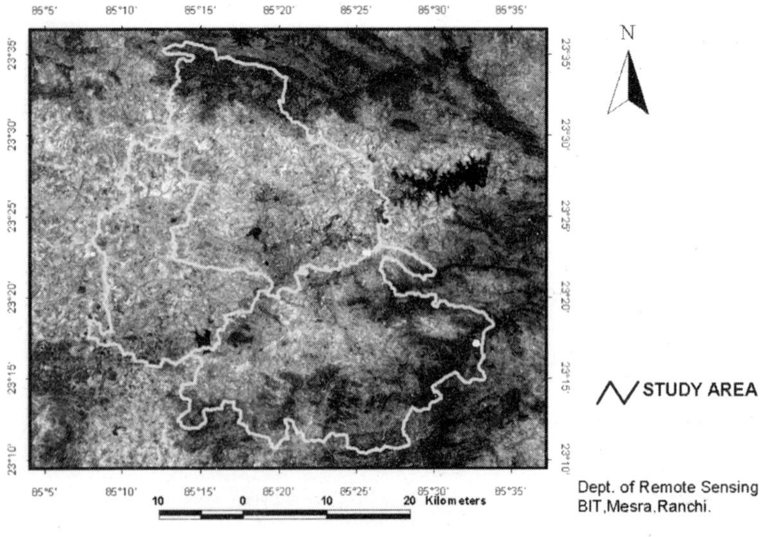

Kanke, Ratu and Namkum Block Map from Census Dept., Patna, Bihar.

## Software's Used

Geomatica 8.2
Arc-View 3.1
Arc Info 8.1

## Landscape Classification

Ecological processes are spatially heterogeneous across anthropogenic landscapes like farms and villages. We use a hierarchical landscape classification system to stratify this variability into components for measurement and analysis of ecosystem properties. Our classification system is designed to identify *ecotopes,* defined as "the smallest homogeneous ecosystem units within landscapes". The ecotope classification system has four levels: Landscape, Landscape System, Land Facet and Ecotope. This Classification was first performed using the base data and then using the satellite data. There was a major change in both the natural ecosystem and the anthropogenic ecosystem between a time gap of 28 years.This statistics of both the time periods are given in Table 1 and 2.

Plate 3 shows the settlement area in 1976 which proves that urbanization was not at all a prominent feature then, however Plate 4 shows the ecotopic map prepared using 2004 data. Plate 5 depicts the Built-up subsystem in totality, and over here a comparison can be made regarding the amount of change that has occurred in anthropogenic system.

## Analysis

— There has been an overall loss of the forest subsystem which has surely lead to degradation of the ecosystem followed by climatic changes. The evidence is the slow and gradual increase in the temperature of Ranchi. The city was earlier known as the summer capital of Bihar for its beautiful landscape and climate. But since the last decade it has witnessed massive changes in urbanization pattern.

**Table 1**
*Ecotope Classification (1976)*

| *Landscape System* | *Land Facet* | *Ecotope Level* | *Symbol* |
|---|---|---|---|
| 1. Natural System | Forest Subsystem | Fairly Dense Sal | FDS |
| | | Dense Sal | DS |
| | | Eucalyptus only | EO |
| | | Eucalyptus and Sal | ES |
| | Plantation Subsystem | Trees | T |
| | | Shrubs | S |
| | | Bamboo | B |
| | | Agriculture | AF |
| | | Other Plantation | OP |
| | Water Subsystem | Fish ponds | FP |
| | | Other ponds | OP |
| | | Lakes | LA |
| | | Tanks | TA |
| | | Rivers | RI |
| | | Reservoirs | RE |
| | Hill Subsystem | Residual Hills | RH |
| | | Stony Mounds | SM |
| 2. Anthropogenic System | Built-up System | RESIDENTIAL | |
| | | High Density | HDR |
| | | Medium Density | MDRV |
| | | Low Density | LDRV |
| | | Rural | RRS |
| | | INDUSTRY | |
| | | Stone Quarry | ISQ |
| | | Industrial Unit | IIU |

**Table 2**
*Ecotope Classification According to 2004 Imagery*

| *Landscape System* | *Land Facet* | *Ecotope* | *Symbol* | *Area km$^2$* |
|---|---|---|---|---|
| Natural System | Forest Subsystem (FSS) | Fairly Dense Sal | FDS | 31.01 |
| | | Dense Sal | S | 44.46 |
| | | Eucalyptus | E | 1.4 |
| | | Eucalyptus and Sal | ES | 2.4 |

Contd...

Contd...

| | | | | |
|---|---|---|---|---|
| | | Forest Blanks | FB | 6.48 |
| | | Uplands with Sal | US | 29.56 |
| | | Degraded Sal | DSF | 28.12 |
| | Plantation Subsystem (PSS) | Other trees | OT | 17.00 |
| | | Shrubs | SH | 2.87 |
| | | Bamboo | B | 0.05 |
| | | Acacia | AC | 1.8 |
| | | Acacia and Eucalyptus | AE | 2.5 |
| | | Agriculture | A | 246.7 |
| | | Grasslands | GL | 120.0 |
| | | Gulmohar | GUL | 0.19 |
| | | Eucalyptus and Pine | EP | 0.33 |
| | | Fallow Land | FL | 22.45 |
| | Water Subsystem (WSS) | Ponds | P | 5.89 |
| | | Reservoirs | RE | 4.34 |
| | | Lakes | L | 0.18 |
| | | Rivers | RI | 69.63 |
| | | Fish ponds | FP | 0.011 |
| | Hill Subsystem (HSS) | Residual Hill | RH | 8.75 |
| | | Stony Mounds | SM | 35.88 |
| | | Hilly Plantation | HP | 2.87 |
| Anthropogenic System | Built-up System (BUS) | Open/Other lands | OL | 94.95 |
| | | Residential: | | |
| | | Rural Residential | RRS | 25.96 |
| | | High Density | HDR | 30.53 |
| | | Medium density | MDRV | 60.27 |
| | | Low Density | LDRV | 51.72 |
| | | Industrial: | | |
| | | Industrial Units | IIU | 2.02 |
| | | Stone Quarry | ISQ | 1.6 |
| | | Railways/Roads | IR | 3.63 |

## Land Facet Change

| *Sl.No* | *Major Land Facet* | *1976 Statistics* | *2004 Statistics* | *Change* |
|---|---|---|---|---|
| 1. | Forest Subsystem | 152.59 $km^2$ | 143.18 $km^2$ | – 9.42 $km^2$ |
| 2. | Builtup Subsystem | 96.34 $km^2$ | 181.99 $km^2$ | + 85.61 $km^2$ |

## Plate 3
*Settlement Map, 1976*

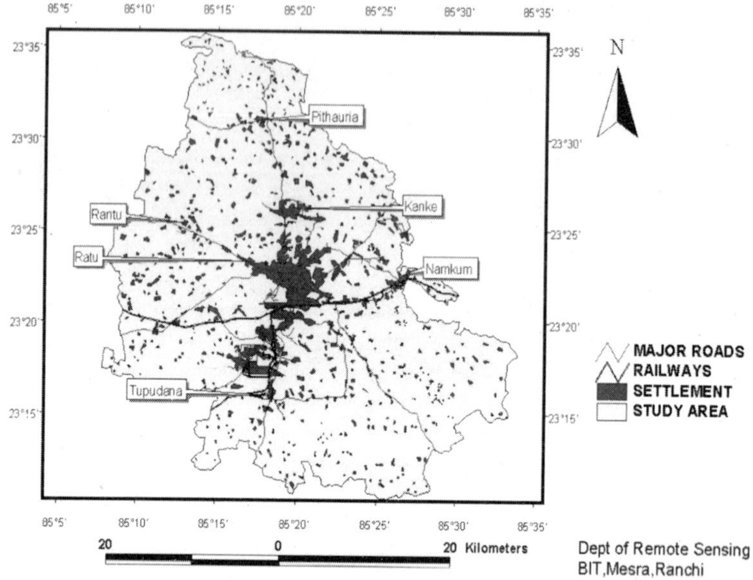


## Plate 4
*Ecotope Classification Map*

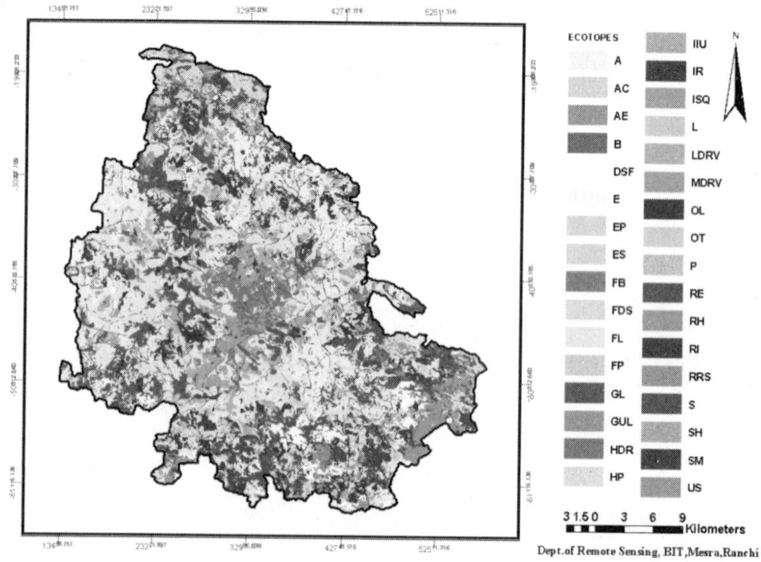

**Plate 5**

*Built-up Subsystem*

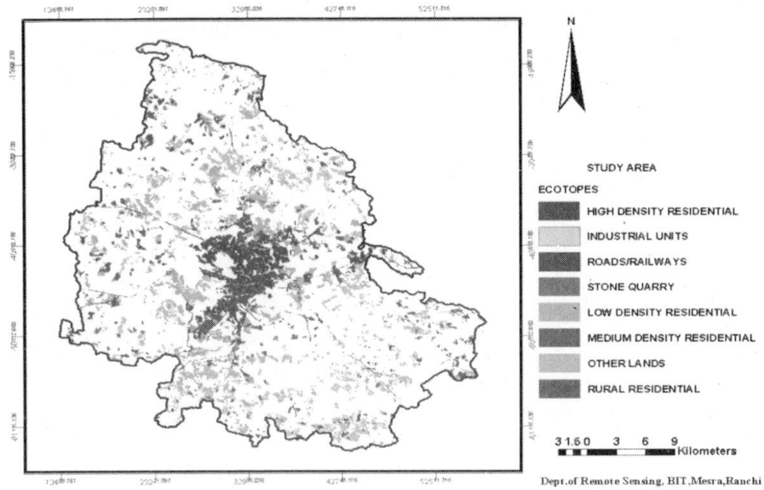


- — Many unplanned settlements have grown over time and this has lead to natural changes via degradation of the overall ecology of the area. The small settlements of the earlier times have grown severely leading to a lot of pressure on the natural bodies of landscape
- — Plantation system has surely increased with time as some areas have witnessed an increase in Acacia plantations. With Ranchi turning into the capital city there has been several changes in the urban lifestyle and to cope up with the emerging pressures the forest department has surely taken steps to preserve the natural beauty of the area by planting trees all over the area and along roads.
- — Agriculture can be seen as small and large patches in the entire area mainly due to mixed population system. Along the periphery of the urban area there are agriculture based population and hence a mixture of agriculture and fallow land can be observed.

— Residual hills are prominent features of the landscape as discussed earlier. They are observed mainly in the Kanke block which has an overall higher elevation. These structures stand as an evidence that the landscape existing is of polycyclic type.

— Plantation system has witnessed an increase in the ecotope type mainly due to an increase in planned and unplanned settlement. There has been an increase in Acacia plantation along linear patterns such as roads and railways. Acacia is a noise absorbent tree and hence they are generally planted along road sides and airports. Gulmohar plantation is restricted to only planned areas where they exist as small patches which cannot be recognized in the imagery. However, in the HEC township they are planted along road sides for beauty and shelter which can be easily interpreted in the imagery as a linear fashion of trees.

## Conclusion

Along with a decrease in the forest cover there has been an increase in the built-up subsystem. There has been a major change in the high density residential area due to a gradual growth of the city. However, along with HDR, medium and low density residential area with plantation has also increased. The remarkable increase in the built-up area has lead to various changes in the landscape. The overall infrastructure of the area has changed. Kanke block has witnessed the highest demographic changes. The changes has reduced the plantation subsystem area, and hence the naturalness has also changed. There has been a significant reduction in the forest cover with overall decrease in the area by 17.84 km$^2$. Main reduction has been in the Sal forest cover. However, there has been comparatively less reduction in Eucalyptus plantation. This may be due to the reason that Sal wood is required for woodworks and fire by local residents. Sal being a very hard and high quality wood is preferred. The loss has surely lead to several natural disturbances like the recent elephant attack in the Ranchi Municipal

Cooperation. Sakua a cheaper variety of Sal species has witnessed maximum threat by local population.

## Select References

Campbell, James B., *Introduction to Remote Sensing*, Third Ed., Taylor and Francis Publications, 1996.

Canter, Larry, *Environmental Impact Assessment*, McGraw Hill Publication, 2nd Ed., p. 563.

Cornelius, Heywood, *An Introduction to Geographical Information System*, Carver, Longman Publications, 1998.

Editorial, *Indian Journal of Landscape Ecology*, Vol. 1, No 1 and 2, June-December 1987.

Human Impact Assessment on Landscape Ecology Using Remote Sensing and GIS. (Unpublished M.Tech Thesis) Kirti Avishek, BIT, Mesra, Ranchi.

Manual of Nationwide Land use/land cover mapping using Satellite Imagery, National Remote Sensing Agency, Department of Space, GoI, Hyderabad, 1989.

Singh, Savinder, *Geomorphology*, Prayag Pustak Bhawan, 1998, p. 517.

Young, Roy H., David Green, Steven Cousin, Taylor and Francis (Eds.), *Landscape Ecology and Geographical Information System*, 1993.

# III

# Geoinformatics for Rural and Urban Planning

# 16

# Role of Geoinformatics in Biomass Management for Resolving Energy Crisis in Rural Sector: A Case Study From Singhbhum, Jharkhand

Deepak Srivastava

The need of biomass augmentation is well established. The paper recognizes the need of biomass augmentations for the reasons already identified (Srivastava, 2003). However, it is suggested in the present paper to incorporate the issues of biomass augmentation and examines the possibilities and implications of using the geoinformatics opportunities for not only augmentation of biomass resources but utilize this resource for further societal benefits in an eco-friendly manner.

A working model for biomass management for river restoration has been established (ibid.). However, in the present context it has been identified that the resources could be utilized for energy and rural development resources too (Srivastava, 2000).

The model suggested for biomass management requires to be integrated with a GIS system for optimum use. It is also necessary that the remote sensing data be utilized for the biomass resource monitoring for different reasons, such as the issues of carbon sequestration, monitoring and arresting illegal biomass resource mobilization, optimization of resource for local development.

A development model must have societal benefits and linkages with sustainable livelihood systems. To achieve this end the paper not only identifies the needs for geoinformatics but indicates the possible areas of creation of sustainable livelihood systems.

## Impacts of Biomass Augmentation

Biomass augmentation through specific programmes can help achieve the following (Srivastava 2000, Srivastava 2003, Valdiya, 1987, Mishra, 1999):

- Slope stabilization
- Soil conservation
- Arrest erosion, reduce sediment load in the streams and the rivers
- Increase carbon sequestration opportunity
- Contribute significantly in climatologic restoration
- Create opportunity for fuel generation for power, industrial, commercial and domestic sectors
- Increase growth opportunity for timber industry
- Create facilities for agro-food processing industries
- Create direct and indirect livelihood system either through the plantation projects or the dependent enterprises
- Assist in management of biodiversity
- Create market opportunity for agricultural waste (biomass is currently not in use for human/animal consumption), in turn support local economy and agriculture

## The Need for Geoinformatics

It is obvious that the development initiatives demands the understanding the needs of local availability of resources of different types of renewable and non-renewable resources, in particular the natural resources such as soil, slope, water resources (surface and subsurface), land use, agriculture, available infrastructure, etc. Additionally, market opportunities, ownership issues of land, census and demographic perceptive considerations have to be qualitatively and quantitatively analysed. This analysis of resources and the appropriate linkages apart from the standard spatial relations of the resources and the temporal variations have also to be considered. The temporal variations may impact the biomass resources either due to triggering of development initiates or it may already be in the cycle of qualitative or quantitative fluctuation, either due to seasonal factors or economic considerations.

It is for the above stated reasons that the intervention of geoinformatic can play a critical role in the management of the biomass resources for sustainable development. The paper does not discuss further the roles and the possibilities for geoinformatics but emphasizes the need of utilizing this tool or opportunity of geoinformatics for biomass management.

The geoinformatic can critically provide the spatial, temporal variations and current biomass status on near real time basis for various economic and market analysis. The facility of utilizing the geoinformatics with the network and a chain of

**Figure 1**

*Illustration on Application of Remote Sensing*

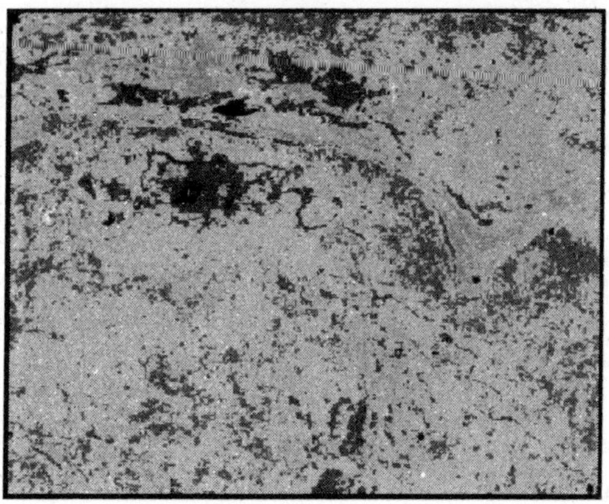

*Source:* IRS data utilized for generating supervised classification for biomass study inputs, legend not shown.

management and decision-making, monitoring systems could very well be established.

The observation on the basis of working for a series of feasibility analysis for setting up decentralized biomass gasification system based on power plant studies, it is suggested that these end development products of development are established either in conjunction with small/micro-industries or local micro-enterprises or for local agricultural or domestic utilities. Normally these projects are executed through NGOs or other such local bodies, such as Panchayat. It has been observed that the competence of use of computers and popular software are common at district, block level or to a certain extent at project levels.

(a) No or very poor expertise exists at project decision-making/implementation levels regarding geoinformatics.
(b) Looking at the present costs, the cost of the software proposed to be utilised for the geoinformatics is too high for the local bodies/agencies/individuals, which in turn could impact the project economics both in recurring and non-recurring manner.
(c) User friendly software package and popularization of IT as such being low are also major impediments in utilizing the available tools.

The author suggests that for developing biomass based commercial enterprises it is advisable to have independent information system. In an effort to develop an independent system a stand-alone software (trial version) has been developed by the author. It is necessary to provide cheap or even free of cost software which are user friendly and that promotes biomass uses such that rural commercial enterprises develop, needs to be promoted and popularized.

With the above perspective in mind it has been suggested to propagate a user friendly low cost software in public interest and in the interest of environmental considerations, particularly the 'Clean Development Mechanism' and other related issues of United Nations and international understandings/agreements.

**Figure 2**

*Biomass and Its Costing (Srivastava 2003)*

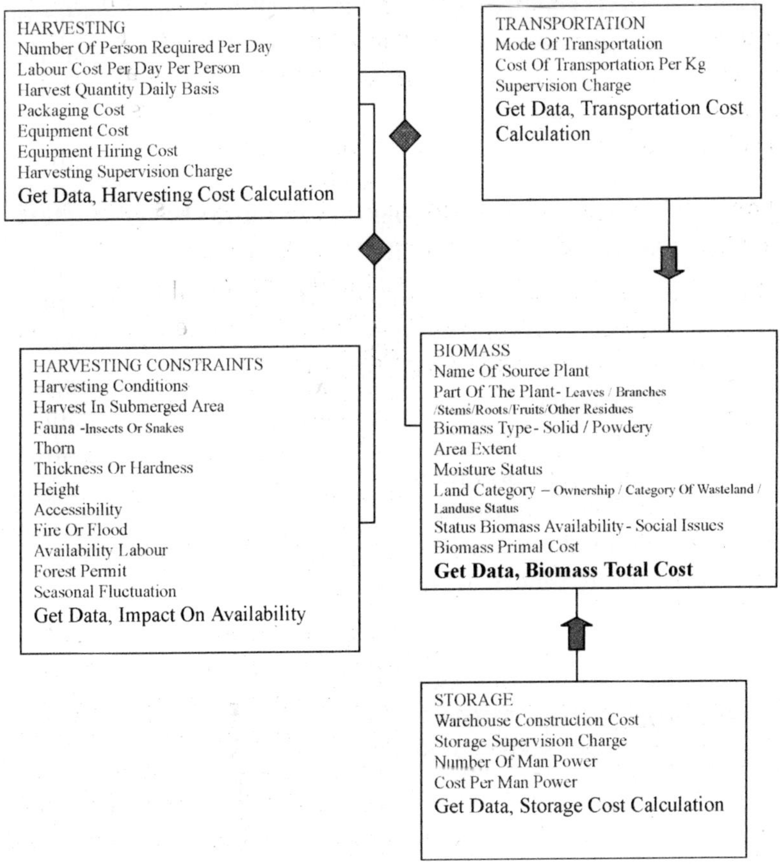


**Figure 3**

*Biomass and Its Link with the Enterprises After: Srivastava 2003*

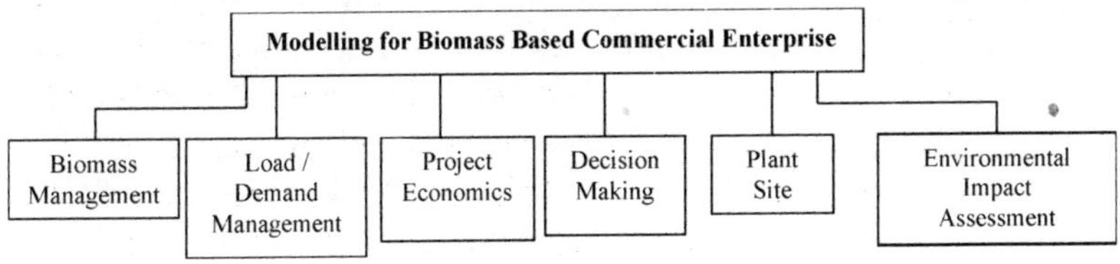

## The Biomass Management

It is quite evident that for biomass management geoinformatics is of critical need. The present software developed is planned to be utilized for biomass management for decentralized biomass based gasifiers. The idea of linking the biomass management issues with an enterprise is essential from the perspective of applications, utility vis-à-vis livelihood systems.

## The Software for Using Biomass for Gasifier Based Systems for Decentralized Power Plants

The package takes care of various needs that would be required for developing a decentarlized power plant based on biomass gasification systems. The critical areas are:

(a) Biomass details
(b) Load details
(c) Site/Location information
(d) Issues related to project economics
(e) Environmental impact assessment factors
(f) A feasibility report

**Figure 4**

*The Biomass Input Sheet*

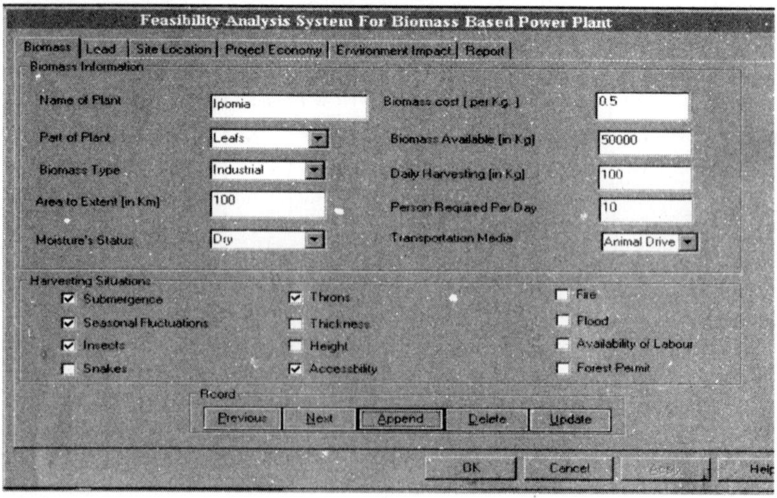

**Figure 5**

*The Report, as Generated by the Software*

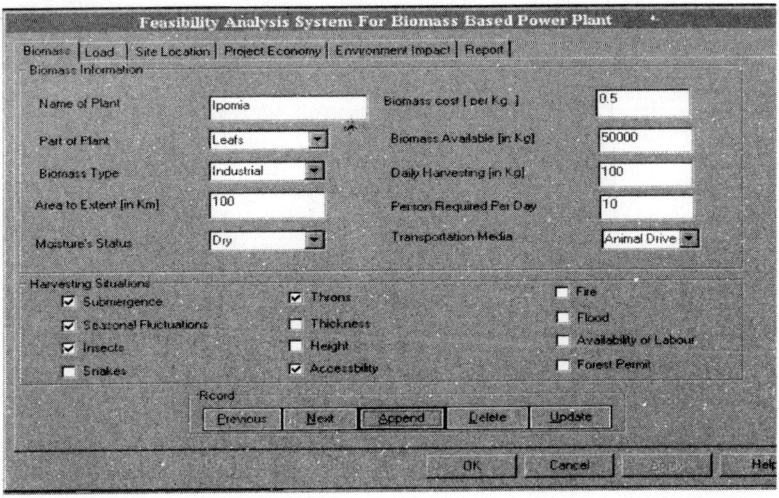


The software developed takes all the above listed critical area for developing the biomass based power plant. The linkages using this software between the different levels of managers, decision makers, financial controllers and other key project controllers, environmentalists can be established and thereby provide a curial development tool for rural uplift. The figures above indicate the different input and report generation capabilities.

## A Case Study for Validations

The study has been completed. The area selected was near Galudi, East Singhbhum. The following data have been collected and analysed:

1. Load, 2. BIOMASS, 3. Social Aspects. The data have been analysed and the feasibility for a 50 and 100 kW power plant has been worked out.

The environmental aspects studied include impacts on land, air, water and displacements.

The standardizations developed have been successfully tested. The results are encouraging.

Standardization for load assessments, biomass assessment, social aspects and costing and project economics for feasibility has been completed and tested in the field.

A software, indicated in the above sections, on Excel based platforms has been prepared for project's economic feasibility assessment and this estimation technique has been field tested. A VB based prototype software has been developed for trial (patent/copy write awaited).

## Conclusion

Based on the above tools and studies it has been concluded that:

(a) A 100kW biomass based gasifier power plant can be set up at Galudi, East Singhbhum, Jharkhand.

(b) It will energize irrigation, domestic sectors, rural micro-enterprises.

(c) It will create 20 persons' direct employment round the year.

(d) It will create indirect employment for 100 persons round the year.

(e) An example of use of Remote Sensing and GIS technologies, however, stand-alone systems are recommended.

The above software can be utilized for feasibility analysis and other decision-making issues related to setting up decentralized power plants based on biomass gasification technology. This model will promote the use of biomass resources and augmentation of biomass resources in the State of Jharkhand. The application of this concept would be useful in creation of sustainable livelihood systems and resolve energy crisis in Jharkhand.

This is a replicable geoinformatics model. As a spin off of the resources generated through biomass related/dependent projects, social development aspects such as health and education sector can be benefited. It can be used as a tool for women empowerment, through appropriate project linkages. Women can be encouraged to take up the projects and support the enterprises and thus create economic development and empowerment opportunities.

## Select References

Anonymous, Ministry of Water Resources Performance Budget 2001-2002 Chapter V; Available at: http://Performance Budget 2001-2002.Htm

Hektor Bo., Socio-Economic Management Models for the Bioenergy Sector; IEA Bioenergy Task 29 'Socio-Economic Aspects of Bioenergy Systems.' Available at: http://IEA bioenergy

Hooda, R.S. and Dye, D.G. 1996, Estimating Carbon Fixation in India, Based on Remote Sensing Data-Acrs, Proceedings, GIS Development.

Mcfarland, Jim, Howard Herzog, John Reilly and Henry Jacoby, 2002, Economic Modeling of Carbon Capture and Sequestration Technologies, Available at: Http://Massachusetts Institute Of Technology

Mac Dicken, K.G., 1997 Forest Carbon Monitoring Programme; Report, Winrock International Institutes for Agriculture Development.

Mishra, D.K., North Bihar Floods and the Proposed Dams in Nepal; Copyright © 1999, 2000 The World Commission on Dams Serial No: Opt028, Available at: http://North Gandak_Bihar Floods And The Proposed Dams in Nepal.Htm

Pathak, M., Srivastava, L. and Sharma, S., 2000, India: CDM Opportunities and Benefits, Available at http://wwo.uri.ors

Singh, N., 2001, *The Sinking Debate*, Vol. 9, No. 19, *Down to Earth*, p. 36.

Srivastava, Deepak, 2000, Water Resource Management and Sustainable Development – A Case Study Along Gandak Sub Basin in Bihar, India. (Published by, River Basin Initiatives, Case Studies for South East Asia, Ramsar Convention).

Srivastava, Deepak, 2003, Biomass Commercial Management Model for River Restoration, Indo-Nepalese Context, *Geospatial Today*, March 2003.

Valdiya, K.S., 1987, *Environmental Geology Indian Context*, Tata McGraw Hill Publishing Company Ltd.

# 17

# GIS/LIS in Land Use Planning and Rural Development

Hira Lal Yadav

Land being the most important resource is regarded as the foundation of all forms of human activities crucial for the economic, social and environmental advancement of all counties. Today every aspect of land development has to be taken into account by policy makers, planners, land administrators and lastly the individual citizens. Knowledge about existing land use and trend of its changes is essential if any nation is to tackle the problems associated with haphazard and uncontrolled growth. The last fifty years have witnessed drastic changes in land use in India caused both by increasing population and natural calamities. Successful interpretation of land and their use entails evaluation of man's adaptation to environment. A systematic framework is needed for updating land use and cover maps that will be timely, relatively inexpensive and appropriate for different needs at national and state levels. Remote sensing techniques can be used effectively to complement surveys based on ground observation and to provide an accurate inventory of current use of the nation's land resources. Therefore, for a standard land use classification using land satellite data and GIS, whic! ill be compatible with the existing land use classification development by various departments is very much desirable. However, the following points must be taken into consideration for integration of land use pattern for area's sustainable development:

1. Establishment of sound database.
2. Suitability and soil capability related to vegetation and production.
3. Legislative control to prevent misuse.
4. Development of cropping pattern to suit the land.
5. People's participation for appropriate land use.

Information about land is a basic requisite for making any policy decisions related to land investment. Therefore, accurate information of land resources—about its physical characteristics, biological conditions, cultural development and proper record are essential for their rational use. Hence, it is essential to have an information system dealing with land related activities. In this background, the concept of Land Information System (LIS) is being used as a tool for effective and efficient storage, analysis and modelling of remotely sensed, spatial and non-spatial data for both policy-oriented information extraction and management decision on scientific basis. GIS/LIS is a computer-based system that provides planners, researchers, resource managers and decision makers also a wide range of applications such as land registration and administration, urban and regional planning and natural resource mapping and management.

## The Concept of LIS

A land information system is essentially a computerization process, which assists in the conventional stages of a land records system. LIS forms the macro-level land management tool. In contrast, GIS are a macro-level tool for decision-making applications. In some GIS spatial/logical database is an assimilation of LIS database. LIS and GIS can work hand in glove for effective data management and analysis. Micro-level planning for rural development generally involves ground study and evaluation of data of a specific area of interest. For example, rural development schemes, irrigation, groundwater development, town planning, mining, etc., require a study of maps (cadastral) and data pertaining to small zones.

## Need of Computerization for Land Records

The system of creation and updating of land records is very elaborate. Maps depicting land parcels (cadastral maps) are required to be updated every 30 years through the process of survey and settlement operations. Most states have not done any survey and settlement operations after independence, as a consequence, updating of records has suffered and they no longer represent the ground realities relating to township and possession. In a primarily agrarian economy with a distorted social structure, it has serious implication in terms of its impact on the execution of all welfare economy development activities.

Land management has been a key focus area for revenue departments throughout the country. The government of India has laid major emphasis in the last decade on land reforms and modernization of land records. LIS thus assist in the following stages of land record system:

1. Preparation of field maps depicting boundaries and measurements and compilation of village maps from these data.
2. Settlement and compilation of revenue assessment on land based on classification, etc.
3. Creation of identifiable field boundary framework on ground by erecting designated marks on land junctions.
4. Updating land maps and land registers incorporating changes in boundaries, titles, etc.
5. Redress of grievances by showing obliterated boundaries, resolving boundary dispute and safeguarding titles and safeguarding titles and arresting encroachments.
6. Revision of land records if changes on ground are too numerous to keep pace with.

Since the First Five-Year Plan, planners have been advocating proper maintenance of land records as the basis for good administration, aimed at social justice through better implementation of rural development programmes. The Sixth Plan has envisaged the completion and updating of land records from 1980 to 1985. The Eighth Plan (1992-1997) and Ninth Plan (1997-2002) have also envisaged the fulfilment of all five-year principle of National Land Records Policies, that is the abolition of intermediaries, tenancy reforms with security to actual cultivators, redistribution of ceilings, surplus land, consolidation of holdings and updating of land records. The general theme underlying the content of all plan documents have emphasized that land is an asset, which provides the primary and secondary needs of the people. Most of the problems of the people in the villages are due to land related issues. Planning and maintaining land records is a prerequisite before any land reforms policies can be successfully implemented to achieve this; latest information technology (IT) is a must for quicker computerization of land records. Therefore, we have to realize the goal of decentralized planning and administration as envisaged in the 73rd Amendment of the Constitution of India.

## Designing a National LIS

The land records of the country, in the present form, have evolved almost a century ago, i.e., during the British rule, and have remained the same without any significant change in their content or application. The existing conventional LIS for Indian villages consist of (1) descriptive land register with the title and other related particulars in respect of each land holding in the village and (2) graphical record consisting of field measurement sketches of the holdings and the village map or, at times, only the village maps. The present day land records have hardly undergone any change during the last seven to eight decades. Consequently, they are unable to serve the contemporary requirements due to their inherent limitation. Since these records presently constitute many shortcomings, there is not much that can be done to overcome them, unless the entire system is reorganized and turned to the needs of the present day. In order to get rid of the inherent deficiencies and to fully cater to the present as well as future requirements of the landholders on the one side and the government departments and organizations involved in land matter on the other, the substitution of the existing and outmoded records with a modern and dynamic system is essential. Scientific and economic generation of up-to-date spatial database through modern cadastral surveys is central to the creation of a modern land system.

# 18

# Geoinformatics System of Uttaranchal Using Indigenous GIS Tool: Preliminary Results

J.S. Rawat, S. Naithani, G. Rawat, C.M.S. Adhikari and V.S. Rawat

In recent years, there has been an increased need for more timely responses to promote effective administration, planning, decision-making and development process. The complexity and huge volume of timely data requirement for developmental planning have proved inadequate in the traditional methods of paper-based data handling, spatial analysis and presentation. Preparation of resource profile at different levels (i.e., national, state and district or below) using GIS should be our main objective. The GIS provides a systematic spatial and attribute database which is a prerequisite for implementing development and research project; for drawing development strategies that are sustainable, area specific and take into account the local need; and to facilitate the process of decentralized (to smaller unit, i.e., district or below) planning, and for more timely response to promote effective administration, planning, decision-making and development process. The process of planning is being decentralized to smaller units, i.e., (state/district or below) to facilitate drawing developmental strategies that are sustainable, area specific and takes into account the local needs. With increasing stress on decentralized planning there is an urgent need to develop a state/district or below level mechanism for effective natural and human resources database system which could help the decision makers, planners, researchers and administrators in implementing there projects.

In view of the significance of the systematic database under GIS environment for development of natural resources, a Natural Resource Database Mangement System (NRDMS) Centre at Almora in the Department of Geography, Kumaun University SSJ Campus, Almora has been establish under a DST, GoI research project.

## Almora the 46th NRDMS Centre of India

The Department of Science and Technology (DST), GoI has recently awarded a research project on "Setting up of Natural Resource Digital Database Centre at Almora, Uttranchal and Creation of an Integrated Database for Decentralized Planning" (Rawat, 2002). Presently under the NRDMS programme of the DST, Government of India, 46 pilot district database centres are operational in different parts of our country. With the help of above projects, 46th NRDMS Centre of our country was established in the Department of Geography, Kumaun University SSJ Campus, Almora on 16, June 2003.

## Objectives of the Centre

The fundamental objectives (as approved by the DST, GoI) of the NRDMS Centre Almora are:

- Development of integrated database as per the DST NRDMS database design.
- Generation of information and sectorial plans needed by District Line Departments.
- Development of technology base/basket, specific to the need of Line Departments.
- Training of the users in District Line Departments of Almora.

- To provide assistance to the researchers and State/District authorities as and when required.

## Study Area

Uttaranchal State (area 53586.53 km$^2$) one of the newly emerged state of India is drained by main perennial and non-perennial river systems, i.e., Ganga, Yamuna, Ram Ganga, Kali, Saruyu, and Kosi, Suyal and many others flows in Bhabhar and Tarai areas respectively. Presently it has two Divisions (viz., Garhwal and Kumaun), 13 Districts (viz., Almora, Bageshwar, Champawat, Chamouli, Dehradun, Haridwar, Nainital, Pauri, Pithoragarh, Rudrapur, Tehri, Udam Singh Nagar and Uttarkashi).

## Methodology

### *GIS Software*

The Centre is using indigenous GIS software, i.e., GRAM++ (Geo Referenced Area Management System) and Vec-Viewer developed by the DST, Government of India through CSRE, IIT, Mumbai. The DST Government of India launched a project entitled Natural Resources Data Management System (NRDMS) to generate computer compatible, spatially oriented database of natural resources and socio-economic parameters to facilitate area specific micro-level planning. The project has emphasized the development of comprehensive spatial database at district level encompassing all the resources. This system (GRAM++) operates on a small and relatively low budget configuration and can be used as a viable tool by local level planners (district administrators, district line departments, block development officers and even village panchyat representatives) for decentralized planning. The important characteristics of GRAM++ is that it can handle both vector and raster data and can be used in different natural resources assessment, village and district or rural and urban planning, resource conservation, watershed management, demographic analysis, trafficability studies, event mapping, marketing and business, alignment studies, environmental impact management and so on.

Thus, GRAM++ is a tool which can collect, analyse and retrieve the spatial and non-spatial data speedily. The GRAM++ has capability to spatial quarries for the maps generated in its domain. The tool contains all the data in particular Geo-referenced mode. The facility of digitization, attribute assignment, registration, joining and analysis are being performed very well in this indigenous GIS package. Vec-Viewer is the supporting tool of the GRAM++. It has the capacity to present the data in spatial form at any scale. It helps the user for generating thematic layers, the database queries and map production generated in the GRAM++.

## Information Need Assessment

For the preparation of the GIS database for Almora District—an "Information Need Assessment Workshop" (Rawat, 2000) was organized to prepare the list of the information needs of the stakeholders of the local level planning, i.e., the different line departments (forest, agriculture, horticulture, soil conservation, "jal nigam", "jal sansthan", public works department, irrigation, tourism, education and electricity) of the district. A list of information needs of these line departments was worked out to prepare the GIS of natural and human resources of Almora District. Digitization of maps was started as per the list of the information needs of the stakeholders of local-level planning.

## Sensitization of GIS Technique

It is difficult, if not impossible to sensitize the GIS technique among the local-level planners, i.e., district or below, who are not aware about this latest GIS intervention in planning. To sensitize the GIS technique among the local-level planners, a District Coordination Committee is constituted in Almora District. The committee comprises of the district administrators (i.e., District Magistrate, Sub-Divisional Magistrate, Chief Development Officer), and Heads of the line departments (i.e., Public Works

Department, Forest, Agriculture, Horticulture, Soil Conservation, Irrigation, Tourism, Jal Nigam, Jal Sasthan, Education and Health, etc.). This committee is headed by the District Magistrate, Almora and the Principal Investigator of the NRDMS Centre, Almora is the Coordinator. Through this committee a Nodal Officer from each line department for NRDMS activities has been appointed which is responsible to depute two officers from his department to bring the maps/data for digitization at the NRDMS Centre Almora. The deputed officers learn the use of GIS and practically do digitization work of their department information/maps under the supervision of NRDMS Centre, Almora. They used to work for two to three weeks at the Centre for learning and doing digitization work of their department and then go back to their department with the soft copy of digital data for their own use. In this way the Centre is sensitizing the use of GIS technique among the stakeholders of the local-level planning and systematically preparing GIS database for Almora District as per user's needs.

## GIS Data Layers

With close association with the district administrators and nodal officers of the line departments as many as 51 spatial raster data maps at 1: 50, 000 scale (see section – Preliminary Outputs) have been prepared so far by the Centre. The data for the study area have primarily been collected from the topographical maps available from Government Departments.

Apart from above, integration of socio-economic data in the digitized District Tehsil, Development Block, and Nyay Panchyat map; and also in road, medical and education facilities maps, have been on Vec Viewer environment.

## Preliminary Outputs

Besides data collection, sensitization and digitization with simultaneous training of the government officials, the Centre has created (a) fifty-one Vector maps (at 1:50,000 scale) in the GRAM++., (b) six Master Database (MDB) files, and (c) nine

Integrated Database on Vec-Viewer Environment. Details of above maps/database is given below.

### Vector Maps Digitized on GRAM++

1. District Outline Map of District Almora with integrated Socio-Economic Data.
2. Tehsil Map of District Almora with integrated socio-economic data.
3. Development Block Map of District Almora with integrated socio-economic data.
4. Nyay Panchayat Map of District Almora with integrated socio-economic data.
5. Administrative Map of District Almora.
6. Village Location Map of District Almora.
7. State Highways Map of District Almora.
8. Major District Roads of District Almora.
9. Other District Road of District Almora.
10. Village Roads (metalled) of District Almora.
11. Village Roads (unmetalled) of District Almora.
12. Light Vehicle Roads(metalled) of District Almora.
13. Light Vehicle Roads(unmetalled) of District Almora.
14. Roads under construction of District Almora.
15. Proposed roads of District Almora.
16. Sanction roads of District Almora.
17. Prime Minister Roads Under Construction of District Almora.
18. Child and Maternity Welfare Centres of District Almora.
19. Ayurvedic hospitals of District Almora.
20. Allopathic hospitals of District Almora.
21. Homeopathic hospitals of District Almora.
22. Community Health Centres of District Almora.
23. Public Health Centres of District Almora.
24. Additional Health Centres of District Almora.
25. Base hospital of District Almora.
26. District hospital of District Almora.
27. Civil hospital of District Almora.

28. Women hospital of District Almora.
29. Government inter-colleges of District Almora.
30. Government Girls inter-colleges of District Almora.
31. Private inter-colleges of District Almora.
32. Government Higher Secondary Schools of District Almora.
33. Government Girls Higher Secondary Schools of District Almora.
34. Private Higher Secondary Schools of District Almora.
35. Professotional Education Centres of District Almora.
36. Central schools in District Almora.
37. Degree colleges in District Almora.
38. Perennial water bodies (streams and springs) of District Almora.
39. Non-perennial warter bodies of District Almora.
40. Forest cover of District Almora (1963AD).
41. Agriculture Land of District Almora (1963AD).
42. Barren Land of District Almora (1963AD).
43. Rain-gauge Network of District Almora.
44. Temperature-gauge Network of District Almora.
45. Sunshine Recorder Network of District Almora.
46. Wind Vane Network of District Almora.
47. Evaporation Tank Network of District Almora.
48. Landslide Scars in District Almora.
49. Floodplains in District Almora.
50. Spot Heights in District Almora.
51. Waterfalls in District Almora.
52. Polling Areas of District Almora.
53. Polling Booths of District Almora.

### Master Database (MDB) File creation

Following MDB files have been created in MS-Access:

1. District Level Database of District Almora (42 fields).
2. Tehsil Level Database of District Almora (42 fields).
3. Block Level Database of District Almora (42 fields).
4. Nyay Panchayat Level Database of District Almora (42 fields).

5. Weather data of district Almora.
6. Land use map (1963) of district Almora.

## Integrated Database Prepared under Vec Viewer Environment

1. District level
2. Tehsils level
3. Blocks level
4. Nyay Panchayat level
5. Road network
6. Educational facilities
7. Medical facilities
8. Land use (1963)
9. Weather data

The software Vec-Viewer having above mentioned database has been installed in the computers of following Line Departments of district Almora for their use: 1. District Magistrate, 2. The Chief Engineer Public Works Department, 3. Superintendent Engineer Public Works Department, 4. Chief Development Officer, 5. Executive Engineer, Uttranchal Pey Jal Nigam, 6. Chief Medical Officer, 7. Uttrakhand Sewa Nidhi (NGO).

## Expected Outputs from the Project

The following outputs are expected from this project:

- Integrated District (Almora) Database.
- GIS-based Digital Database of Natural Resources and Socio-economic aspects.
- Development of Sustainable Plans for Selected Sectors.
- Generation of Action Plans for Selected Sectors
- Technology Baskets Specific to the Needs of the Line Departments.
- Improved Planning Tools and Methodologies.
- Trained Human Resource.
- Technology Base/Basket Specific to Resources.
- Fully equipped and well established NRDMS Centre for District Almora.

*Acknowledgement*

This paper constitutes a part of DST, Government of India project, on setting-up of Natural Resource Digital Database Centre at Almora, Uttranchal and creation of an Integrated Database for Developmental Planning.The authors are thankful to Dr. D. Dutta, Department of Science and Technology, New Delhi; Dr. B.K. Mohan, Mrs. Suri and Mr. Mayur, CSRE, IIT Mumbai, Dr. A.N. Singh and Late Dr. Ravi Sharma, Dr. Ravinder Kaur, IARI, New Delhi, and Prof. A.K. Gosain, Department, of Civil Engineering, IIT Delhi, for their technical guidance; Prof. H.S. Bisht, Head of the Department, Geography, Kumaun University, SSJ Campus, Almora for providing departmental facilities and Prof. L.S. Bisht, Dean, Faculty of Arts, SSJ Campus, Almora and Shri K.K.S.Bisht, Scientist, Vivekanand Parvatiya Krishi Anusandhan Sansthan, Almora, District Magistrate Almora and Mr. M.L. Kanojia for their logistic support and cooperation to conduct this project. We are extremely thankful to all the line departments who have helped us in digitization work of the base maps.

## Select References

Rawat, J.S., (2000): "Information Need Assessment Workshop on Land and Water Resources in Almora District", Sponsored by DST, New Delhi and Kumaun Univeristy Campus, Almora in association with the District Administration, Almora, Techniocal Report, Department of Geography, Kumaun Univeristy Campus, Almora, pp. 1-42.

Rawat, J.S., (2002): "Setting-up of Natural Resource Digital Database Centre at Almora and Creation of an Integrated Database for Developmental Planning", DST Project, Department of Geography, Kumaun University Campus, Almora.

# 19

# Use of GIS for Natural Resources and Environmental Planning for Sustainable Rural Development: A Case Study

Sujit Choudhury and G.S. Chattopadhyay

In a larger perspective of the world, human development process is in conflict with earth's natural system, which results in global warming, ozone layer depletion, loss of biodiversity, etc., at a larger scale. The cumulative effect of a large number of micro-economic activities, which are in conflict with local ecology, is having its impact on the global environment. India has also its own share of development versus environmental conflicts. The sharp population increase in last four decades, coupled with lack of alternative livelihood exerts pressure on existing natural resources, mainly land, resulting in continuous loss of topsoil, shrinking of forestland, lowering of water table and many other environmental problems and hazards. It is a vicious cycle; more loss of natural resources causes more exploitation of natural resources and ultimately the local ecology becomes irrevocably destroyed. These numerous micro-issues create macro-environmental problem in the long run.

The study[1] of the natural resource management and creation of the target area model in GIS for the sustainable rural development programme of Patratu Block, Hazaribagh district, Jharkhand, India, has given the insight of the impact of economic and developmental dynamics on local ecology and the need for formulating an environment friendly strategy for sustainable rural development. This paper deals with the environmental management perspective of the target region for rural development with the help of GIS model. The study covers geology, geomorphology, hydrology, soil classification and vegetation pattern with special emphasis on maximum utilization of natural potential.

## Environment and Rural Development

There is always a profound balance being maintained between biosphere and the ecosystem by Nature. Habitation in large parts of rural India, particularly the tribal areas, lived harmoniously with nature a few decades ago. But, with faster population growth coupled with external economic models and change of socio-economic pattern this interaction is becoming more conflicting. The external development initiatives, many a times lack proper knowledge of the area, understanding and technical know-how. Such developments are often creating an imbalance in the overall ecosystem and thus directly or indirectly responsible for creating environmental problems and hazards. There are numerous evidences of man-made hazards seen in the Indian countryside.

For a sustainable rural development, man has to live and work with nature, and not blindly oppose her, particularly in rural areas where people are primarily depending on natural resources. The local ecology provides the livelihood to the rural people. All the earth's processes such as glacial, fluvial, marine, aeolian and other related processes are not active in all parts of the country at a time. Some of them are more pronounced in some areas, which may be totally absent in other areas. But, all the processes are so interrelated with each other that one has to take knowledge of the effects of all of them in order to assess the overall environmental problems that are introduced by them individually or collectively. Environmental management is essentially dependent on scientific inquiries where interdisciplinary approach is the real objective.

Before adopting any developmental planning procedure, one needs to have a thorough knowledge and understanding of the environmental components, namely the geological and geo-morphological, hydro-geological, pedological, ecological and other related factors for a particular area.

## Resources and Environmental Planning Model

It was necessary to create a model of the area with different natural and socio-economic component as separate entities in different layers. This GIS Model is the framework for organizing knowledge of this particular area with its different variables known as entities, in space. The model is data intensive cross-disciplinary, dynamic and complex. To deal with such model one has to keep in mind that "Environmental models often include modules or sub-models backed by rigour of physical theory, but being strongly cross-disciplinary and synthetic, environmental science (and modelling) is by and large a theory-poor domain".[2] Natural systems like biological, geological, and ecological systems are very complex and it is practically impossible to model such systems perfectly. Therefore, these systems require simplification and approximation. At the time of creation and analysis of this model this basic concept has to be kept in mind.

The goal of this model is to understand the relationship and the interlinking between various natural elements existing within the target area which may be a watershed boundary. It is also important that the terrain model incorporates socio-economic dynamics with respect to the available infrastructure.

The structure of this model includes several aspects:

(1) The extension and the geometry of the domain to which the model is applied.

**Figure 1**
*Interdependence of Factors*

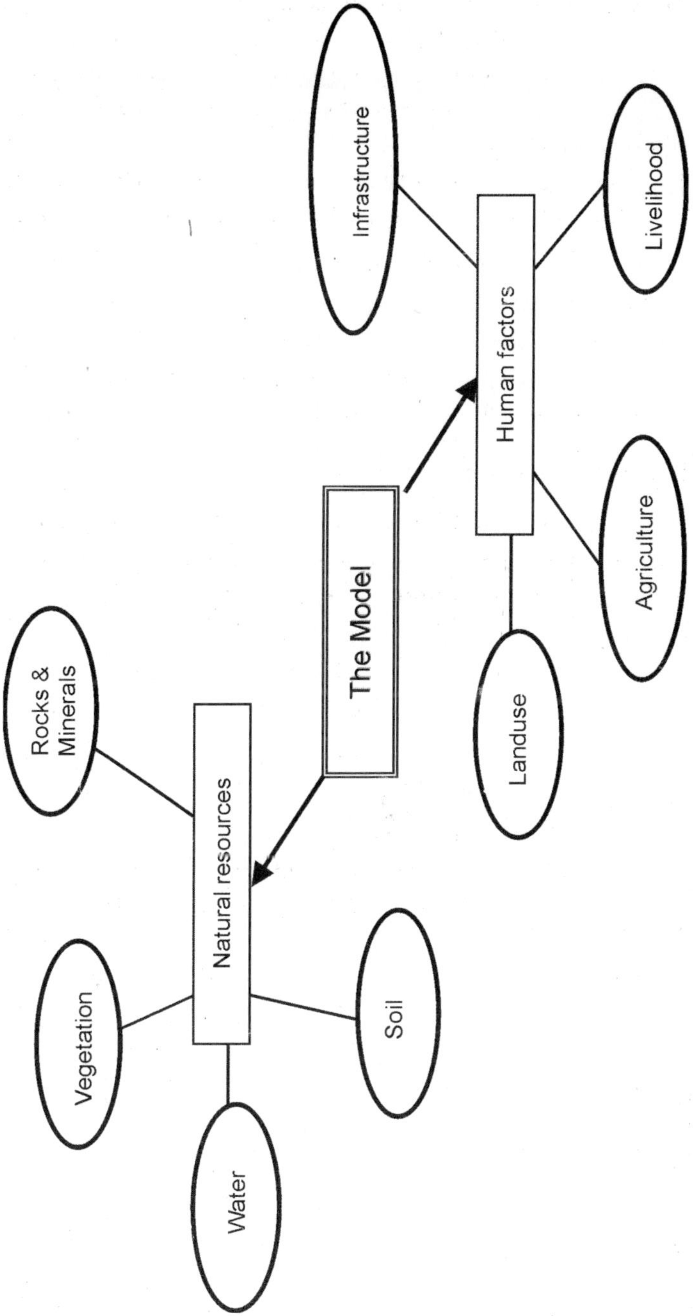

(2) It is a conceptual model, with schematic representation of the essential features of the physical system. For instance, this corresponds to geomorphology with drainage system, groundwater flow model, subsurface lithology and soil classification.
(3) The boundary conditions exhibit characteristics of interactions between the target domain and the rest of the world.

The objective of the model is to depict the existing human interaction with natural and environmental components spread over different domains of physical elements of ecosystems like atmosphere, hydrosphere and lithosphere and ultimately affects the community living in immediate vicinity of this area.

## The Study Region

The area situated at Patratu Block of Hazaribagh district in Jharkhand has been taken as a model to illustrate the requirement of environmental planning in sustainable rural development.

**Map 1**
*The Region Under Study*

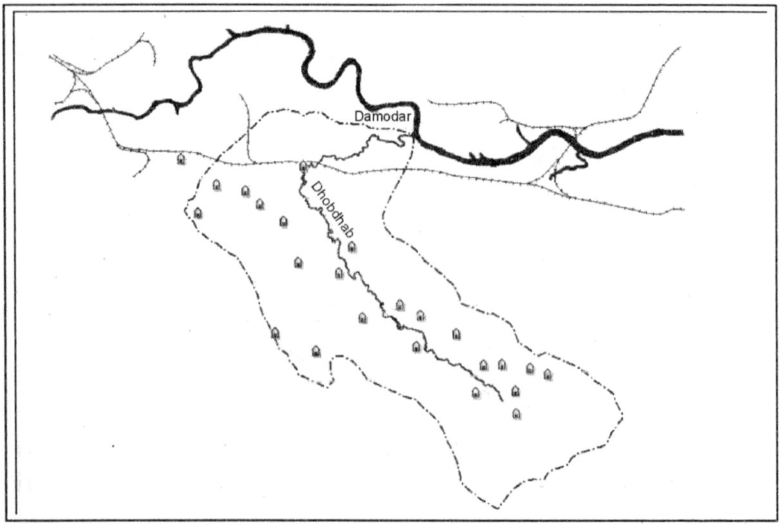

The total area is around 64 square kilometres with 18 villages and a population of around 19,000. The area (as shown in Map 1) under review is characterized by an undulating and rugged topographical terrain, traversed by a number of nalah cuttings with its trunk stream Dhobdhab Nalah, running from south-east to north-west direction, more or less sub-parallel to the hills on northern and southern boundaries. Principal natural potential resources of the area are: (i) rocks and minerals, (ii) vegetation, (iii) surface water and groundwater, and (iv) soil, which are being utilized by rural people as means of livelihood.

## Livelihood, Environmental Planning and the GIS Model

The GIS model with different entities in several organized layers provides an idea about the terrain in question. The Digital Elevation Model (DEM), as shown in Figure 2 depicts the terrain surface with altitudes. The project area is lying between two hill ranges running parallel to NE-SW direction, closed at the south-eastern corner. Dhobdhab Nalah flows between them and

**Figure 2**

*Digital Elevation Model of the Region*

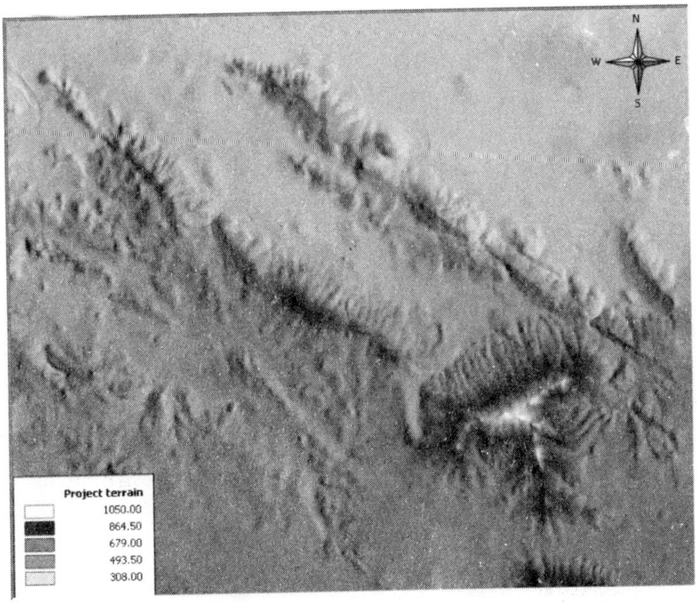

meets Damodar River at the North-West, (see Map 1). This depicts the geomorphology of the terrain with important features like drainage pattern and piedmont at the foothills region and the difference in soil morphology, which in turn helps to understand the variations of vegetation growth in the area.

This model alone raises numerous scientific queries. When various layers like geology, soil, land use, drainage and ultimately the cadastral maps are overlaid on the model terrain, the interrelations of various natural components and human interaction become clear. This also helps to understand the cause and effect of various human activities.

### *Geology and Geomorphology*

The GIS model provides a clear demarcation of various rock bodies depicting effects of environmental agencies on the terrain morphology. The high hill range at southern side along with the highest hill at the south-eastern corner, (Figure 2), consists of amphibolites, hornblende schist rocks and the northern lower hill range consists of two parallel rock bodies, i.e., (i) the northern side coarse-grained porphyroblastic granite, (ii) southern side with hard quartzite bodies. In the main project area, the valley consists of bands of subordinate metasedimentary rocks such as quartzite, mica schist, phyllites, and limestone/calc silicate.

During the later part of Pleistocene epoch, the landform in this terrain has undergone enormous weathering effects converting it into different platform sectors. Varied land practices suitable for that particular type of landform and keeping numerous options for future optimistic approaches with appropriate technical methodology that has not been adopted so far. These are mainly due to non-availability of traditional knowledge and lack of scientific approaches resulting in mishandling of the natural resources.

The model gives the exact location of different land platforms along with existing land uses, which comes from post- and pre-monsoon LISS-PAN merged data.

The characteristics of these platforms are different for different land uses. Piedmont deposits with assorted sedimentary materials in the upper sector near the hills are less suitable for field crop based extensive agricultural practices, but more suitable for horticulture and agro-forestry practices. The next platforms with lower breaks in slopes constitute finer sediments lying along the flood plain of the trunk channel makes it suitable for intensive field crop agricultural practices. The existing land use practices show large-scale deviation from the above practices. It has also been noted that few remote places like two-hill top villages the traditional practices are in accordance with above practice.

Hills and mounds consisting of granite gneiss occur mostly as domes with intense weathering effects due to gullying through all around drainage channels causing enormous soil erosion and breaking up of the rock masses along exfoliation planes making them vulnerable for further erosional hazards. Among other rocks like harder amphibolite group of rocks are comparatively less prone to erosion and the gullying pattern is different from the northern hills. The model not only provides a very clear picture of different drainage patterns in different hill ranges but also helps to make quantitative analysis on drainage morphometry.

### *Soil*

The model shows the distribution of different types of soil in different parts of the terrain in conjunction with geology and geomorphology. The soil analysis data of different soil sample collected from the different parts of the area in the model demarcates the suitability of different crops. It has been noted that in many villages the cropping practices are not in conformity with land suitability, which leads to lower productivity as well as loss of soil fertility.

The model shows two different aspects of soil formation and demarcates their boundary:

(i) The areas with high run-off result in great loss of soil from the original surface, and
(ii) The area where this soil is being accumulated over the piedmont sector of vast pediplain.

Thus, the model shows the necessity of the various conservation measures to be adopted in different parts of the region.

To estimate the soil loss, universal soil loss equation[3] has been used;

$$A = R\,K\,L\,S\,C\,P$$

where

A = estimated gross soil erosion
R = rainfall erosivity factor
K = soil erodibility factor
L = slope length factor
S = slope gradient factor
C = crop management factor
P = supporting conservation practice factor

Now except R, the erosive power of rain, all other factors (KLSCP) combined is the erodibility of land surface. The GIS model provides the required data for calculation of erodibility of land surface for different segment of the project area. The model estimates different quantities of soil losses in different conservation practices to choose the best option.

## Water Potentials

The existing drainage pattern interpreted from satellite imagery incorporated in the model shows their relationship with the geology, geomorphology and soil. The tributaries supply large amount of sediments with run-offs after precipitation, due to high gradient can be identified for measures like gully plugging and contour bundings.

A groundwater contour of the terrain in the model shows the groundwater potential at different parts of the area. The groundwater gradient in the model helps to identify the recharge zone and its characteristics.

The land use practices depend on the availability of the water resources and the surface and groundwater in the watershed region are deeply interrelated. Any change of land use must consider this interrelationship factor. The model shows the region with intensive multi-cropping agricultural practices with large number of wells used for irrigation purposes causing water scarcity in other parts of the watershed region. It also shows the locations of various water harvesting measures required for conservation of soil and water.

### *Vegetation*

Vegetation in the area possesses two very important aspects, such as (i) forestry and (ii) agriculture. The dense forestry in the two ridges, North and South of Dhobdhab Nalah, is a valuable asset to protect the ecological balance. The model provides the existing condition of the forests at different parts of the project area. The foliage cover in the southern hill range is in much better condition, whereas northern lower hills are stripped of the forest cover in most parts. The vegetation map in the model shows: (i) natural growth of vegetation like mahua (*Madhuca Indica*), polash (*Butea Monosperma*) in the vast pediplain zone, (ii) growth of deciduous trees, such as mango *(Mangifera Indica),* jackfruit *(Artocarpus Indica)* and neem in the foothill areas admixed with colluvial slope wash materials. It also locates the suitable area for horticulture practices and agro-forestry to conserve the ecology and provide livelihood to the local people.

In the model it is easier to identify the land with alluvial soil suitable for agriculture and the soil map provides the basis for selection of cropping pattern.

## Conclusion

For any rural development of such magnitude involves large number of people from different spheres of life – the specialists, funding agencies, government bodies and lastly the stake-holders. They need to interact regularly from planning stage to completion of the project. A model with quantification capability provides a powerful interaction platform amongst all the members during different stages of the project. It helps the stake-holders to understand the crucial aspect of the interrelationship of the natural elements on which the livelihood of the local population depends. The model provides the basis on which the future livelihood planning of different spheres of stakeholders can be outlined. The development of this model needs very high level of technical competence with an in-depth understanding of the rural development dynamics of the target region supported by extensive field experience. The utilities of such digital model are many and some of them are:

1. For the analysis of interrelationship of various components of the natural resource base of the target region.
2. For the understanding of human dependence on various components on natural resources, segment-wise.
3. As a common platform for a meaningful participatory planning.
4. For overall understanding about the future impact of the planned/suggested recommendations.

Lastly, it is necessary to realize that the use of such a model in the present development context where rural urban divide is widening becomes essential.

# 20

# Geo-Information Strategy for Rural Development: A Case Study of Aurangabad District, Bihar

Kumar Ramjee Prasad Singh

## The Study Region

The study region is a political unit officially termed as Aurangabad district of Bihar state in the socialist sovereign republic of India. The study region is spread over an area of 3305 sq.kms. Geographically the study area is situated from 24°28′ N lat. to 25°08′ N lat. and from 84°00 East long. to 84°45′ East long. It lies in the southern part of the middle Ganga plain, although its southern part is characterized by the environmental features of the Chotanagpur plateau. The district consists 11 community development blocks, viz., Aurangabad, Deo, Kutumba, Nabinagar, Barun, Obra, Doudnagar, Haspura, Goh, Rafiganj and Madanpur blocks (Figure 1).

## Methodology and Analysis

(1) Geo-informations have been put into GIS (Geographical Information System) of ARC/INFO software.

**Figure 1**

*Aurangabad District: Distribution of Settlements*

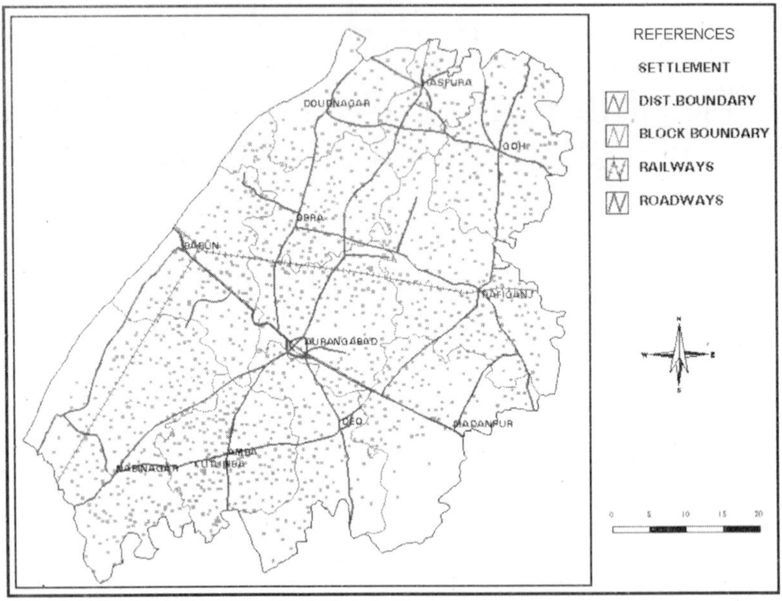


(2) Physical Geo-informations like geology, relief, groundwater potentiality, drainage, soil, land cover/land use (Figure 1,1a) etc., have been prepared with the help of remote sensing technique in which digital analysis have been done using LISS-III – IC 104 – 54 PATH and ROW scene of two season's digital data with the help of Erdas imagine software of image processing.

(3) Information of social provisions like, hospital, primary health centres, dispensary, maternity child welfare centres, post offices, post and telegraph offices, high schools, colleges, training and technical institute—their location have been digitized from information received by district centres, SoI-Toposheet 72-D, (Scale 1: 250,000) and District Census handbook have also been incorporated.

(4) Economic provisions like canal networks, transport networks, etc., have also been digitized and were placed in GIS environment.

**Figure 1a**

*Aurangabad District: Spatial Aspect of Land Use/Land Cover*

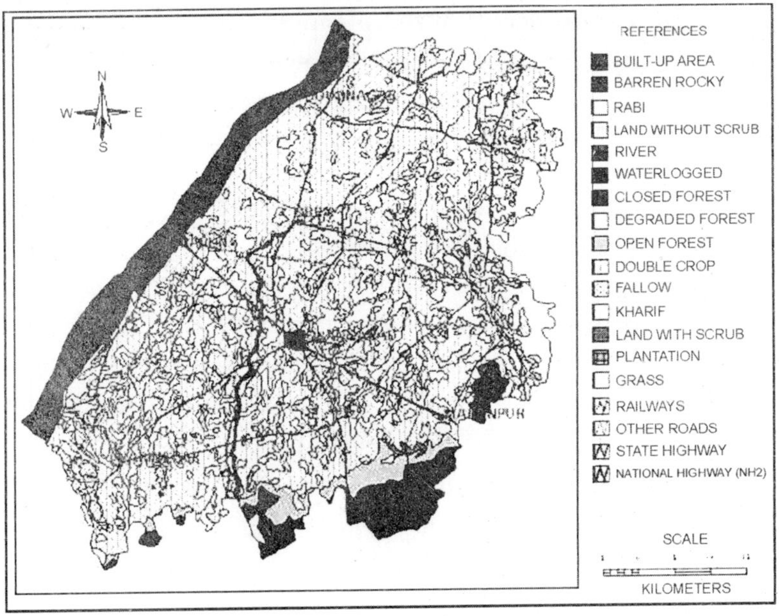


(5) Identification of central places: Those settlements which are predominately concerned with serving the needs of their surrounding areas are called central places. Central places play a very crucial role in the regional development. Their role in the regional development is not only because central places function as economic service centres to its surrounding region, they also spread ideas of modern technology and modern values towards their hinterland, which also helps in better development and prosperity.

The planning processes for decentralized planning has also identified the role of central places. After Fourth Five-Years Plan it was felt that sectoral priorities and planning from top level is of little help in providing better results as far as socio-economic development to the people is concerned. Since then the sectoral approach has been changed into integrated area development in which central places either urban or rural service centres are being considered as focal points for the decentralized planning. This is really a basic approach towards assessing the regional or local resources and evaluating their role in the regional development.

Therefore in the present study central places have been identified, as these places are closely related with the influence areas. Generally degree of development decreases as we go away from central places.

(6) For the identification of central places and service centres, all those settlements having more than 1,000 population have been chosen at first stage. There are total 1,884 villages (Figure 1) out of that only 1,749 villages are inhabited with 20, 04960 persons (2001) in the study area.

Settlement were stratified under the following groups: (Table 1)

**Table 1**

| *Sr. No.* | *Population Size* | *No of Settlements* |
|---|---|---|
| 1. | 1001-1500 | 138 |
| 2. | 1501-2000 | 70 |
| 3. | 2001-2500 | 33 |
| 4. | 2501-3000 | 15 |
| 5. | 3001-3500 | 8 |
| 6. | above-3500 | 23 |
| | Total | 287 |

All 287 settlement were digitized and were kept in GIS environment with their central functions. Later on 71 settlements were selected through GIS query system based on reasoning that these centres are performing more than 7 services ranging from VII group services. Services considered have been listed in Table 2.

**Table 2**

*List of Service Group and Central Services*

| *Service Groups* | | *Services* | |
|---|---|---|---|
| (i) | Administrative Provisions | 1. | District headquarter |
| | | 2. | Sub-division office |
| | | 3. | Police stations |
| | | 4. | C.D. Block |
| | | 5. | Gram Panchayat office |
| | | 6. | Municipal office |
| (ii) | Educational Provision | 7. | Primary school |
| | | 8. | Middle school |
| | | 9. | High school |
| | | 10. | Intermediate school |
| | | 11. | Degree college |
| | | 12. | Traning and Technical Institute |
| (iii) | Medical Provision | 13. | Hospital |
| | | 14. | Health centre |
| | | 15. | State dispensary |
| | | 16. | Nursing home |
| | | 17. | Veterinary centres |
| | | 18. | Veterinary sub-centres |
| (iv) | Communication | 19. | Post office |
| | | 20. | Post and telegraph office |
| | | 21. | Telephone services |
| | | 22. | Bus Stand |
| | | 23. | Railway stations |
| (v) | Commercial Provisions | 24. | Permanent market |
| | | 25. | Periodical hats |
| | | 26. | Commercial bank |
| | | 27. | Electrification |
| | | 28. | Large Industrial unit |
| | | 29. | Small Industrial unit |
| | | 30. | Tiny Industrial unit |
| (vi) | Agricultural Provisions | 31. | Biscouman depot |
| | | 32. | Co-oprative Bank |
| | | 33. | Land Development Bank |

| | |
|---|---|
| | 34. Fertilizer depot |
| | 35. H.Y.V. seed depot |
| | 36. Vaypar Mandal |
| | 37. Agricultural implements repairing centres |
| (vii) Other Services: | 38. Tap water |
| | 39. Urban centres (NAC) |

On the very basis all total 71 settlements (Figure 2) have been identified as central places. There are 66 rural central places and only 5 urban central places. Except 5 urban centres all selected 66 rural central places are dominant in primary service functions. Some secondary functions which are related with primary functions like agricultural implements repairing centres, etc.

Hierarchy of selected central places have been recognized to provide priority in planning at grass-root level. Weighting technique has been applied to provide relative difference of

**Figure 2**

*Aurangabad District: Distribution of Central Places*

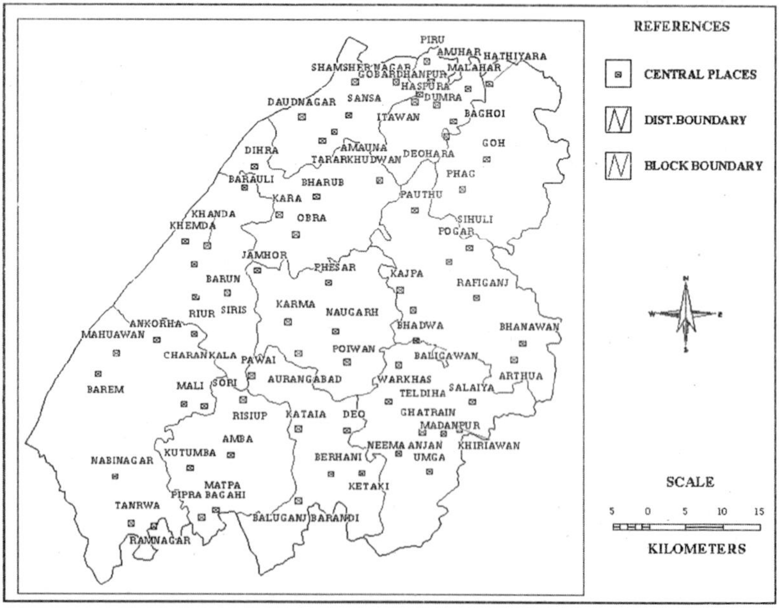

**Table 3**

| *Centres* | *TOT_POPU* | *CEN_FUN* | *TOT_SER* | *TOT_VAL* | *CEN_NDX* |
|---|---|---|---|---|---|
| Aurangabad | 47565 | 36 | 119 | 8071.34 | 14.71 |
| Daudnagar | 30348 | 31 | 79 | 3319.22 | 6.05 |
| Barun | 4485 | 30 | 65 | 2582.40 | 4.70 |
| Nabinagar | 14879 | 30 | 51 | 2303.50 | 4.20 |
| Rafiganj | 18531 | 31 | 57 | 2183.47 | 3.98 |
| Madanpur | 5337 | 28 | 45 | 1635.53 | 2.98 |
| Goh | 7454 | 27 | 42 | 1510.92 | 2.75 |
| Hanspura | 4555 | 26 | 40 | 1489.70 | 2.71 |
| Obra | 7935 | 28 | 43 | 1461.75 | 2.66 |
| Deo | 10066 | 26 | 47 | 1448.41 | 2.64 |
| Jamhaur | 6729 | 24 | 28 | 1447.69 | 2.64 |
| Kutumba | 4800 | 25 | 35 | 1322.70 | 2.41 |
| Amba | 3328 | 20 | 26 | 704.56 | 1.28 |
| Ketaki | 12237 | 17 | 27 | 543.87 | 0.99 |
| Pauthu | 3121 | 15 | 20 | 409.15 | 0.75 |
| Pawai | 3856 | 16 | 22 | 378.40 | 0.69 |
| Khemda | 3992 | 15 | 17 | 351.17 | 0.64 |
| Siris | 1589 | 14 | 17 | 283.69 | 0.52 |

Contd...

Contd...

| | | | | | |
|---|---|---|---|---|---|
| Risiap | 2373 | 14 | 17 | 282.61 | 0.51 |
| Tanrwa | 2473 | 14 | 17 | 257.80 | 0.47 |
| Kataia | 1954 | 10 | 11 | 220.38 | 0.40 |
| Dihara | 5137 | 11 | 12 | 217.18 | 0.40 |
| Tarar | 6374 | 14 | 14 | 216.88 | 0.40 |
| Ankorha | 1259 | 11 | 13 | 211.32 | 0.38 |
| Kara | 4494 | 13 | 15 | 195.51 | 0.36 |
| Baghoi | 2835 | 11 | 14 | 195.12 | 0.36 |
| Khiriawan | 4106 | 10 | 14 | 189.08 | 0.34 |
| Kajpa | 2052 | 10 | 11 | 188.78 | 0.34 |
| Khudwan | 3293 | 13 | 15 | 188.21 | 0.34 |
| Deohara | 3621 | 13 | 15 | 188.21 | 0.34 |
| Baligawan | 2045 | 11 | 14 | 183.33 | 0.33 |
| Phag | 1893 | 11 | 13 | 183.05 | 0.33 |
| Charankala | 1159 | 13 | 15 | 181.03 | 0.33 |
| Bhadwa | 2766 | 11 | 15 | 166.72 | 0.30 |
| Poiwan | 1747 | 11 | 14 | 161.11 | 0.29 |
| Riur | 4485 | 10 | 13 | 158.49 | 0.29 |
| Salayia | 3550 | 10 | 13 | 156.70 | 0.29 |

Contd...

Contd...

| | | | | | |
|---|---|---|---|---|---|
| Ghatrain | 1169 | 11 | 17 | 156.68 | 0.29 |
| Mahuawan | 1430 | 11 | 15 | 154.32 | 0.28 |
| Matpa | 2608 | 10 | 11 | 151.62 | 0.28 |
| Phesar | 1753 | 8 | 10 | 149.36 | 0.27 |
| Shamsher nagar | 7710 | 11 | 15 | 148.57 | 0.27 |
| Mali | 1381 | 12 | 13 | 141.07 | 0.26 |
| Naugarh | 1291 | 9 | 11 | 140.29 | 0.26 |
| Malhara | 1880 | 9 | 12 | 137.35 | 0.25 |
| Pipra bagahi | 2689 | 10 | 11 | 137.14 | 0.25 |
| Sansa | 5947 | 10 | 15 | 136.71 | 0.25 |
| Bharule | 4320 | 11 | 12 | 133.94 | 0.24 |
| Piru | 5705 | 11 | 16 | 128.99 | 0.24 |
| Sihuli | 2184 | 9 | 11 | 128.29 | 0.23 |
| Gobardhanpur kanap | 2054 | 10 | 15 | 126.64 | 0.23 |
| Baluganj barandi | 1181 | 10 | 15 | 122.10 | 0.22 |
| Khanda | 1654 | 10 | 11 | 117.39 | 0.21 |
| Arthua | 2310 | 11 | 13 | 111.22 | 0.20 |
| Amjhar | 4802 | 10 | 13 | 110.25 | 0.20 |
| Warkhas | 2167 | 11 | 12 | 110.04 | 0.20 |

Contd...

Contd...

| | | | | | |
|---|---|---|---|---|---|
| Dhanwan | 1455 | 8 | 9 | 96.08 | 0.18 |
| Karma | 3392 | 10 | 11 | 93.50 | 0.17 |
| Barem | 1739 | 10 | 11 | 93.50 | 0.17 |
| Sori | 1500 | 10 | 11 | 93.50 | 0.17 |
| Umga | 13996 | 10 | 13 | 92.54 | 0.17 |
| Nima anjan | 5973 | 9 | 11 | 87.55 | 0.16 |
| Dumra | 3534 | 9 | 10 | 86.37 | 0.16 |
| Teldiha | 1924 | 9 | 10 | 86.28 | 0.16 |
| Barauli | 1778 | 9 | 11 | 78.13 | 0.14 |
| Itawan | 2928 | 9 | 10 | 76.95 | 0.14 |
| Hathiyara | 4114 | 8 | 12 | 73.27 | 0.13 |
| Berhani | 2653 | 8 | 10 | 71.00 | 0.13 |
| Amauna | 2403 | 8 | 10 | 71.00 | 0.13 |
| Ramnagar | 1908 | 8 | 9 | 69.82 | 0.13 |
| Pogar | 2902 | 8 | 9 | 69.82 | 0.13 |

**Table 4**

| | | | *Hierarchy Level* | |
|---|---|---|---|---|
| *Order of hierarchy* | *Name of hierarchy* | *Centrality index value* | *Number of centre places* | *Name of central places* |
| 1st order | Growth centre | >7 | 1 | Aurangabad |
| 2nd order | Growth point | 5-7 | 1 | Daudnagar |
| 3rd order | Service centre | 3-5 | 3 | Barun, Nabinagar, Rafiganj |
| 4th order | Rural service centre | 3-1 | 8 | Madanpur, Goh, Haspura, Obra, Jamhaur, Kutumba, Amba |
| 5th order | Village service centre | < 1 | 58 | Ketaki, Pauthu, Pawai, Khemda, Siris, Risiap, Tanrwa, Kataia, Dihra, Tarar, Ankora, Kara, Baghoi, Kiriawan, Kajpa, Khudwan, Deohara, Baligaon, Phag, Charankala, Bhadwa, Poiwan, Riur, Salayia, Ghatrain, Mahuwan, Matpa, Phesar, Samser nagar, Mali, Naugarh, Malhara, Pipra, Bghoi, Sansa, Bharuli, Piru, Sihuli, Gobardhanpur, Kamnpi, Baluganjbrandi, Khanda, Arathua, Amjhar, Warkhas, Dhanwan, Karma, Barem, Sori, Umga, Nimaajan, Dumra, Teldiha, Baruli, Itwan, Hathiyara, Beshani, Amauna, Ramngar and Pogar. |

**Table 5**

| *Sr.No.* | *Name of Growth Centres* | *Percentage of Working Population in Secondary and Tertiary Activities* | *Existing Important Industrial Works* | *Proposed Industrial Works* |
|---|---|---|---|---|
| 1. | Doudnagar | 64.67 | Brass and pottery works, carpet weaving, blanket weaving, lac works, brick klin, stone crushing, cloth weaving, handloom, oil and flour mills, rice mills. | Glass industry, sugar industry, food processing industry, dairy industry, woolen clothes industry, etc. |
| 2. | Barun | 21.14 | Vanaspati ghee, glass, smokeless fuel, stone crushing, rice clin, oil and flour mills, etc. | Iron and steel works, food processing, hush based paper industry, sugar industry, etc. |
| 3. | Nabinagar | 48.75 | Cloth weaving, handloom works, biri making, stone crushing, rice mills, pottery making, basket making, lac works. | Agro-based industries like sugar.food processing, dairy, etc.. Forest based industries like paper, brick klin, glass industry, etc. |
| 4. | Rafiganj | 79.33 | Cloth weaving, handloom, biri, stone crushing, pottery making, basket making, lac works. | Food processing, stone chips, dairy industry, cloth industry, readymade garments industry, etc. |
| 5. | Madanpur | 55.03 | Rice and flour mills, brick clean, cloth weaving, blanket weaving, stone crushing, lac works, etc | Food processing, paper industry, lac industry, ready made garments, mat weaving, etc. |
| 6. | Goh | 41.53 | Rice, flour and oil mills, stone crushing, biri making, handloom and pottery works. | Food processing, packaging, sugar, glass industries, etc |

Contd...

Contd...

| | | | | |
|---|---|---|---|---|
| 7. | Haspura | 73.65 | Cloth weaving, handloom, pottery works, brick clin, rice and flour mills, etc. | Readymade garments, dairy, food processing, hush-based paper and power industries |
| 8. | Obra | 71.68 | Carpet weaving, dyeing, cloth making, blanket weaving, carpet weaving, handloom and powerloom industries, etc. | Carpet weaving, readymade garments, blanket weaving, dairy and food processing, dyeing industries, sugar industry, etc. |
| 9. | Deo | 54.01 | Blanket weaving, handloom and power loom, brass works, earthen tile, pottery making, rice and flour mills. | Food processing and dairy industries, cement, carpet weaving, textile, readymade garments, fertilizer, tourism industries, etc. |
| 10. | Jamhaur | 46.23 | Rice and flour mills, handloom and powerlooms, blanket industry, brick klin, pottery and earthenware. | Cement, dairy, carpet weaving, readymade garments, textile, fertilizer, cardboard industries, etc. |
| 11. | Kutumba | 13.33 | Stone crushing, cloth, handloom blanket weaving, rice and flour mills. | Rice and flour mills, paper and hardboard industries, packing, cork, glass, cement, plywood and building industries, etc. |
| 12. | Amba | 52.57 | Carpet and blanket weaving, rice, flour, lac works, wooden works, earthen pottery and tile industries, etc. | Furniture, sugar, food processing, dairy, glass, hardboard, paper, plywood, lac works, glass industries, etc. |
| 13. | Pauthu | 14.34 | Rice and flour mills, sugar, rope weaving, earthen tile and pottery making, etc | Hush based paper power, dairy, iron and steel works food processing, ready made garments industries, etc. |

services to measure the relative importance of services and central places. Finally, centrality has been determined.

The principle involved adopting the technique is that the scarcity increases the importance of function in terms of centrality, as such the weight of that function has been kept high.

The used technique for weighting the functions are as follows:

$$W_i = \sqrt{\frac{P}{N}}$$

where $W_i$ = awarded score value (centrality value with reference to the ith function)

P = Total Population.

N = No of occurrence of a particular service in the study area.

Population has been rooted over to get relative value at the same time the awarded score value will not be very high.

Table 3 shows selected central places, population, central function, total services, total score value, and centrality index.

With above weightage technique five orders of central places have been recognized (Table 4), when compared with national strategy some deviation have been established. At the national level there are only four hierarchical order, while in the present study there are five orders without taking growth pole. In the present study growth centre is in the highest order, followed by growth point, service centres, rural service centres and village service centres.

(7) Those central places have been defined as rural growth centres which qualify as growth point under the growth pole strategies. All proposed rural growth centres have been kept on geo-information layers in GIS environment and a humble attempt has been taken in the recommendations of probable industrial growth. Keeping local raw materials nearness to market transportations facilities and other geo-informations, etc. (Table 5), one more centre place Pauthu has been included as it is having the 3rd largest size of population of block.

**Figure 3**

*Aurangabad District: Composite Hinterlands of Service Centres*

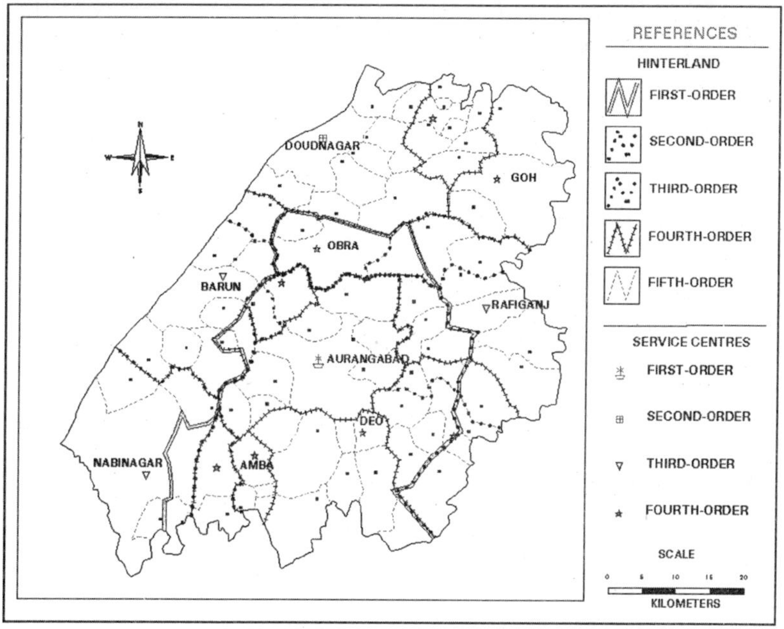


The above study shows that Geoinformatics can play a decisive role in the rural development as it provides better visual aid shares and query system to reach on certain decision-making for the decentralized planning.

## Select References

Bhat, L.S. 1976, *Micro-Level Planning: A Case Study of Karual Area,* K.B. Publication, New Delhi.

Brush, J.E. 1953, The Hierarchy of Central Places in South-western Misconsin, *Geographical* Review No. 43.

Burrough, P.A. 1986, *Principles of Geographic Information System for Land Resources Assessment,* Oxford: Clarendon Press:

Canon, C. 1937, *An Enquiry into the Nature and Causes of the Wealth of Nation,* New York Publication.

Christaller, W. 1933, *The Central Places in Southern Germany,* translated by C.W. Baskin, New Jersy, 1966.

Mark, D.M. et al., 1994, Determining spatial relations between lives and regions in ARC/INFO using the intersection Model; in proceedings of the 14th Annual ESRI user conference.

Morehouse, S. 1985, ARC/INFO: A georelational model for spatial information, in proceedings of AUTOCARTO 8. Falls Church.

Ozemoy, V.M. et al., Evaluating Computerized Geographic Information System using Decision Analysis Interfaces V.II.

Sen L.K. et al. 1971, Planning Rural Growth Centres for Integrated Area Development. A Study in Hiryalguda Taluka, National Institute of Rural Development, Hyderabad.

Singh, R.B.P. 1985, *Social Welfare for Rural Development A Case Study,* Inter India Publications, New Delhi.

Tiwary, P.C., 1988, Regional Development and Planning in India, New Delhi.

# 21

# Land Information System for Rural Development

Rajiv Narayan

Land is the foundation of all forms of human activity; from it we obtain the food we eat, the shelter we need and the space to work. The land is man's most valuable and precious resource. Very recently the need for thoughtful and careful stewardship of the land, together with the more intensive use and management of its resources has emerged as a matter of major global concern. Utilization of land for agricultural purposes and the dependence on agriculture for the political, social and economic stability will continue to a large degree in the rural sectors of the developing countries of the world. Though it is a very narrow concept but the rural development is very much interconnected with development of education facilities, health care, etc. But only land development in the rural areas can be one of the effective measures for solving the problems of a high rate of population growth and rising unemployment. The socio-economic problem of inequitable distribution of wealth evidenced by the disparity of income levels between the urban and rural areas can be solved with land development in this part of the world.

Rural development is advocated today as a basic strategy for economic development all over the world. The sheer largeness of numbers who continue to live and work in the rural areas in the developing countries compels our attention to its important sector of the economy. Conceptually rural development is taken as development of the rural areas. Often, rural development has meant the extension of irrigation facilities, expansion of electricity, improvements in the techniques of cultivation, construction of school buildings and provision of education facilities, health care, etc. But only land development in the rural areas can be one of the effective measures for solving the problems of a high rate of population growth and rising employment in the developing countries. The socio-economic problem of inequitable distribution of wealth evidenced by the disparity of income levels between the urban and rural areas can also be overcome through planned and economically viable land development projects.

The importance of land development has led to a revaluation both for information about land and of the strategies and programmes that may provide it. Increasingly it has been recognized that policy makers, planners, land administrators and individual citizens all have a need for information about the land and make significant use of spatial data. Accurate knowledge of natural resources and accurate descriptions and record of such knowledge are the first essentials to their rational use and conservation.

The land information is a prime requisite for making decisions related to land investment, development and management. The information value and the effectiveness of it in decision-making process are directly related to the quality of the information and the manner in which it is made available.

The responsibility for providing the information is being taken on by complex, diverse group of individuals and institutions who make up what may be described as the "land information management community". It is a community which includes geographers, surveyors, cartographers, foresters, valuers as well as system engineers, computer scientists, record managers, land planners, lawyers and resource specialists. It is a community which is increasingly interested not only in the technology for gathering and processing information and in the design and development of land information systems but also in the policies and strategies for their effective use.

## Land Information System

An information system can be defined as a combination of human and technical resources, together with a set of organizing procedures that produces information in support of some managerial requirement.

Data are raw collection of facts. Data relating to land may be in alphanumeric (cadastral record) or graphical (maps) or digital (using modern procedure) forms. To become information, raw data must be processed so that they can be understood by a decision maker. A land information system gives support to land management by providing information about the land, the resources upon it and the improvements that can be made to it. The operation of a land information system includes acquisition and assembly of data, their processing, storage, maintenance, retrieval analysis and dissemination. The usefulness of such a system depends upon its updatedness, accuracy, completeness and accessibility and also upon the extent to which the system is designed for the benefit of the user rather than for the producer of the information. The land information systems include environmental information, infrastructure information, cadastral information and socio-economic information.

They may be designed to serve one primary function or can be multifunctional. They may have been for strategic planning or for management control or operational control. Each requirement needs a special set of information for cadastral information systems to handle data of land parcels. The main function of this information system is to provide data on land ownership, value and use. Cadastral information system is developed using the cadastral map showing the land parcels and the land records data.

Other referencing systems such as co-ordinates can be added to facilitate data manipulation and the exchange of information with other systems. Important uses of cadastral information systems are rural land development, upgradation of wasteland and strategies for land use management.

Computerized land information systems (LIS) have become useful tool for public and private organizations which perform a wide variety of geographical analysis and land management functions. Parcel based land information system are a special type of LIS which focus on the inventory and management of well-defined units of land ownership. Interest in land management is not a recent concern. Agencies involved with deed registries and tax assessment, as well as private interests concerned with development potential, have long maintained system pertaining to the cadastre, defined as "a record of interest in land, encompassing both nature and extent of these interest" (McLaughlin, 1984: 82). These records, however are often maintained by separate jurisdictions at separate locations, resulting in duplication and error. Those who need access to this information do not always know where it resides. Furthermore, information may need to be compiled for several sources are not always directly compatible. So, over the time, there has emerged a movement to modernize and automate land records. Though obviously there is some disagreement in the terminology referring the automation and modernization of land records, McLaughlin and Nichols (1987) advocate a taxonomy of information systems in which a multi-purpose cadastre (MPC) is considered a special case of a LIS, which is a special case of GIS. Lodwick and Feuchtwanger (1987) consider LIS and GIS as essentially the same in the point of view of analytical functions and other operations performed on the data. Meltz (1988) offers the useful distinction that the principle of a LIS is on the land parcel while the database architecture of a GIS is organized around mapable features, but also notes that in terms of functionality, a LIS can effectively serve as a GIS.

## Case Studies

1. In Thailand a 20-year project titled "The Land Titling Project" has been undertaken in 1985 for the comprehensive reform of its land record system, focused on the issuing of title deeds to landowners and improvements in land administration. This project aims to establish the following:
   (i) A system of survey and mapping control, based on the national geodetic network, with positions expressed in the Universal Transverse Mercator (UTM) Co-ordinate system.
   (ii) A set of new base maps, in the form of photomaps, produced by the rectification. In rural areas the maps will be at a scale of 1:4000, and in the urban areas at 1:1000. There will be approximately 90, 000 new rural maps and 15, 000 new urban maps.
   (iii) A complete new set of cadastral maps, based on the photomaps and at the same scales – a total over 100,000 maps.
   (iv) A highly decentralized system of land offices, arranged for the convenience of landholders who register land transaction there.
   (v) A system of land records in the land offices, which will be complete, secure, up-to-date and effective.
   (vi) Land valuation information, which is being built up rapidly and will eventually include valuation rolls for all local authority areas in Thailand.
2. In 1985 Govt. of Tanzania had started digital database for development proficiency. Main social and technical purpose of this project are following:
   (i) Establishment of land inventory system to assist in a variety of long range planning and infrastructure.
   (ii) Helping in the levying equitable property tax.
   (iii) Supporting in the introduction in the land consolidation when it becomes necessary.
   (iv) Introduction of fast and more comprehensive storing and retrieving land records.

(v) Effective maintenance and security of records.
(vi) Maintenance of data quality standard.
(vii) To enable on-line access by the public to the cadastral record.

3. In the Upper Komering watershed of southern Sumatra a study was conducted to analyse the present patterns of land use and current dynamics of resources utilization for the development of information system management of natural resource system. The first part of the study deals with analysis of the farming system in the watershed and dynamic tendency over time. The second part deals mainly with the matching data from different sources to extrapolate "Farm survey information system" to obtain data requirement for the projection of future land use.
4. A wide assorted data was combined with the help of GIS for the assessment of present and potential agriculture production in rural Argentina. The study area is south of Rojas in the northern part of Buenos Aire province. The database includes topography and soil type to identify erosion hazard zones. Satellite imageries were interpreted and field surveys were conducted to identify present land use system. Each land parcel was assessed in terms of over and under utilization. Finally, this information system helps to determine the potential profit for farmers by identifying the proximity to grain storage facility and availability of easy transport facilities. This study is important for the rural development.

In Indian context, the project Natural Resources Data Management Systems conceived by the Department of Science and Technology, Government of India aimed towards the development of spatial information system and the decision support tool for integrated planning and management of resources at the rural level. Adopting the concept of GIS methodology, resources profiles encompassing both natural resources and socio-economic sectors have been built in a number of districts in the country. In order to provide software support for data analysis, modelling and applications, an indigenous GIS package GRAM has been developed. Studies have been carried out for rural development in the sectors of groundwater development, infrastructure management, wasteland reclamation, crop production, etc. The National Afforestation and Eco-Development Board, Ministry of Environment and Forests, Govt. of India has supported multi-institutional project on the line of GIS technology for the wasteland development at the rural level. Under this project, land information systems have been developed using cadastral records at the village level and strategies for integrated wasteland development have been suggested. Also the district level planners in our country are exploring the possibility of building cadastral information systems for rural land management.

## References

Lodwick, G. and M. Feuchtwanger (1987). Land Related Information Systems, UCSE Report No. 10010 (Calgary: Department of Surveying Engineering, University of Calgary).

McLaughlin, J. (1984). The multi-purpose cadastral concept: Current status, future prospects. In Seminar on the Multi-purpose Cadastre, edited by B. Neimann (Madison, Wisconsin Institute for Environmental Studies), pp. 82-93.

McLaughlin, J. and S. Nichols (1987). Parcel-based Land Information Systems, *Surveying and Mapping*, 48, 35.

Meltz, S. (1988). An Approach for Implementing Land Information Systems for Rural and Urbanizing Counties, *Surveying and Mapping*, 48, 35.

Merolla, Silvia, et al. (1994). A GIS Application for Assessing Agricultural Land. *ITC Journal*, Vol. 3, pp. 264-269.

Pauli Sadki, J. (1989). Development of a Cadastral Database System for the Mbeya Region of Tanzania, *ITC Journal*, Vol. 3 and 4, pp. 217-220.

Zabel, Peter et al. (1988). Farming Systems in the Upper Komering Watershed, *ITC Journal*, Vol. 1, pp. 87-90.

# 22

# Application of Remote Sensing and GIS in Rural Planning: A Case Study of Ranchi District

Jitendra Shukla

Modern paradigm of development and planning is largely influenced by mechanistic, 'system' thinking where man and nature are seen as separate entities: man as a creative originator and consumer and nature as a source of commodity. Spatial planning is concerned with identification and choice of location in a region from where a wide range of socio-economic infrastructure and service facilities are to be provided effectively. For sustainable development of district/block planes, information on all respect of resources, its utilization level, growth tends and future prospects and social structure and preparation of well defined rural land use/land cover map is requisite. The advance techniques of aerial photographs, remote sensing and GIS provide quick, reliable, accurate and unbiased information in spatio-temporal analysis.

Space is common denominator in all human activities and therefore spatial planning cuts across and integrates all human activities (Mishra, et al., 1974). The resources managers and planners need detail, reliable, timely and accurate data of a region and comprehensive land use planning can help in solving the loss of agricultural land and degrading environmental quality. The function of land use planning is to guide decisions on land use in such a way that the resources of the environment are put to the most beneficial use for man, whilst at the same time conserving those resources for the future. The remote sensing technology has the potential to make most significant contribution in the area of land use/land cover data collection specially in agriculture, urban area analysis, wasteland, vegetation coverage, etc.

## Objectives of the Study

To demonstrate the use of satellite remote sensing and geographical information system in the development and planning in rural areas, focus on specific problems and solutions and to evolve strategy for sustainable development of natural resources and protection of degraded ecosystem. An attempt, therefore, has been made to:

— assess the availability of natural resources with emphasis on land use/land cover and water potential of Ranchi District,
— identify graded and waste land areas and study the changes in land use over a period of twenty-eight years,
— estimate the present infrastructure facilities for development and trends of utilization of resources,
— evaluate any gap in resource identification and utilization, and
— suggest suitable measures for micro-level and area specific plan.

## Study Area

Situated in the middle of Chotanagpur plateau, Ranchi District consists of 22°5′ N to 23.6°N and 84°9′ E to 85°9′ E, having 7,689 sq. km. It is inhabitated by population of 3,783,577 according to 2001 census. Out of total geographical area of 7,698 sq. km., 184.55 sq. km is occupied by urban land use. As per 2001 census the study area is inhabitated by 1,806,682 rural population.

## Methodology and Data Base

For the compilation of land use/land cover variables the methodology as shown in Figure 1 has been adopted.

**Figure 1**

*Methodology and Data Base*

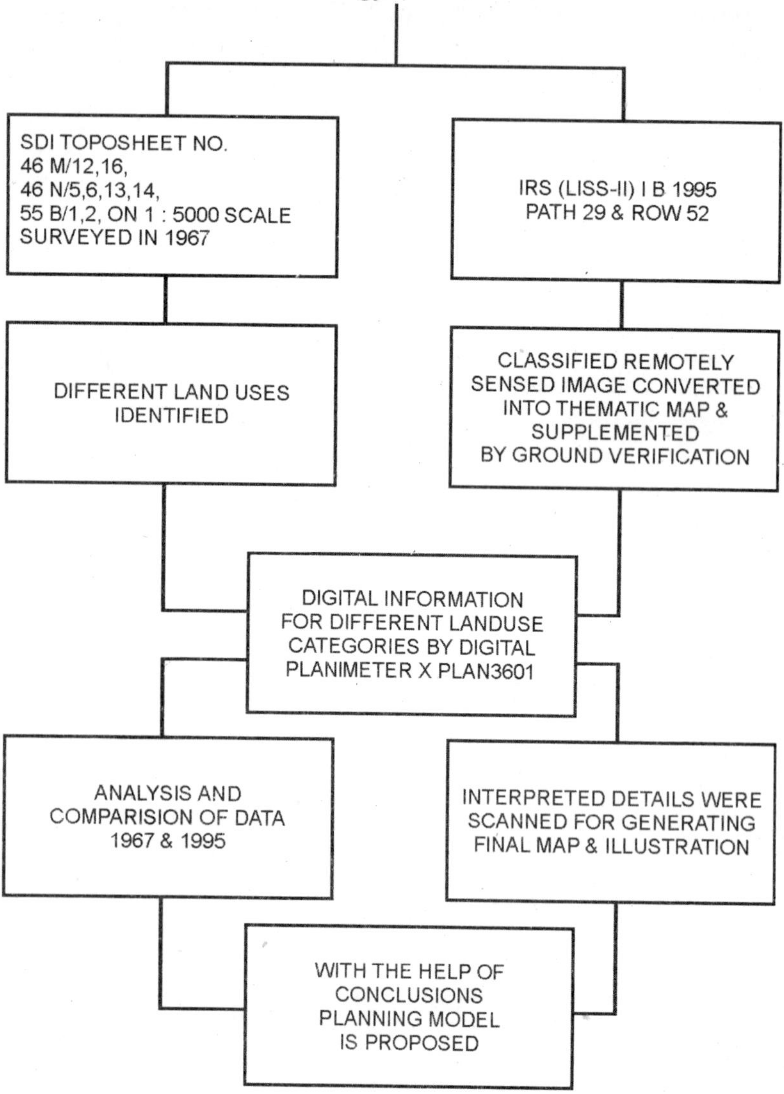

## Results and Discussions

Sustainable developments in its simplest form is defined as development of natural resources to meet the immediate needs of the present population and requirements of future generations without anyway, and endangering the ecology and environment (Anonymous, 1991). For integrated development of rural infrastructure the resource utilization and management is basic on 'integrated area development' approach in which the evaluation and assessment of natural resources have been based on optimal utilization, multiple use, no wastage, conservation and management of natural resources. The present land use/ land cover systems must be borne in the mind before planning strategy for the development of the area is suggested:

1. Since Ranchi is a sprawling city, there appears a drastic change in urban expansion. The percentage increase of 716.53 in urban land has been noticed within a span of 28 years. The urban area of 184.455 sq. km however, includes township Ratu, Kanke, and Ormanjhi also. They contribute 28.5 and 19.35 sq. km respectively. As per 1981 census they were rural in character. Thus, out of a total area of 948.77 sq.km of Ranchi District approximately 20 per cent area is contributed by urban land use. This not only suggests intensive use of land but also inefficiency of existing land use, because urban sprawl has taken place at the cost of fertile agricultural land around the metropolis.
2. The city sprawl also confirms the sprawling population of urban areas which is over 27 lacs according to the 2001 census. This also suggests high degree of organization in study area whilst rural population (1.60 lacs) lives in 154 villages and percentage of increase in rural population is very small.
3. The agricultural land has been reduced. This loss of agricultural land (22.88%) may be attributed to urban sprawl. M.Northam Ray (1979) has repeatedly emphasized that the sprawling city reduces the efficiency of land use.

# Table 1

| | Land use Categories | Area $km^2$ | Area in% | | | Area $km^2$ | Area in% | Absolute change in area 67-95 | Variation in% |
|---|---|---|---|---|---|---|---|---|---|
| 1. | Agricultural Area | 2665.34 | 73.13 | Agricultural | Irrigated Non-Irrigated | 116.50 | 12.28 | +152.203 | –22.08 |
| 2. | Urban Area | 522.59 | 2.38 | | Ranchi | 136.637 | 41.80 | | |
| | | | | Urban Area | Ratu | 28.50 | 3.00 | | |
| | | | | | Kanke | | | | |
| 3. | Rural Area | 755.91 | 2.98 | Rural Area | | 30.25 | 3.19 | –2.00 | +7.08 |
| 4. | Hill Forest and Reserve Forest Area | 2151.74 | 15.99 | Hill Forest and Reserve Forest Area | | 160.128 | 16.88 | –8.388 | +5.53 |
| 5. | Water Bodies | 212.50 | 1.32 | Water Bodies | | 12.25 | 1.29 | +0.25 | -2.00 |
| 6. | Wasteland Area including Stony waste | 431.15 | 3.28 | Wastland Area | | 12.05 | 1.27 | +19.10 | -48.39 |
| 7. | Open Scrub Area | 7698.00 | 237.20 | 3.92 | Open Scrub Area | 36.50 | 3.85 | +0.70 | -1.88 |

*Sources:* SoI Toposheet IRS (LISS II) IB 1995 No. 73 E/6, E/7, E/4

**Figure 2**

*Planning Strategy for Agricultural Areas*

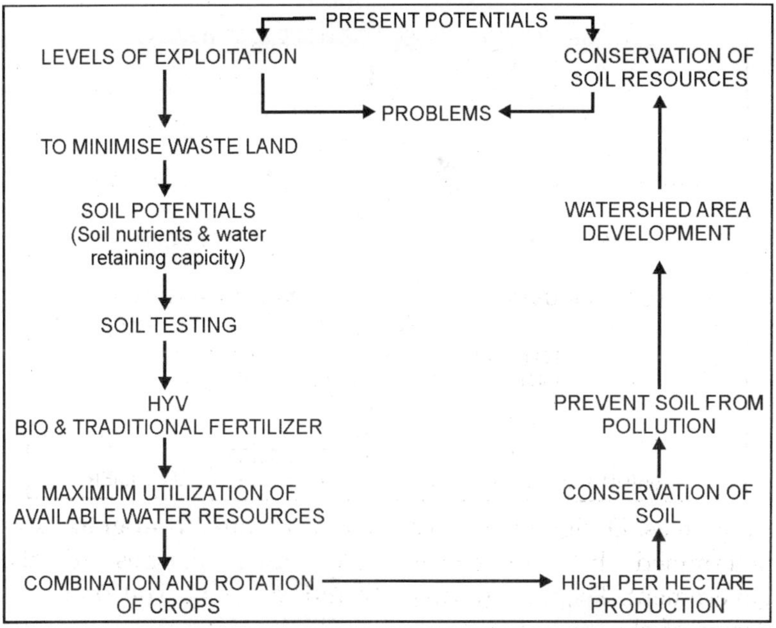


4. The contraction of wasteland by 48.39 per cent suggests the intensity of agricultural land use. D. Stamps (1964) has also emphasized that there should be no land which is not serving some useful purpose for the community.
5. It is interesting to note that a positive percentage variation of 5.53 has been observed in forest coverage of Ranchi District.
6. Insignificant areal changes have been recorded in rural settlements (plus 7.08%), water bodies (minus 2.00%), and open scrubs (minus 1.88%).

Integrated development of rural infrastructure: physical, economic and social is another approach that has engaged attention of planners and policy makers. Since agricultural development must be taken as a key to sell sustaining growth of rural areas, a rural development model can be designed around agricultural development strategy which will definitely be beneficial for all farmers having large or small landholdings and which can provide the basis for other development activities in

**Figure 3**
*Subsystem Model*

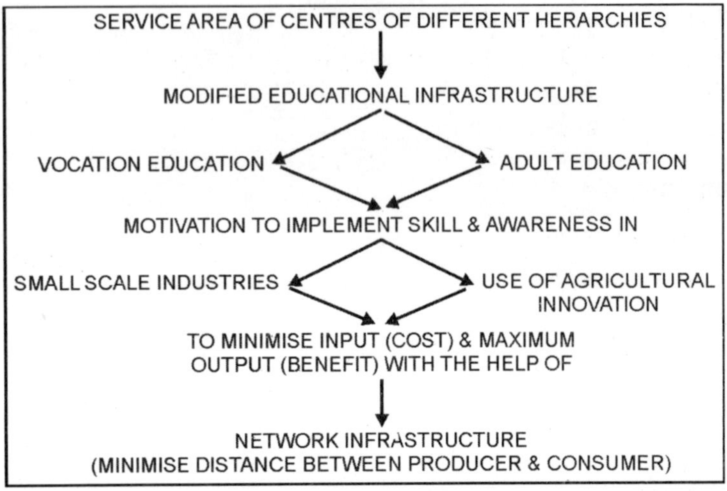


rural areas. Design of an integrated development programme is determined by existing local potential policies of the government, keeping in view the future needs. The proposed design of planner may be:

Agricultural development, although basic to rural development, cannot make headway without intervention in the sphere of education, training and perception towards innovation and raising level of aspiration of the farmers. Simultaneously, introduction of small scale industries cannot proceed too far without giving stress on awareness of environment ecology, occupational training, infrastructure and linkage network to markets and consumers. In addition to this, another subsystem model for harmonized development in rural sector is proposed. Subsystems can be rural settlement pattern service area of service centres of different hierarchies, more effective infrastructure network, employment pattern, rural organization and institution systems.

Specific programes for increasing productivity, improving level of understanding, motivation and perception is needed. Though design of an integrated rural development is determined by existing local potentials and policies of the government but their must be coordination between plan and policy maker, and rural population of the area to know what their problems are? Whether integrated development plan meets their problems and needs? What is their level of perception? Stages of implementation of plan and follow-up programmes for evaluation of planning model.

## Conclusion

The study has resulted in number of significant conclusions:

— With the use of multi-date, multi-type, large scale data, it was possible to prepare land use map with all the information needed for planning the afforestation and soil conservation activities. Mapping of rural environment provides the basic platform for developmental plans and factorial design.

— To prepare detailed report of eco-degraded area with the use of remote sensing data.

— All planning design and execution should be based on scientificly established database. The local planners and administrators must be aware that their efforts can be more effective if a systematic analysis of the total rural environment is undertaken by interdisciplinary approach.

— Planner can use scientific modelling techniques in understanding cause and effect relationship of land uses, landuse allocation to minimize soil degradation and pollution.

— For the better planning and favourable results of the planning of the study area, the participation of local people may be sought so that grass-root approach may be persued. After implementation of the planning strategy the feedback programme is to be sought after regular interval so that the viability of the proposed strategy can be evaluated and can be supplemented by a modified course of action.

## Select References

Stamps, D. (1964). *Applied Geography,* Penguin Book, London, pp. 65, 100-101.

Food and Agriculture Organisation (1977). *A Framework for Land Evaluation,* International Institute for Land Reclamation and Improvement, Netherland, No. 22.

Ray, M. Northam (1979). *Urban Geography,* John Wiley and Sons Inc. New York, p. 469.

Mishra, R.P. et al. (1974). *Regional Development Planning in India: A New Strategy,* Vikas Publishing House Pvt. Ltd., New Delhi, p. 40.

# 23

# GPS-GIS Integrated System for Dynamic Route Guidance

U.C. Sahoo

Though GPS and GIS are no more new terms to the world, but its use in the field of transportation engineering is becoming more popular nowadays. As far as land transportation or vehicle navigation is concerned, GPS can be applied to all types of Intelligent Transportation Systems (ITS), where positioning requirement is there.

Transportation data is usually associated with spatial data, like traffic counts from particular sites, the traffic volumes along particular roads or links, etc. Geographical Information System (GIS) can be used as a database for storing these transportation data. The primary advantage of using GIS as a database for transportation data is the fact that GIS can integrate the spatial data and display the attribute data in a user-chosen format. The Global Positioning System (GPS) is widely being used as a tool for collecting the spatial data. Systems which chiefly use GPS as a spatial data source for a GIS are called as GPS-GIS integrated systems.

## Applications of GPS-GIS Integrated System

The various applications of GPS-GIS integrated system in the field of ITS requiring real-time traffic information are:

1. Advanced Traveller Information System (ATIS)
2. Automatic Vehicle Location (AVL)
3. Dynamic Route Guidance (DRG)

Apart from these above applications, GPS/GIS can also be used in the following areas of ITS:

- Advanced Traffic Management System (ATMS)
- Advanced Public Transport System (APTS)
- Commercial Vehicle Operation (CVO)
- Electronic Toll collection and Traffic Management (ETTM)

## Overview of GPS and GIS

### *Global Positioning System (GPS)*

Development work on Global Positioning System (GPS) commenced within the Department of Defense (DOD) in 1973, the motivation being to develop an all-weather, 24-hour positioning system to support the positioning requirements for the armed forces of United States and allies, civil aviation, maritime, surveying and mapping, construction, mining, earth

**Figure 1**

*GPS-GIS Integrated System for Different ITS Applications*

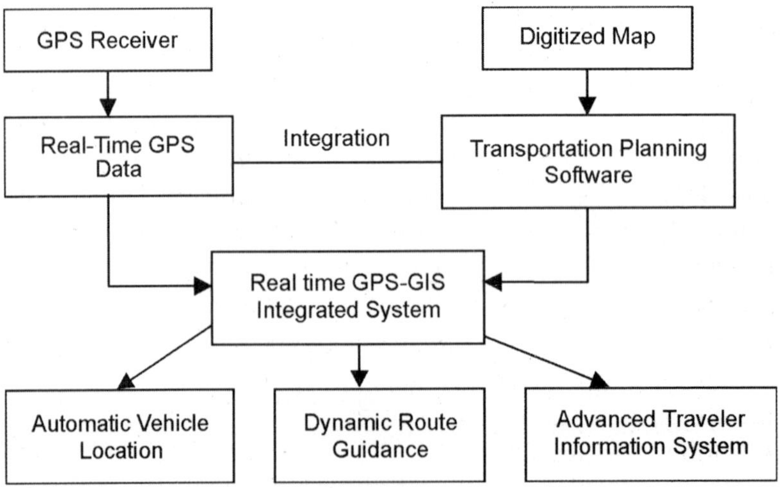

sciences, electric power systems, telecommunication and outdoor recreational activities. GPS is a satellite based radio navigation system, which provides practical and affordable means of determining position, velocity, and time around the globe. Civilian access is guaranteed through an agreement between the DOD and Department of Transportation (DOT).

GPS consists of three main parts:

1. Space segment (satellites)
2. User segment (receivers)
3. Control segment (management and control)

The space segment consists of the constellation of satellites and the signals broadcast by them that allow the users to determine the position, velocity and time. The user segment is the receiver part, which is a combination of hardware, software and operational procedures. From the location and navigation point of view, these two segments are of our interest. The observation of at least four satellites simultaneously permit determination of the 3D co-ordinates of receiver and the time offset between the receiver and GPS time.

### *Geographical Information System (GIS)*

Geographic Information Systems (GIS) provides a current, spatial, visual representation of transit operations. It is a special type of computerized database management system in which geographic databases are related to one via a common set of location coordinates. This allows information to be developed and displayed to assist operators, dispatchers, and street supervisors to make on-the-spot decisions, and to assist planners in service assessment, restructuring and development. GIS is most often used for:

- Transportation planning and modelling;
- Demographic analysis;
- Route planning, analysis and restructuring;
- Bus dispatch and scheduling;
- Bus stop and facility inventory;
- Rider ship analysis;

- AVL and monitoring;
- Para transit scheduling and routing; and
- Accident reporting and analysis

Spatial data is the crucial component of a GIS. The important sources of spatial data are the already existing digital files, maps, which can be digitized, and more recently GPS (mapping type receivers).

## Basic Modules of Dynamic Navigation System

### *Digital Map Database Module*

A digital map database is an indispensable module for any vehicle and navigation system that involves map related functions. Without a map, it is very difficult for a traveller to explore an unfamiliar area and make correct decisions concerning the route. With a map as medium, complex information can be communicated very easily. In general, there are two methods by which a computer can present a map to a traveller. One is to digitize a paper map (or photographs) using a scanner so that it can be stored in and retrieved from computer memory as a digitized image or a raster-encoded structure (raster encoding). The other is to convert a paper map into a data structure or a vector encoded structure so that it can be stored in and retrieved from computer memory easily, and presented in different ways to the user based on the functions or features required (vector encoding). Both the methods require the use of GIS.

### *Positioning Module*

Positioning involves determination of the co-ordinates of a vehicle on the surface of the earth. Location involves the placement of the vehicle relative to landmarks or other terrain features such as roads. These tasks are accomplished by the positioning module in conjunction with other modules. Three positioning technologies are most commonly used: stand alone, satellite based, and terrestrial radio based. A common satellite-based technology involves equipping a vehicle with a global positioning system (GPS) receiver.

### *Map- Matching Module*

To provide drivers with proper manoeuvring instructions or to correct display the vehicle on a map in an error-free fashion, the vehicle location must be precisely known. Therefore, an accurate vehicle location is considered a prerequisite for good system performance. The positioning system can track the position of a vehicle relative to another position, such as an origin. Typically, vehicle heading and distance traversed are used to determine incremental changes in the position of the vehicle relative to the origin. These sensory systems have errors associated with their operation. The result is more uncertainty in the true position of the vehicle. To deal with this uncertainty, map-matching systems have been developed to match the position given by the positioning system (or trajectory) with a position associated with a location on a route map.

### *Route-Planning Module*

Route planning is a process that helps vehicle drivers plan a route prior to or during a journey. It is widely recognized as a fundamental issue in the field of vehicle navigation. Route planning can be further classified into either multi-vehicle (system-wise) route planning, which plans multi-destination routes for all vehicles on a particular road network, or single-vehicle route for a single vehicle according to the current location and a given destination.

### *Route-Guidance Module*

Route guidance, an integral part of Intelligent Transport System (ITS), is the process of guiding the driver along the route generated by the route-planning module. Guidance can be given either before the trip or in real time while en route. The pre-trip guidance could be presented to a driver as a printout. Such a printout would be similar to travel trips provided by a travel agency, but with detailed turn-by-turn driving instructions, or just like the output generated at a kiosk near the car rental counter. These instructions might include turns, street names, travel distances, and landmarks. On the other hand, en-route guidance would require providing turn-by-turn driving instructions to a driver in real time. It is much more useful, but requires a navigable map database, an accurate positioning module, and demanding real-time software and computational power.

### *Human-Machine Interface Module*

The human-machine interface is a module that provides the user with the means to interact with the location and navigation computer and devices. This is often treated under ergonomics or human factors engineering in the literature. Researchers in the field of ergonomics conduct research and apply information about human behaviour, abilities, limitations, and other characteristics to the design of tools, machines, systems, tasks, jobs, and environments. The objectives are to increase productivity and safety, to maximize performance, and to ensure comfortable and effective human use. To develop a successful human–machine interface, a certain procedure must be followed that may include identification of requirements, determination of functions to be supported, specification of interface type(s), selection of controls and displays, and finally, designing and implementing these interfaces.

### *Wireless Communication Module*

Wireless data applications in Intelligent Transport System (ITS) play a critical role in making the vision of mobile computing a reality. It provides a very valuable opportunity to present relevant information to the vehicle and its occupants as well as to obtain data for transportation management systems. Many quality services can be provided to drivers using communication technology. For example, the drivers and in-vehicle navigation systems can receive updated traffic information to help them manoeuvre through traffic, the traffic control centres can obtain current traffic reports from vehicles for traffic management and travel time predictions, and traffic control centres could even locate and help to navigate in the road network.

## Dynamic Route Guidance

A dynamic navigation system uses real-time traffic information to assist users travelling on the road networks. This technique is also known as dynamic route guidance. The other sensors along with GPS receiver used for route guidance in a vehicle are shown in Figure 2 and described below:

The sensors those are used along with GPS for dynamic route guidance are as follows:

**Figure 2**

*Generic GPS Aided Vehicle Navigation System*

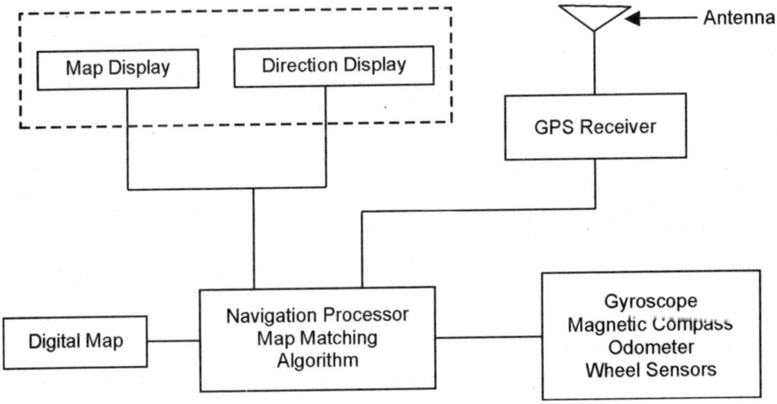


*Odometer:* It is a distance sensor that may be mounted singly or in pairs on to either the wheel or the transmission of the vehicle.

*Magnetic Compass:* It measures the heading of a vehicle. The most popular electronic compass technology for land vehicle application uses the flux-gate principle.

*Tilt Sensor:* It gives information about the pitch and role angles of the vehicle. It may include one or more inclinometers.

*Gyroscope:* It measures the rate of change of heading of the vehicle

*Digital Maps:* These are used through GIS to relate mathematical co-ordinates to locations on the street segments and intersections. In turn, the stored co-ordinates of the map features provide a means in co-ordinating space, and hence allow the digital map contribute the navigation function. En-route guidance of vehicles is done using the position data received by a GPS receiver. Once the position is known, proper signals and instructions can be prepared for the driver after comparison of the position against the planned route. This planned route consists of a sequence of road segments stored by digitizing the map through GIS.

At any point during the trip, the vehicle should be travelling on a segment contained in the planned route. The route following task closely monitors the position of the vehicle on this segment to determine when to take proper guidance action. Those actions are prescribed to the driver through the display units. The display information may be in terms of a series of voice announces to warn the driver of the approaching manoeuvre.

### *Centralized Dynamic Navigation System*

This system shown in Figure 3 relies upon a multi-vehicle route-planning module to guide vehicles on the road based on real-time traffic information. In this system a host server (i.e., traffic management centre) uses the position data from vehicles through wireless communication and then using this position data plus digital map database, gives information or guidance to the drivers of the mobile vehicle through wireless.

## Conclusion

GPS has the ability to collect and store large amounts of useful data. If these data could be used in real time, then large number of application can be made in the field of ITS. The dynamic navigation system is very much useful for travellers having little information about the road network in a given city. GPS

**Figure 3**

*Generic Centralized Dynamic Navigation System*

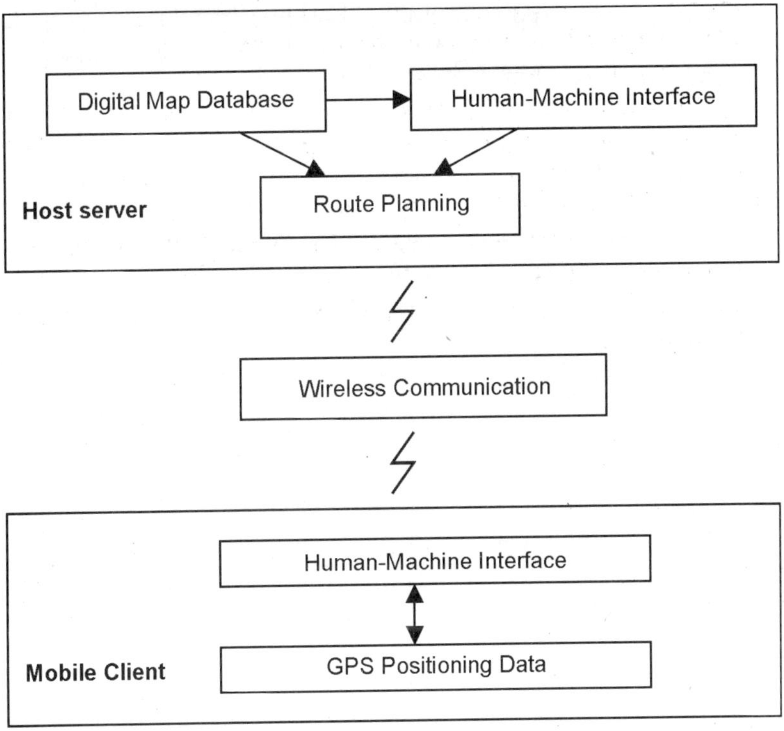


integrated with GIS is growing its network of application as smart traffic assistance tools. But there is still a lot to be done to apply these technologies in the developing countries like India for mixed traffic flow conditions.

## Select References

Kulkarni, M.N. and Choudhary, Vijay Singh "Global Positioning System: A Useful Tool for Intelligent Vehicle-Highway Systems (IVHS)" GISDevelopment.net.

Kumar, Praveen and Sahoo, U.C., "Use of GPS in Transportation Engineering", *Indian Highways*, August 2004.

McNally, M.G., Marca, J.E., Rindt, C.R., and Koos, A.M., "GPS/GIS Technologies for Traffic Surveillance and Management" Draft Interim Final Report, Institute of Transportation Studies and Department of Civil and Environmental Engineering, University of California, Irvine, March 2002.

Rao, V. Trinadha, Bhardwaj, Alok and Rao, Ch.V. Subba, "Intelligent Transportation Systems (ITS): A Technology Solution for 21st Century Public Transport System Management", csi big 2003.

Sivaram, CMSL and Kulkarni, M.N., 2001, "GPS-GIS Integrated system for transportation engineering", Map India 2001 International Conference, CSDMS, NOIDA, India.

Zaho, Yilin, "Vehicle Location and Navigation Systems", Artech House, London, 1997.

# 24

# Geoinformatics in Decentralized Planning: A Case Study

P.S. Rathore

## Importance of Decentralized Planning in Indian Scenario

In our country the greatest limitation to well guided district planning exercise has been the absence of an integrated database of physical features and natural resources either at macro or micro-level. The conventional methods of obtaining essential data on physical, geographical, natural resources and land use data and maps were extremely slow, wherein the rapid changes could not be brought out. In India the importance of district planning arose primarily out of the following pragmatic considerations:

- Block Development Programme (BDP), which failed to make an impact on grass-root development, though block development officers knew the problems of the people.
- Officials at grass-root level lacked the planning perspective, authority (both administrative and financial) and also the technical competence to formulate plans and implement them on their own.
- Subsequently, Community Development Programme (CDP), which established a coordinated staffing pattern, did not have a plan for integrated resource development.

- Macro-plans for planning commission also failed to make the requisite impact at the grass-root level because they are too far removed from the local resources and realities.

Thus, the idea of decentralised planning (district planning) has emerged as an attempt for development from below with an emphasis on the removal of disparities. Conceptually, districts are more homogeneous unit areas than states from planning point of view. Planning at district level plays a vital role in the economic growth and the development of the area, more so, the degree of development is a function of natural resources occurrence, its utilization and sustenance for future population needs and demands. Today, to meet the growing demands of teaming millions these resources are being restored to intensive utilization, resulting in decrease in vegetative cover, loss of forest wealth, transformation of production agricultural land into degraded land dwindling of water resources, environmental pollution due to rapid urbanization and industrialization diminishing per capita availability of land, famine and drought, migration of labour, etc. Up-to-date and reliable information about the use of various natural resources and socio-economic features is essential for the proper management, regulation and planning of such resources. If the country is to overcome the problems associated with haphazard and uncontrolled growth there is an urgent need for a sustained effort to reverse the situation to normalcy through optimal district level planning and scientific management of all the natural or physical endowments.

## Objectives

The objective of the present work is to design and prepare a developmental plan for District Rajsamand of Rajasthan with the view to support the decision makers at the district level involved in the district planning exercise. A computerized information system prepared with the extensive use of Satellite Remote Sensing and Geographic Information System (GIS) techniques would be aimed at the broad objectives mentioned below:

(1) Development of geographical information system for district level planning, integrating the information on natural resources and socio-economic features.
(2) To develop a systematic procedure of data integration for development around a GIS package.
(3) Assess the utility of Satellite Remote Sensing data and prepare different thematic maps on 1:50,000 and 1:250,000 scale.
(4) Identification of resources management strategies to meet the basic needs of the people from backward group.
(5) Identify and suggest measures to increase productivity for the economic growth of the district.
(6) Recommending a developmental plan keeping in view the intricate ecological balance.

## Data Integration Using GIS

The process of unifying existing data sources into a single framework is called database integration; it is one of the most sophisticated and powerful approaches to data interpretability, where a complexity is made manageable. Spatial information in the form of resources maps, viz., Land Use/Land Cover, Soil, Hydrogeomorphology and Ground Water Prospect, were prepared using satellite remote sensing data using Visual Interpretation Techniques. Other thematic maps, viz., Settlement Location and Transport Network, Administrative Boundaries, Drainage and Surface Water Bodies were prepared using the base details from Survey of India topographical maps on 1:50,000 and 1:250,000 scale. A slope map was prepared using the contours and spot height of SoI topographical maps, utilizing the capabilities of ArcInfo software. There are 989 inhabited villages in district Rajsamand as per 1991 census. A village boundary map at the scale of 1:50,000 was prepared for all the blocks of the district; to this the census maps has been scanned and digitized using AutoCAD MAP and Arc/Info. The non-spatial data used for these villages include demographic and socio-economic data.

## *Data-base Design, Creation and Organization*

In order to meet the objectives, the whole exercise calls for a planned and rationally structured database. To this it becomes imperative to segregate the database into two parts, namely Spatial Database and Non-Spatial Database. The spatial database contains a store of all map information derived from various sources, whilst the non-spatial database basically stores alphanumeric and all other forms of data, which did not get covered in the above. As spatial analysis forms the backbone of this whole exercise, the data storage facility also required to be up-to-date and efficient for quick data manipulation.

### *Database Design*

The very first task in the process of data base design is the identification of complete set of primary input data elements. This was done keeping in view the non-redundancy in the data input as well as multiple usage of single data element for different purposes. Input data elements are available in two forms, viz., spatial and non-spatial. Design of the database for district planning had to recognize the fact that the system has to support information retrieval in terms of spatial units, which are generally used by planners at district level. The spatial inputs are generally referred to SoI topographic maps. An administrative unit may include partially or fully, more than one map sheet. Tabular data is referred to either point locations or village/taluk names. Keeping this in view the database has been designed in such a way that it facilitates basically physical (schema) set, which is adequately supported by logical set-up.

### *Physical Schema*

It is the system view providing the basic framework for database creation and management. In this definition the actual data elements are defined uniquely. Physical schema consists of graphic and tabular components. Graphic components include a set of thematic layers used for different purposes. Each layer has a set of common ticks, registration marks, which are the intersection points of latitude and longitude lattices. This is essential for bringing the different thematic layers onto same co-ordinate database creation from multiple map sheets by providing the flexibility of individual map inputs. Tabular data elements include the description tables for the graphic features stored as a part of graphic layers as well as data on socio-economic aspects.

### *Database Creation and Organization*

In order to have the database ready for use, the main tasks carried out are:

- Collection and/or preparation of input data sets.
- Database creation and its organization into logical and physical schema as detailed earlier.
- Transformation/analysis of primary input data in order to generate desired information and organization of information into different set-up.

The data sets collected/prepared through multiple sources and methods were put to GIS environment. To this, in the present study, Arc/Info GIS package has been used as the core of the spatial database. The data entry was via scanning and digitization (in case of map inputs using the capabilities of Arc/Info and AutoCAD MAP) and direct entry through alphanumeric terminal in case of tabular inputs.

### ***Creation of Spatial Framework***

The most significant task prior to map scanning or digitization due to the fact that entire district is covered in multiple map sheets and inputs were on single map sheet basis. It involved:

- Definition of a co-ordinate system with origin lying at lower left corner of the lower left map sheet.
- Identification of registration tick marks at latitude-longitude crossings.
- Design on co-ordinate units chosen as true distance in kms on ground.

- Calculation of co-ordinate values for selected registration points using polyconic projection with appropriate central meridian and standard parallel.

An important aspect of integrated area-level planning is the combined analysis of the thematic natural resource data and the tabular socio-economic data. For this, a criteria based analysis of the integrated databases has been carried out in GIS environment. User interactive programme has been developed in GIS using Arc Macro Language (AML) of Arc/Info GIS Software.

## Socio-Economic Data Input and Analysis

In order to demonstrate the over all socio-economic status of district Rajsamand, different demographic and economic indicators were shown in map form which would further help for better understanding of the current phenomena in the district.

### *Derivation of Spatial Patterns*

Demonstration of the utility of a GIS to represent the non-spatial data spatially, different non-spatial parameters have been used to derive spatial patterns. These spatial patterns highlight the low development areas in the district and also potential in showing the trend of development which is some of the pertinent spatial patterns that are generated in this study. These spatial patterns have been represented on a village-wise basis.

## Integration of Spatial and Non-Spatial Data

The development in a district is never uniform. For the formulation of a development plan of a district, it would be necessary to make a critical assessment of present scenario, on the basis of available resources (natural/physical) and potentials, nature of local needs and problems, the priorities assigned to different socio-economic activities and the trends of development. Owing to the difference in the levels of economic development and provision of various facilities in a district, it is essential to carry out intra-district disparity analysis for prioritizing the developmental programme within the district. The purpose of the present study therefore is to examine the distribution of infrastructural and socio-economic facilities in the district and to categorize various development blocks of the district, based upon overall level of development.

Through systematic analysis of land use/land cover, slope, hydrogeomorphology and soil coupled with agro-climatic and socio-economic profiles, suitable combination of land use practices have been recommended for land resource management on a long-term basis. The basis of these development packages can be grouped based on different geophysical and environmental factors, viz., landform, slope, soils, rainfall and other climatic factors and human induced factors like land use, irrigation practices, infrastructure development and other socio-economic factors. Keeping in view the current scenario of the district as mentioned previously, the developmental plan for land resource management was developed recommending best practices: management of forest land, reclamation measures for waste land, crop land management, allied agricultural and forestry activities.

## Results and Conclusions

It is generally felt that one of the reasons of our failure to achieve the basic goals is that the plans are formulated at higher levels, namely national and state levels, which are quite distant from the grass-roots realities. Thus, there is a wide gap between those who plan (at higher levels) and those who implement it (at the local level). Planning at the lower levels can reduce this gap. It is perceived that lower the level or unit of planning, smaller will be the gap between planning and implementation. This is one of the strongest justifications of decentralised planning. In addition to this, there are many advantages.

In the present study Remote Sensing is found to be a wonderful technique for thematic map preparation, on the other hand, GIS is extremely useful to establish links amongst all the facilities as well as to compute the various indices quickly on the basis of different weightage assigned to different facilities. The database is quite flexible in its design such that it can be easily updated and any complex transformation equation can be used involving all the facilities. Here it would be apt to mention that efforts are made to port the entire database on a web enabled GIS (Autodesk MapGuide) for wider application needs.

# 25

# Application of Radar Data and Interferometry with Optical Data in Urban Growth Sprawl Measurement in and around Kolkata, West Bengal

Tanima Sarkar and M.S. Nathawat

In recent years, cities all over the world have been experiencing rapid growth because of the rapid increase in world population and the irreversible flow of people from rural to urban areas. Specifically, in the larger towns and cities of the developing world the rate of population increase has become a matter of grave concern. Many of them are now facing unplanned and uncontrolled settlement growth at the densely populated sites or fringes. India, like many of the developing countries has the problem with urban expansion and growth of population in the main cities. For example, over the last decades, Kolkata has significantly expanded due to different developmental activities and migration of people from rural and other sites. Various changes have been and are occurring in the city but there are no regularly updated maps to indicate those changes. For this reason, Kolkata is taken up for this study of precise mapping of the urban sprawl and the advantages or merits of using radar data along with optical data for the purpose.

Multispectral remote sensing data has been widely used for this study. The mechanism of ground object recognition is based mainly on its spectral reflectance characteristics. However, in many cases, spectral information may not provide sufficient information on surface texture and surface roughness, which are important for land cover classification. SAR remote sensing data on the other hand supplies information on surface, texture and even surface roughness. Independent application of these data leads to certain limitation of accuracy in ground objects recognition. To achieve the best result it is required to combine both the data sets. In this work, multispectral, optical and SAR remote sensing data have been used separately and synergetically for urban study. The datasets used are Landsat TM data of May 14, 1987 and March 9, 1998 and ERS SAR tandem pair of 1996 (orbits 25214 and 5541). First attempts have been made by digitally enhancing the multispectral optical data, viz., addition of textural information with raw multispectral optical data for highlighting settlement area. Subsequently, useful information from SAR data, viz., amplitude and coherence have been merged with multispectral optical data in a manner which highlights settlement class in general. It is observed that the built-up area not only gets highlighted as a single class, but according to different density and structure, different subclasses within the settlement class can be easily identified.

## Fusion of SAR Data with Multispectral Optical Data for Highlighting Built-up Areas

The Landsat TM data is used for the fusion with SAR data to produce a high resolution multispectral image without distorting the spectral properties of the fused image. The fused multispectral images are then input to the multitemporal classification scheme that allows the differentiation of constant and variable land use types.

There are several approaches of image combination, for example, data fusion, overlying, etc. In this study both the techniques are used. The post classification approach is divided into following steps:

— Maximum likelihood classification of land cover
— Extraction of urban area in SAR image using optical image mask
— Level slicing of the SAR image according to backscattered energy
— Overlay of the level sliced SAR image on the ML classified image to create final built-up map of Kolkata city.

On Figure 1 is shown level sliced SAR image and its overlay on the TM classified image. The housing and built-up area has been classified into 2 classes: low density and high density of construction. These classes are coloured in light green and light pink respectively. Instead of only one pink colour (one class) in left image for urban settlement, on the right we could see better structure of urban construction by following different colour shades (pink and green). High density construction is located mainly along old streets, while the low density construction dominates newly developed parts. The colour distribution has enabled to enhance urban structure that is impossible to recognize on the optical (TM) classified image. According to the level slicing, the dense settlement and the low density settlement can be highly differentiated.

The data fusion technique is also used in this study in the different possible ways. Since the settlement class in the Envisat data has well separatibility in the ERS amplitude information and fairly well separatibility from other land use/land cover classes in the ERS amplitude image of orbit 25214, these two images are used for the purpose of better separation of settlement class from other classes. The ERS 25214 and the Envisat images are subjected to Principal Component Transformation and the first principal component image or the PC1 is the image, which contains all the important variations of the ERS and the Envisat images. In other way, it can be said that the PC1 image is the image, which definitely contains the distinct signature of the urban settlement area. The PC1 image derived from ERS and Envisat data was

Figure 1

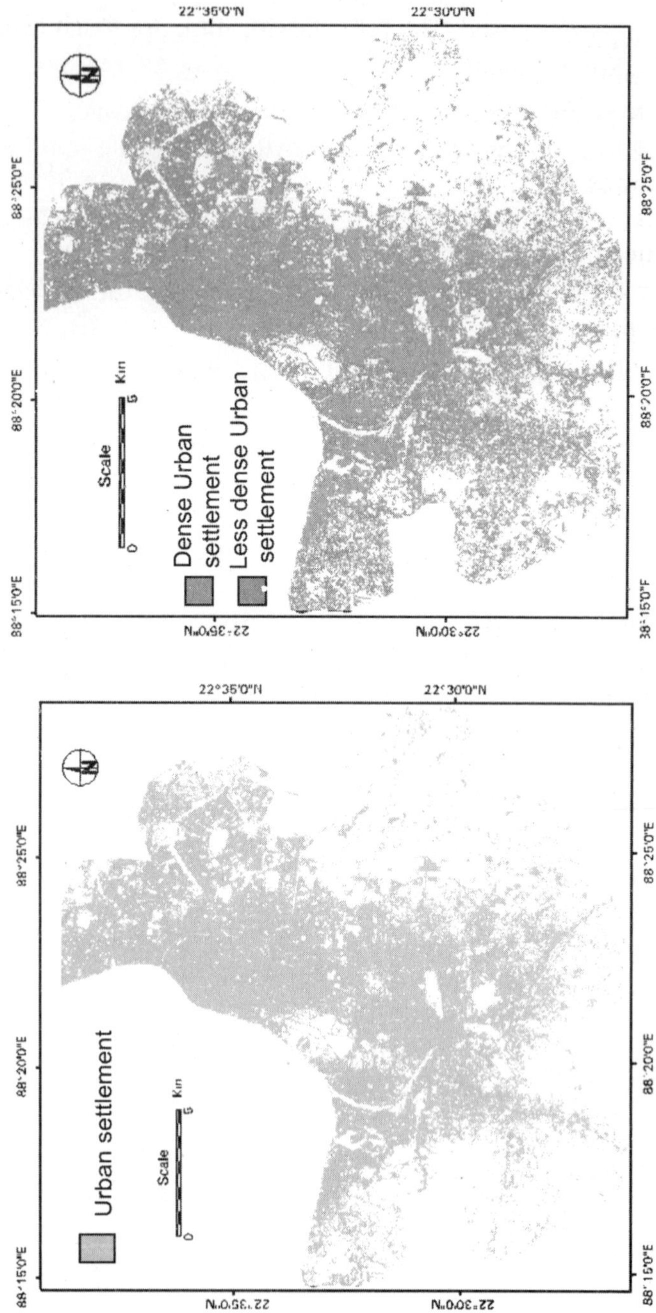
Dense Urban settlement
Less dense Urban settlement
Scale
0
5
Km
88°15'0"E
88°20'0"E
88°25'0"E
22°35'0"N
22°30'0"N
Urban settlement

fused with Landsat TM data. First, the RGB Image of the Landsat TM (432 band combination) is transformed to IHS image and then from this IHS image the intensity, hue and saturation components are separated individually. The intensity band is histogram matched with already derived PC1 image of ERS and Envisat amplitude information and finally combined back with the hue and saturation components.

In another type of optical and radar data fusion type for the highlighting of the settlement area, the amplitude information from ERS and Envisat data is added as the intensity channel whereas the coherence information is added as saturation channel with the hue component obtained from Landsat TM 98 FCC. The final IHS Image is classified by automated technique to have an estimate of the total settlement area in general and different classes of settlements in specific.

In a different approach, 6 optical bands of Landsat TM data subjected to principal component transformation. The first principal component image (PC1) is used as one channel and stacked together with Envisat amplitude image and coherence image of ERS couple of orbits 25214 and 5541. Besides, the unsupervised classified image has also been fused with the Envisat amplitude image and coherence image of ERS tandem pair.

The classified images obtained from Landsat TM Data and from the fused product of principal component image of SAR amplitude and Envisat image, coherence with TM Data have been studied for settlement mapping. In the case of this fused product it was observed that different classes of settlements, viz., dense settlement and less dense settlement can be identified by the difference in colour shades which are deep brown and orangish brown respectively. It enables to enhance the separatibility between different types of settlements, which is otherwise difficult from Landsat TM data alone. High density settlement and low density settlement classes can be differentiated and their spatial coverage can be determined.

## Results

Demarcation of urban and suburban settlements with land vegetation is possible in all the fusion combinations, but the best result is obtained by the fusion of ERS and Envisat amplitude, ERS tandem pair coherence and Landsat TM Data (Figure 2) One of the merits of these fused products is that the data appears free from haze as observed in the optical image. The roads and the boundaries between the different objects become very clear. One can very easily interpret from the images that the growth of the urban city is along the roads, and the suburban settlement growth is along the boundaries of the dense settlement area.

**Figure 2**

*IHS merged image of First Principal Component Image of ERS and Envisat, hue channel of Landsat TM 98 data and average backscattered coherence image as the intensity, hue and saturation component.*

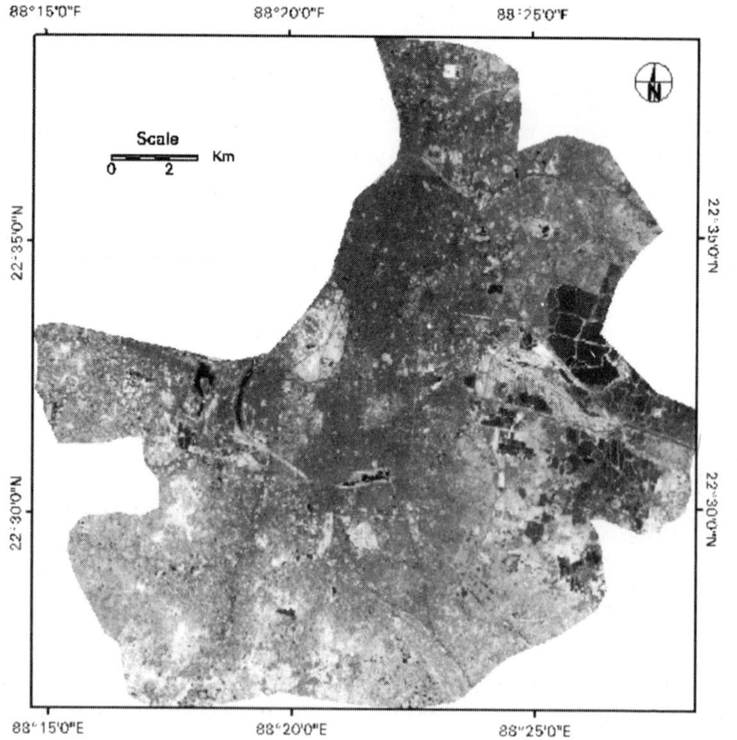

The fused product of Envisat amplitude and ERS tandem pair coherence image with the PC1 of 6 optical bands of Landsat TM data image or with unsupervised classified image is also good in terms of broad demarcation of the settlement area. The dense settlement area having very high brightness values when added with other images, the high brightness value is contributed to the image for the settlement area. The suburban settlement area is only fairly demarcable. The said fused products are therefore good enough for separating high and low density settlements, but can be used for broad delineation of the total settlement area.

## Refrences


Banerjee, Ajit Kr., (1983), importance of evolving a Management Plan for Groundwater Development in the Calcutta Region of the Bengal Basin in Eastern India, Proc. International Symposium convened by UNESCO in cooperation with the National Committee of the Federal Republic of Germany for the International Hydrological Programme, pp. 45-54.

Banerjee, Sanhita, (2003), Application of Remote Sensing for mapping and monitoring of mangroves of Sagar Island, Sunderbans, West Bengal, India, Unpub. thesis P.G. Diploma in Remote Sensing and GIS, IIRS.

Biswas, A.B., and Saha, A.K., (1985), Environmental hazards of the recession of piezometric surface of the Groundwater under Calcutta, Proc. Ind. Nat. Sci.Acad., Vol. 51 A, No. 3, pp. 610-662.

Borgeaud, M. and Wegmuller, U., (1996), On the use of ERS SAR interferometry for the retrieval of geo-and bio-geophysical information, Fringe 96, Zurich, Switzerland.

Graham, L.C., (1974), Synthetic interferometer radar for topographic mapping, Proc. IEEE, Vol. (62)6, pp. 763-768.

Jensen, J.R., (1996), *Introductory Digital image Processing: A Remote Sensing Perspective,* Prentice Hall Englewood Cliffs, New Jersey, pp. 316.

Sikdar, P.K., Biswas A.B. and Saha A.K. (1996), A Study on the possible land subsidence in Calcutta and Howrah Cities due to Groundwater Overdraft, *Ind. Jour. of Geol.,* Vol. 68, No. 3, pp. 193-200.

# 26

# GIS-Based Drug Survey: A Case Study within Ranchi Municipal Limits

Subrata Ghosh and Richa N.K. Sharma

## Introduction

GIS has made a tremendous impact in many fields of applications. The present paper has tried to explore its feasibility in mapping a popular drug.[1]

1. Distribution of $SO_2$, $NO_x$ and $CO_2$ emissions from fuel combustion and industrial activities can be estimated using GIS.[2]
2. Analysis of regional hazards (landslide) can be done using GIS.[3]
3. GIS has applications to sales and marketing. GIS can answer queries like which locations in the city are suitable for certain retail outlet, what route should distribution trucks take to minimize delivery cost and time, which are the potential rural areas having great scope for marketing a particular product etc.[4]

Coronary artery disease (CAD) is considered to be India's original sin. Studies that compare the risks between different ethnic groups show just how much more danger Indians face. The findings of Dr. E.A. Enas, director of the non-profit Coronary

Artery Disease in Asian Indians Research Foundation, published in the Indian Heart Journal and based on studies in Canada, England, Singapore and South Africa (all countries with significant ethnic South Asian populations), show that Indian men and women consistently suffered higher CAD mortality rates than Whites, Chinese, Blacks and Malays. The incidence of CAD is between two and four times as high for Indians than it is for Americans, Britons, and other Asians like the Chinese. A whopping 25 per cent of all heart attacks affecting people from the Indian subcontinent hit people under 40, a rarity in any other population. Diabetes, high blood pressure, tobacco usage, central obesity, high LDL levels, high triglyceride levels, high saturated fat diet and sedentary lifestyle are the major factors responsible for this. India adds 45 per cent to the world coronary burden now and it could cross 50 per cent by 2020.[5]

Atenolol chemically is 4 -2 -(hydroxy-3-iso -propyl-amino-propoxy) – phenylacetamide.[6]

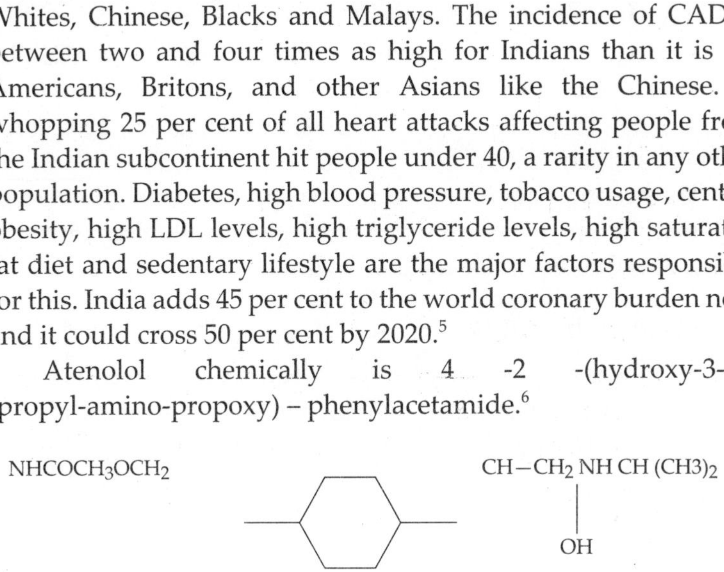


Mol formula: $C_{14} H_{22}N_2 0_3$ Mol. Wt. : 266.34

It is official in 1P[6], BP [7] and EP[8]. It is a white or almost white powder, sparingly soluble in water, soluble in ethanol, slightly soluble in methylene chloride and practically insoluble in ether.[8] Atenolol is a β–adrenceptor blocking agent. It is a cardio selective $\beta_1$ receptor (found mainly in the heart) blocker having no intrinsic sympathomimetic activity. Its cardio selectivity is dose related.[9]

β–receptor blocking drugs were introduced way back in 1966 to treat cardiovascular diseases. The synthesis of atenolol was first reported in 1970 and the first pharmacological and clinical studies were made in 1973.[10]

Atenolol is used in the management of hypertension, angina pectoris, cardiac arrhythmia and myocardial infarction. Atenolol reduces resting and exercise-induced heart rate as also myocardial contractility. It reduces systolic and diastolic blood pressures. The reduction in blood pressure and heart rate results in reduced myocardial work and oxygen requirement. Anginal attacks are reduced in intensity and frequency and exercise tolerance is improved.[11] The dose of atenolol is 50 to 100mg daily in 1 or 2 doses.[6] Tablets of strengths 25mg, 50mg and 100mg are available in the market.[11]

The present paper is based on the application of GIS to sales and marketing of various brands of atenolol tablets (50 mg), which is a drug of choice for the cardiovascular diseases.

## Objective of the Study

The study was carried out with the following objective in mind,

— To survey the locational availability of an essential cardio-active drug.

## Study Area

The Ranchi city was selected for the study.

## Methodology

A market survey was conducted for 38 marketted brands of atenolol tablets (50 mg). In all 75 drug outlets were surveyed.

The available information was plotted on the location map.

Rank correlation coefficient was calculated between the ward development index and number of brands available using

$$r = 1 - 6\Sigma di^2/(n^3 - n)$$

where $n$ = Total no. of observations

$di$ = Difference in the ranks of the two variables in the *i*st observation

## Results

An overlapping image of the ward map, route map and the drug outlets shows that the outlets are more or less located in and around the main roads. Also the concentration of outlets are more around the medical college-hospital. It is seen that the outlets are more concentrated in the densely populated areas

No correlation was obtained between the price and popularity (i.e., lesser the price, more should be the sale). This implied that the popularity of the brand had more to do with its efficiency and marketing rather than its price. A very high correlation(0.914)was found between the socio-economic status of the wards and the number of brands available, i.e., more the developed ward, more is the range of brands available.

## Conclusion

The result of the study showed that a thorough GIS-based drug survey is very necessary so that the same chemical drugs placed at different price can be targeted for the appropriate economical groups with respect to their geographical locations.

The present study can be extended to find out the disease pattern of a particular area by conducting a survey on the types of medicines that are stocked in the drug outlets of that particular area and then trying to correlate it to the socio-economic structure of that area.

## Notes

1. M.K. Gillespie, D.C. Howard and M.J. Ness, *Environmental Monitoring and Assessment,* 39, 385 (1996).
2. H. Akimoto and H. Narita, *Atmospheric Environment,* 28(2), 213 (1994).
3. K.R. Gharat and H.S. Lakra, *ISG Newsletter,* 9(1), 21(2003).
4. *Outlook,* Vol. XLIV, No. 19, May 11-17, 56(2004).
5. *Pharmacopoeia of India,* Vol. I, Government of India, Ministry of Health and Family Welfare, 1996, p. 72.
6. *British Pharmacopoea,* Vol. I, Her Majesty's Stationary Office, London, 1993, p. 55.

7. *European Pharmacopoeia,* 4th Edn., Published by the Directorate for the Quality of Medicine of the Council of Europe, France, 2001, p. 680.
8. *Goodman and Gilman's The Pharmacological Basis of Therapeutics,* 10th Edn., J.G. Hardman and L.E. Limbird, Eds, McGraw Hill Publications: New York, 2001, p. 249.
9. V. Caplar, in *Analytical Profile of Drug Substances,* Vol. 13, K. Florey, Ed., Academic Press: New York, 1984, p. 1.
10. *CIMS Drug Profile,* 85, April 2004 (Update 2), p. 129.
11. Kanungo, A., (2002), "Assessment of land use/land cover change of Ranchi municipal area through dynamics of space and time with special emphasis on Ethenic contribution to development process", 2002, M Tech Dissertation, Deptt. of Remote Sensing, BIT, Mesra, Ranchi.

# 27

# Enterprise GIS for Decentralized Planning

Kamal Jain and Rahul Goel

Information technology is revolutionizing the way in which we live and work. It is manipulating all aspects of our life and lifestyle. The amount of calculation power that is available to mankind is increasing at an exponential rate. Computers and communication are becoming integral parts of our lives (Leon Alexis, 1999). Over the last 10 to 20 years, Geographic Information Systems (GIS) technology has steadfastly provided tools for the collection, storage, management, analysis and reporting of spatial information. GIS technology has provided many governments and other organizations with an opportunity to introduce major improvements in traditional methods and procedures of providing services to citizens and customers. The use of Internet and related web technologies for accessing spatial data as well as for performing basic spatial query and analysis has opened the world of GIS to the masses, instead of keeping its domain to the technically sophisticated Information Technology (IT) personnels and spatial data professionals.

In Geo-business based organizations, practicing GIS, Photogrammetry and GPS solutions, IT has a crucial role to play, both at organizational and at departmental level. At the organizational level, IT should assist in specifying objectives and strategies of the organization. At the departmental level, IT must ensure a smooth flow of data (spatial and non-spatial) and information across departments, and should guide organizations to adopt the most viable business practices with the development and maintenance of enterprise wide geo-database. This geodatabase will eliminate the need of the isolated data islands that existed in each department and make the organization's data accessible across the department boundaries. This enterprise wide data sharing has many benefits like availability of high quality information for better decision-making, faster response times, etc.

## Basics of Enterprise Geographic Information System

In small organizations usually a single person, keeping in mind the overall companies objectives, makes decisions. As the size of organization grows, managing the entire operations becomes impossible for a single person. When the organization becomes larger, each person hires people to assist him/her and the various departments are evolved. The departments become close and watertight as the size of the organization enhances. Although information technology provides the perfect answer, in the haste,

**Figure 1**

*An On-Demand System (Thum Peter G., 2002)*

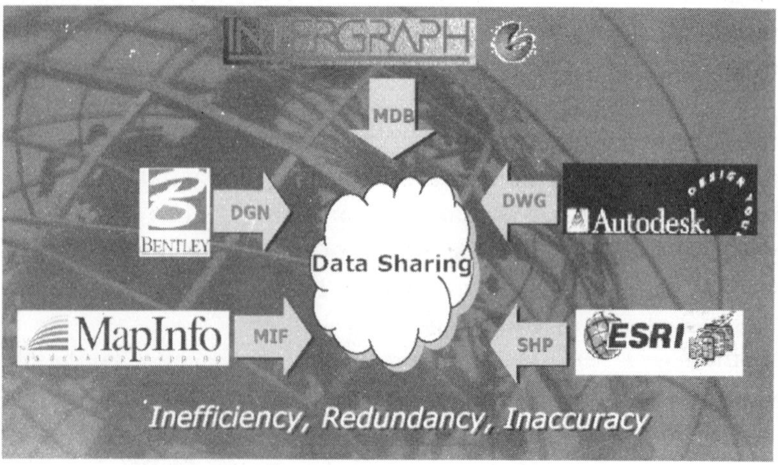

**Figure 2**

*An Integrated System (Thum Peter G., 2002)*

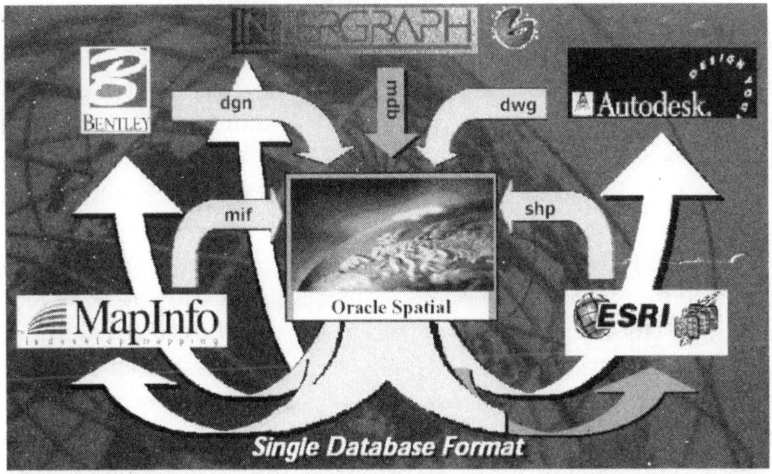


most developers usually end up developing need-based, isolated and non-compatible information systems, as shown in Figure 1.

Most of this happens because IT is not integrated into the corporate strategy. To draw real benefits from a technology as powerful as IT, one has to devise a system with a holistic view of the enterprise. Such a system has to work around all the major activities of the organization, and should facilitate seamless flow of information across departmental barriers as shown in Figure 2.

### *What is an Enterprise?*

An enterprise is a group of people with a common goal, which has certain resources at its disposal to achieve that goal. This group has some key functions to perform in order to achieve that goal (Leon Alexis, 1999).

In other words an enterprise can also be defined as a business organization or an undertaking, especially one of some scope, complication and risk (Teri Boudreau Landrum, 2001).

## Enterprise Geographic Information System

As deregulation takes effect, the business climate has a decidedly Darwinian atmosphere. Survival goes not just to the fittest but also to those companies that can do the best job of fitting information on facilities, customers, and competitors together to provide a competitive edge (Linda Hecht, 2004).

In GIS community, the term Enterprise GIS is defined as follows:

"Enterprise GIS covers the techniques and concepts that uses a shared central geodatabase, that allows integration and dynamic updating of multiple GIS data sources without time-consuming compatibility and translation issues. It is based around industry IT standards and web services, so it enables GIS applications to easily access the functionality of mainstream business applications and IT systems. Moreover it also allows non-GIS applications and systems to easily access GIS functionality." (Crocker Charlie, 2004)

The major benefits of implementing a GIS come from shared information. Enterprise GIS links GIS applications throughout an organization to realize benefits through maintaining data sets corporate-wide. With an enterprise GIS, users of spatial data spend more time on the analytical capabilities and business functions of a GIS and less time searching for, compiling, and integrating the data they require.

### *Geodatabase*

"A geodatabase is a relational database that stores and manages geo-spatial information" (ibid.). Unlike file-based geo-data storage schemes, a geodatabase stores the shapes of features and raster images in a table inside a relational database. By storing geo-data within a relational database, a typical multi-user GIS will see an immediate performance boost, as well as greatly enhanced data security and validity, as shown in Figure 3. Since modern relational databases are designed with robust data back-up and disaster recovery tools, a GIS using a geodatabase enjoys greater availability, uptime, and access to traditional enterprise data.

**Figure 3**

*An Enterprise Geodatabase (Thum Peter G., 2002)*

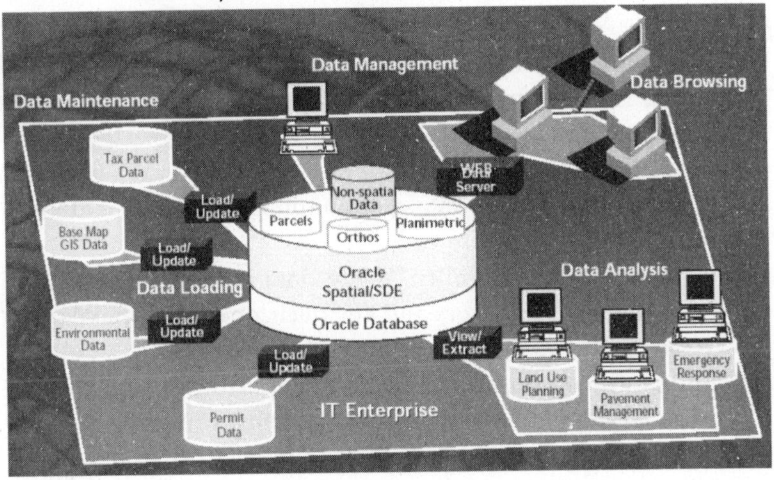


Some of the tangible benefits of modern Enterprise Geodatabase are:

(a) Significantly reduced data development and management costs.
(b) Controlled access to valid and current geo-spatial information.
(c) Support multiple concurrent GIS users and workflows.
(d) Serve GIS data in high demand environments.
(e) Securely store GIS data and protect against unauthorized data access.
(f) Allow long transactions, data versions, and other real-world scenarios.
(g) Store many GIS features in a single, seamless, spatially indexed database.
(h) Enable sophisticated GIS data models, data validation and spatial rules.

### *Benefits of Enterprise GIS*

The use of central integrated geodatabase in an Enterprise GIS system delivers significant benefits and enhances worker productivity and capacity to deliver goods and services to customers (Crocker Charlie, 2004). Some of the tangible benefits of Enterprise GIS are listed as below:

(a) Best tools selection for any job: Each division in an organization can use its preferred software and data types, while working from a shared enterprise Geodatabase.
(b) Productivity enhancement: Eliminate the time wasted to convert or translate data and eliminate out-of-sync data or errors.
(c) Ease of data accessibility: When data is stored in an enterprise database, all users and key decision makers have quick access to the most accurate and up-to-date data.
(d) Enhanced communication among departments: In an enterprise system all independent systems can communicate quickly and effectively, regardless of data format used.
(e) Higher standards of data security: By storing data (spatial and non-spatial) in a central database, i.e., geodatabase, an organization can easily maintain secured data.
(f) Enhanced reliability: Since all the users are accessing central database, so all the changes incorporated by different users will be reflected in the central database. So, data remains reliable when used by any other user in future.
(g) Scalability: An Enterprise GIS is a scalable configuration. At any instance of time, with the expansion of organization, the number of nodes attached can easily be increased in the enterprise GIS configuration.
(h) Easy web-based access: Many different kinds of applications can concurrently access and use an Enterprise GIS view web. The GIS data is available to many different clients, from desktop applications to Internet web based solutions. This in turn enhances data sharing and worldwide community participation.

## Enterprise GIS Solutions Approaches

In the current scenario Enterprise GIS solutions can be delivered through two different approaches, as shown in Figure 4.

**Figure 4**
*Technology Considerations (ESBI Computing, 2003)*

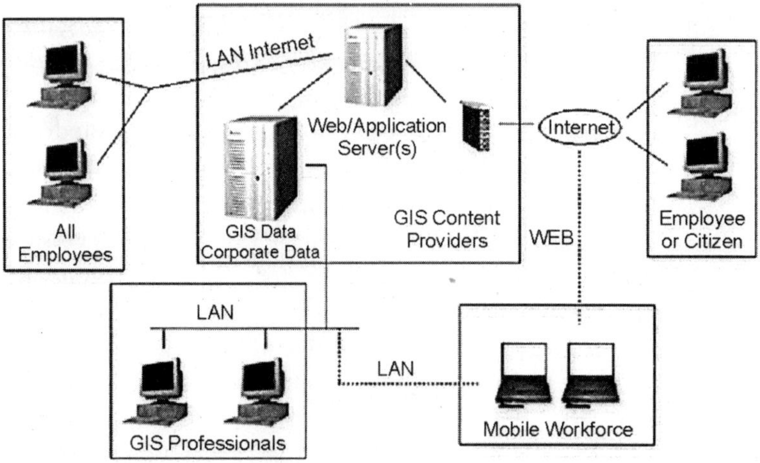


(A) Custom Desktop based solution.
(B) Web (Internet) based solution.

## Custom Desktop Based Solutions

The custom desktop based solution can be implemented in three different ways:

(a) One way is to create a custom application. This is the most popular and feasible solution. Various functions are implemented using the pre-defined objects available and a custom application is developed. These objects are referenced and programmed through a project to produce an application. The application and the objects are bundled together and then can be delivered as a collective package to any desktop in the organization.

The most significant benefit of this type of application is software independency. One need not deploy the complete GIS package on each and every desktop in the organization, when using custom application. So the users who are not proficient in using the regular GIS software can easily work through these custom applications. Moreover, the user can easily integrate and grow its database without usage of any GIS package.

(b) The second approach is to drive the GIS software as a server application. This is typically used when working across applications. In this type of application, an application form is developed which basically handles only the top priority framework window of the GIS software. Through this form the user enters into the conventional GIS software framework with which most of the users are generally familiar. The changes that are performed in the application form are reflected directly into the software package.

(c) The third approach is to build custom commands. Using the GIS software Command Wizard, commands can be created. The wizard registers the command and creates the necessary project files required for producing a DLL. The project files contain boilerplate code that incorporates selected command features. The command is completed by adding the automation code and compiling a DLL. In this case the user once again works inside the GIS software and no software independency exist.

This approach is most beneficial for proficient GIS software users since the component object is embedded inside the regular software only. Most of the software development companies adopt this approach to enhance the feature of their regular software with every new release. The programmers from the research and development section of the company build the custom command utilities and simply embed them prior to every new software release.

## Web (Internet) Base Solution: Web GIS

Geographic Information Systems (GIS) are powerful applications that manage and manipulate spatially referenced data. Web GIS means access to GIS applications/functionality and data via a Web Browser (Ramesh Wahi, Sandeep Kuniel, 2003).

**Figure 5**

*3 Tier Web Architecture (Petersen Jeremy, 2001)*

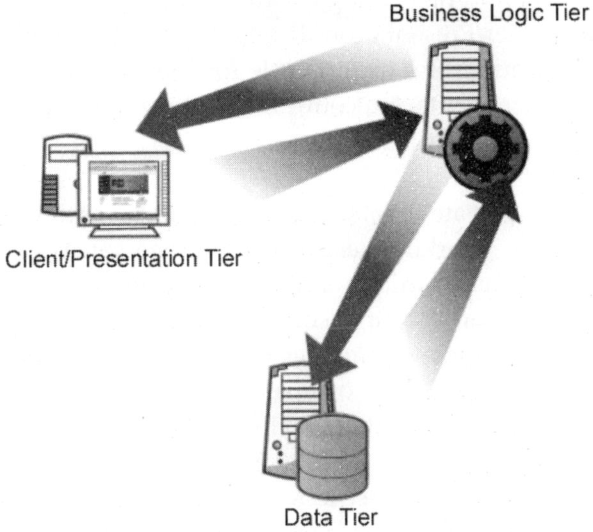


A typical Internet system consists of 3 distinct tiers as shown in Figure 5.

Database/Server Tier: Database plays the most important role in creating an effective web application, since it stores whole data, and the information retrieval depends upon the structure in which data is stored.

Middle/Application/Business Logic Tier: The number of users in an Internet system is exponentially higher than the number of clients in a client-server system. As a result, the middle tier in the Internet system should be equipped to handle mass user authentication. In addition, support for single-sign-on makes the user experience seamless as the user moves from application to application. Another important requirement in an Internet system is the click-response metric and this requires efficient caching of pages in the middle-tier and the ability to farm out these caches over a geographical area to help local performance. Finally, another important mid-tier requirement is a mapping engine. The mapping engine in the mid-tier takes in data like map name, centre location, data source and map size and returns a map image (or a URL for the image) and the minimum-bounding rectangle (MBR) of the map.

Client Tier/Presentation: It implements the "look and feel" of an application. It is responsible for the presentation of data, receiving user events and controlling the user interface.

## Conclusion

As today's private geobusiness companies and government organizations working especially in the area of national planning, transportation, utility management, etc., try to compete in a market that is constantly changing, the visionaries behind each of these companies are trying to look forward into how they can make their company do more with less. Successfully achieving the potential of GIS in a complex organization is a difficult task. How they can match more and more government regulations, yet still use less and less manpower. How to continue to grow into new markets, yet maintain current staffing levels. How to eliminate redundancy within different departments and different parts of the organization. The answer lies in Enterprise Geographic Information System implementation.

Reduced redundancy, improved data sharing, faster response, scalability and greater payback from geospatial investments, all make Enterprise GIS solution a wise choice for any private geobusiness company or government organization.

## References

Crocker Charlie, 2004, "The What and Why of Farallon's Enterprise GIS Approach", *http://www.fargeo.com/enterprise_gis/index.html*

ESBI Computing, 2003, "Enterprise Web GIS", *http://www.esbic.ie/geobusiness/Enterprise_Web_GIS/components.html*

Leon Alexis, 1999, "Enterprise Resource Planning", Tata Mcgraw Hill.

Linda Hecht, 2004, "GIS Helps Utilities Thrive in a Deregulated Environment", *http://www.gis.com/specialty/utilities/utilities.html*

Ramesh Wahi and Sandeep Kuniel, 2003, "Web GIS Applications in Real Estate", *http://www.gisdevelopment.net/application/lis/urban/mi04112.htm*

Teri Boudreau Landrum, 2001, "Enterprise GIS – Building a Strong Foundation", *http://www.gisdevelopment.net/proceedings/gita/2001/lgp/lgp005.shtml*

Thum Peter G., 2002, "Technical Dimensions of Enterprise IS/GIS", *www.geoanalytics.com/publications/technical_enterprise_PT2.pdf.*

# 28

# Urban Land Use Pattern and Change Detection Studies of Jaipur City Using Remote Sensing and GIS

Rupali S. Mathur, M.S. Nathawat and M.P. Punia

Problems related to rapid transformations of land use have reached a level of significant impact; this unabated trend is increasing pressure on the limited land available in the city and its environs. It is, therefore, desirable to plan for the city and its peripheral areas in an integrated manner, so that, in due course, as the city grows, the periphery can merge into the whole. This kind of planning needs knowledge of the composition of the urban land use patterns and the quantitative analysis of the space devoted to each type of land use. Jaipur is the oldest planned city of India and experienced a phenomenal growth during the post-independence period due to the heavy influx of displaced persons. This resulted in the change of the original city structure; many new residential areas were developed. The expansion of the city towards the north and the east was restricted because of physical constraints; development therefore, continued towards the south, manifold in the last few decades both in terms of population and in the extent of area. Keeping in view of the above, the present study has been performed to assess the land transformation patterns by change detection in land use of different time periods during 1975 to 2003 of the Jaipur city.

## Objectives

- To analyse the status of existing land use/land cover patterns and urban environment of the Jaipur city.
- To find out the trends of the urban expansion during 1975 to 2003.

## Study Area

Entire Jaipur city (area covered by old JDA boundary) is considered for the purpose of land transformation and change detection studies. The whole region has experienced an emergence of environmental crises in terms of depletion of groundwater, forests, soil erosion, flora and fauna and climatological changes. Whole area towards east has experienced substantial deforestation. This has further deteriorated due to mining operations for building stones. Devastation of tree and soil cover has turned hilly areas to rocky exposures. Wasteland areas are continuously increasing and degrading the environment. Salinization is prominent in west and the south-west of Jaipur region. All the above phenomena have put heavy pressure on the environment, which in turn has created environmental changes in the ecosystem and their subsystems.

## Location, Topography and Climate

Jaipur is the capital of Rajasthan, the largest state of India. It is located at 26°4′ North latitude and 75°45′ East longitude. It is surrounded by Aravalli hills to the north and east. The elevation varies from 462 m to 382 m above mean sea level.

Jaipur experiences a semi-arid climatic condition with hot and dry summers and cold winters. The mean maximum temperature is 41°C, and the mean minimum temperature is 6.5°C. The average annual rainfall in the area is 620 mm. The month of July experiences highest relative humidity (74%). The predominant wind direction is from north-west and west.

## Data Input

Details of data used in the present study are outlined below:

- IRS 1C LISS III (FCC) data of February 2003, of path 96, row 53.
- IRS 1C LISS II (FCC) data of 1991, of path 29, row 49, scale – 1:50, 000.
- LANDSAT-TM (FCC) data of April 7, 1986 of path 147, row 41, scale – 1:1000, 000.
- LANDSAT-MSS (FCC) data of March 2, 1975 of path 158, row 41, scale – 1:1000, 000.
- SoI Toposheets, bearing no. – 45N/13, 45N/14, 45N/9, 45N/10, 45N/8, 45M/12, 45M/16, 54A/4 and 54B/1 of 1:50000 scales.
- Guide map of Jaipur city on the scale of 1:20, 000.

## Softwares Used

- ARC/INFO (ArcGIS) version-8.02
- Geomatica version – 8.2
- ArcView version – 3.2a
- Image Analyst

## Methodology

In the present study mainly two aspects, land use/land cover and change detection have been demonstrated and mainly remote sensing techniques were used for the purpose.

## Land Use Classification

### *Data Requirements and Collection*

For the purpose of land use classification of different time periods, the regional scale maps have been prepared using remote sensing data as well as other ancillary data, published maps and field data.

## Classification System Adopted

In the present study, a detailed classification system is developed for mapping. The land use/land cover classes developed are:

(a) Built-Up Land
- High density area
- Medium density area
- Low density area

(b) Agriculture
- Cropped area
- Fallow land

(c) Forest Area
- Degraded forest
- Actual forest

(d) Mining/Quarrying Area

(e) Wastelands
- Gullied affected area
- Sandy area
- Saline area

(f) Industrial Area

(g) Parks and Playgrounds

(h) Land Under Transformation

(i) Water Bodies

## Methodology For Preparing Land Use/Land Cover Maps

The broad methodology followed for the preparation of land use/land cover maps, involves the following main steps.

### Preparation of Base Map

The base map has been prepared by SoI toposheets and updated by the base maps prepared by the Jaipur Development Authority (JDA) at the scale of 1:20000 and by IRS 1C LISS III satellite data. The features such as water bodies, national highways, metalled and unmetalled roads, railways, settlement have been included in the base map.

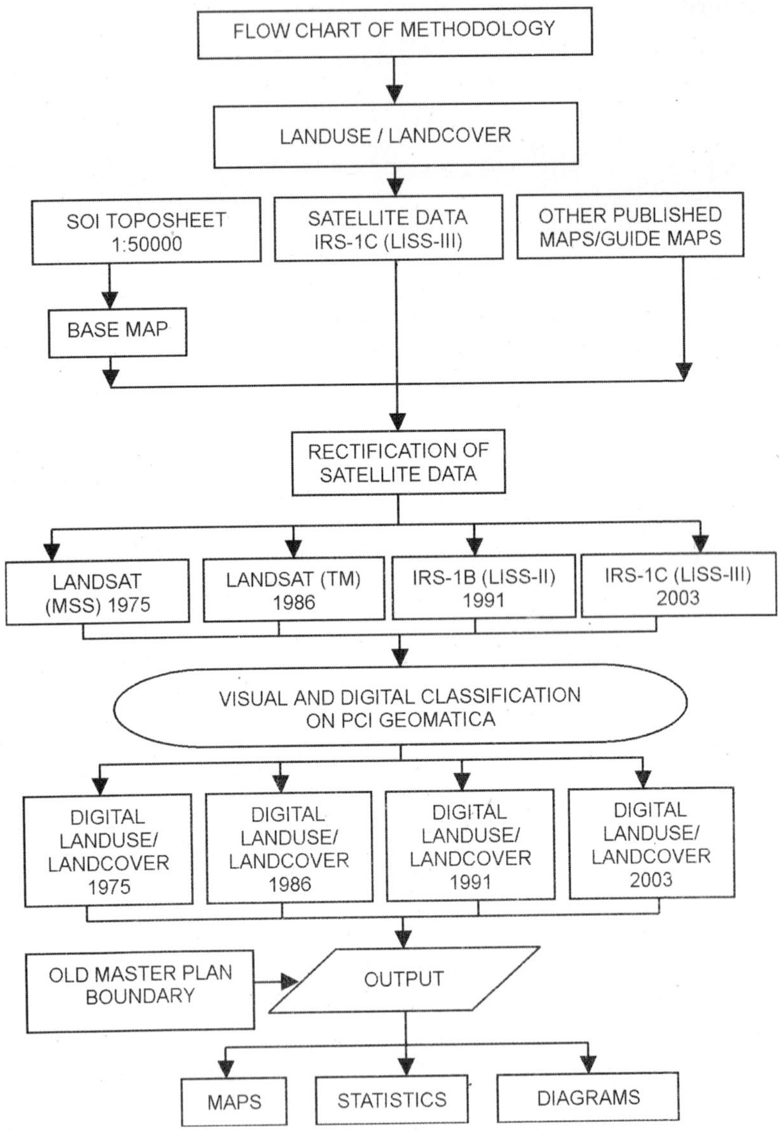
FLOW CHART OF METHODOLOGY
LANDUSE / LANDCOVER
SOI TOPOSHEET 1:50000
SATELLITE DATA IRS-1C (LISS-III)
OTHER PUBLISHED MAPS/GUIDE MAPS
BASE MAP
RECTIFICATION OF SATELLITE DATA
LANDSAT (MSS) 1975
LANDSAT (TM) 1986
IRS-1B (LISS-II) 1991
IRS-1C (LISS-III) 2003
VISUAL AND DIGITAL CLASSIFICATION ON PCI GEOMATICA
DIGITAL LANDUSE/ LANDCOVER 1975
DIGITAL LANDUSE/ LANDCOVER 1986
DIGITAL LANDUSE/ LANDCOVER 1991
DIGITAL LANDUSE/ LANDCOVER 2003
OLD MASTER PLAN BOUNDARY
OUTPUT
MAPS
STATISTICS
DIAGRAMS

## Preparation of Land Use/Land Cover Maps for the Years 1975, 1986 and 1991, Based on Visual/on Screen Visual Interpretation

Land use map of 1975, 1986 were obtained by the LANDSAT MSS and TM respectively, IRS 1B data was used for the preparation of land use/cover map of 1990 of entire Jaipur region and map for 2003 was prepared by digital cum on-screen visual interpretation of IRS 1C LISS (III) data.

## Land Use/Land Cover of 1975

For the preparation of land use/land cover map, old JDA (Jaipur Development Authority) boundary had been considered. This boundary covered almost 44472.27 hectares of land.

*Area Under Land Use/Land Cover in 1975*

| *Category (LU/LC)* | *Area (in hectares)* | *Area (in%)* |
|---|---|---|
| High density built-up land | 1001.66 | 2.25 |
| Medium density built-up land | 1523 | 3.42 |
| Low density built-up land | 2113.69 | 4.75 |
| Croped area | 16385.65 | 38.22 |
| Fallow land | 12308.26 | 27.68 |
| Degraded forest | 2911.33 | 6.55 |
| Mining area | 48.61 | 0.11 |
| Gullied area | 249.52 | 0.56 |
| Sandy area | 4823.54 | 10.85 |
| Saline area | 334.08 | 0.75 |
| Industrial area | 302.92 | 0.68 |
| Land under transformation | 1427.1 | 3.21 |
| Rocky wastelands | 789.75 | 1.78 |
| River beds | 253.15 | 0.57 |
| Total | 44472.27 | 100 |

## Land Use/Land Cover of 1986

After the gap of eleven years since 1975, following observations are made: 41.20 per cent of the total study area was cropland, which has increased to 2.98 per cent. As a result, fallow land was highly decreased. Degraded forest was increased, while due to afforestation 2.53 per cent of area converted to forest class. Areal extent of built-up land clearly indicates the development of the city during this period.

*Area Under Land Use/Land Cover in 1986*

| *Category(Lu/Lc)* | *Area (in hectares)* | *Area (in%)* |
|---|---|---|
| High density built-up land | 1268.12 | 2.85 |
| Medium density built-up land | 4216.78 | 9.48 |
| Low density built-up land | 3086.38 | 6.94 |
| Croped area | 18252.98 | 41.2 |
| Fallow land | 3942.14 | 8.86 |
| Degraded forest | 4147.72 | 9.33 |
| Actual Forest | 1124.58 | 2.53 |
| Mining area | 167.56 | 0.38 |
| Gullied area | 415.35 | 0.93 |
| Sandy area | 1301.88 | 2.93 |
| Saline area | 194.8 | 0.44 |
| Industrial area | 807.67 | 1.82 |
| Land under transformation | 4426.73 | 9.95 |
| Park/Playground | 80.81 | 0.18 |
| Rocky wasteland | 785.61 | 1.77 |
| River beds | 253.15 | 0.57 |
| Total | 44472.27 | 100 |

## Land Use/Land Cover of 1991

During 1991 the built-up land has grown from 19.27 per cent to 25.47 per cent registering an increase of 6.20 per cent area of built-up land. During this period croped area and land under transformation categories cover 35.54 per cent and 13.49 per cent of total geographical area, which indicates that city is in growing stage.

*Area Under Land Use/Land Cover in 1991*

| *Category (Lu/Lc)* | *Area (in hectares)* | *Area (in%)* |
|---|---|---|
| High density built-up land | 2988.92 | 6.72 |
| Medium density built-up land | 4131.99 | 9.29 |
| Low density built-up land | 4209 | 9.46 |
| Croped area | 15803.2 | 35.54 |
| Fallow land | 2960.52 | 6.66 |
| Degraded forest | 5489.28 | 12.34 |
| Actual Forest | 230.92 | 0.52 |
| Mining area | 65.78 | 0.15 |
| Gullied area | 436.9 | 0.98 |
| Sandy area | 475.55 | 1.07 |
| Saline area | 180.75 | 0.41 |
| Industrial area | 376.19 | 0.85 |
| Land under transformation | 6000.88 | 13.49 |
| Park/Playground | 81.73 | 0.18 |
| Rocky wasteland | 789.75 | 1.78 |
| River beds | 250.91 | 0.56 |
| Total | 44472.27 | 100 |

## Land Use/Land Cover of 2003

A remarkable change has been observed in land use during this period. Area statistics shows that city is developing at very fast rate, as built land has increased by 33.98 per cent (Map 3). Crop area has covered 26.62 per cent; 3.47 per cent and 4.36 per cent was under gullied and saline lands respectively.

*Area Under Land Use/Land Cover in 2003*

| *Category(Lu/Lc)* | *Area (in hectares)* | *Area (in%)* |
|---|---|---|
| High density built-up land | 2444.2 | 5.5 |
| Medium density built-up land | 12665.09 | 28.48 |
| Crop area | 11837.84 | 26.62 |
| Fallow land | 2909.38 | 6.54 |
| Degraded forest | 233.05 | 0.52 |
| Actual forest | 5289.06 | 11.89 |
| Mining area | 82.37 | 0.19 |
| Gullied area | 1543.39 | 3.47 |
| Saline area | 1937.84 | 4.36 |
| Park/Playground | 153.16 | 0.34 |
| Rocky wasteland | 12.38 | 0.03 |
| Plantation | 157.88 | 0.36 |
| Scrub | 2653.92 | 5.97 |
| River beds | 250.5 | 0.56 |
| Water body | 5.82 | 0.01 |
| Rail | 143.08 | 0.32 |
| Road | 2153.32 | 4.84 |
| Total | 44472.27 | 100 |

## Land Use/Land Cover of 2003 within New JDA Boundary

The new master development plan for the development of Jaipur was prepared and approved in 1995, which came into force from September 1, 1998. Land Use/Land Cover map of the new area has been prepared by digital image processing techniques of the IRS-LISS III data of the year 2003.

The computations show that out of 149,438 hectare land, 18,180.75 hectare land is under high and medium density built-up lands, which is 11.72 per cent of the total study area. Computations also show that 36.55 per cent area is under crop land, 17.16 per cent area is under fallow land (Map 5), 9.48 per cent area is under scrub land.

Within the new JDA boundary, some small townships like Bagru, Sanganer, Chaksu, Kukas were also considered. The regional appearance of the area is urban and of suburban categories.

*Area Under Land Use/Land Cover in 2003 (New Boundary)*

| *Name* | *Area (in hectares)* | *Area (in%)* |
|---|---|---|
| High density built-up land | 2494.83 | 1.67 |
| Medium density built-up land | 15685.92 | 10.5 |
| Crop area | 54616.66 | 36.55 |
| Fallow land | 26323.31 | 17.61 |
| Degraded forest | 1104.48 | 0.74 |
| Actual forest | 12131.08 | 8.12 |
| Mining area | 82.37 | 0.06 |
| Gullied area | 7675.55 | 5.14 |
| Saline area | 9902.76 | 6.63 |
| Park/Playground | 153.16 | 0.1 |
| Rocky wastelands | 852.54 | 0.57 |
| Plantation | 157.88 | 0.11 |
| Scrub | 14168.74 | 9.48 |
| River beds | 900.69 | 0.6 |
| Water body | 73.84 | 0.05 |
| Rail | 280.97 | 0.19 |
| Road | 2834.15 | 1.9 |
| Total | 149438.94 | 100 |

## Trends of Urban Development

The urban sprawl maps of Jaipur city have been prepared on the basis of multi-date data of Landsat (MSS), Landsat (TM), IRS – IB (LISS–II), IRS–IC (LISS–III) data in conjunction with SoI topographical maps. These maps depict, growth of the city as well as the direction of the growth. After analysing these maps, the following observations are made on the basis of their land Use/land cover classification.

## Change Detection During 1975 to 2003

LANDSAT (MSS) sensor was used for finding the change for the period of 1975 and after 11 years the same study area covered by LANDSAT(TM) in 1986. Again same study area was covered with the help of IRS 1B LISS (II) data in 1991 to find out the change during the gap of 5 years. Likewise, again IRS LISS (III) data has been taken for the year 2003, to find the change since 1991.

To get the changes at periodical intervals one cross matrix has been generated with the help of Complex Logic Program in

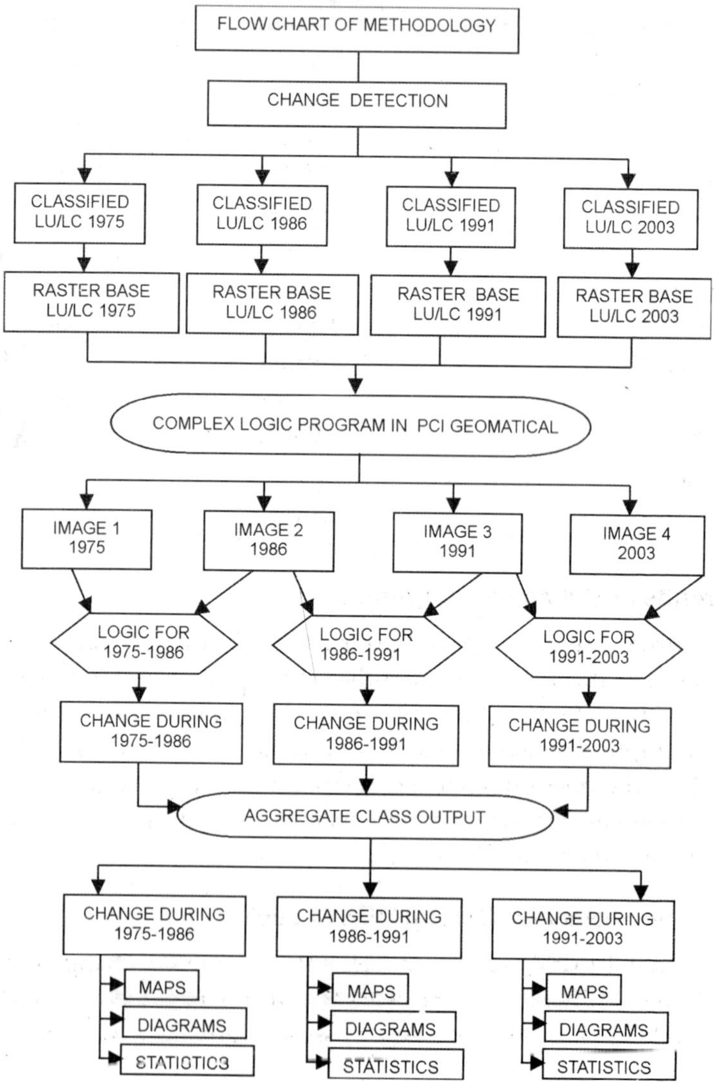

PCI Geomatica, which gives the direction and dimensions of change in land use and land cover during two time periods.

## Change Detection Studies

### *Change During 1975-1986*

Total built-up land in 1975 was 10.4 per cent of total study area, which has become almost double and covered 19 per cent land of the study area. The major changes are observed in southern and western parts where density of houses has increased to high density and open areas have been converted into built-up lands. The mining area in the study area is mainly stone quarrying found near foothills of Jhalana and Nahargarh hill ranges. During 1975-86 period it has increased from 48.61 hectares to 167.56 hectares, showing three-fold increase. Land under transformation category mainly showing the transition phase. Whereas 6979.9 hectares land of this category has remained as open land. This shows that the city is developing outwards and encroaching agricultural lands.

Categorywise changes are shown in the following table:

*Change During 1975 to 1986 (Aggregate Classes)*

| *S.No.* | *Class Name* | *Area (in Hectares.)* | *Area (in%)* |
|---|---|---|---|
| 1. | High density | 1002.41 | 2.25 |
| 2. | Moderate density | 1377.04 | 3.1 |
| 3. | Low density | 1038.64 | 2.34 |
| 4. | Built-up development | 1221.98 | 2.75 |
| 5. | Crop to settlement | 1901.55 | 4.28 |
| 6. | Open area to settlement | 3330.89 | 7.49 |
| 7. | Land trans. to settlement | 913.88 | 2.05 |
| 8. | New settlement | 42.57 | 0.1 |
| 9. | Crop area | 16435.7 | 36.96 |
| 10. | New crop area | 390.01 | 0.88 |
| 11. | Land under transformation | 6979.85 | 15.69 |

Contd...

Contd...

| | | | |
|---|---|---|---|
| 12. | Fallow land | 2212.07 | 4.97 |
| 13. | New plantation | 317.43 | 0.71 |
| 14. | New park | 11.4 | 0.03 |
| 15. | Degraded forest to open area | 126.6 | 0.28 |
| 16. | Degraded forest | 3645.56 | 8.2 |
| 17. | Dense forest | 53.68 | 0.12 |
| 18. | Industrial area | 302.8 | 0.68 |
| 19. | New industrial area | 608.95 | 1.37 |
| 20. | Mining area | 48.5 | 0.11 |
| 21. | New mining area | 150.62 | 0.34 |
| 22. | Wasteland | 1161.45 | 2.61 |
| 23. | New wasteland | 152.76 | 0.34 |
| 24. | Barren rocky | 790.96 | 1.78 |
| 25. | River | 249.64 | 0.56 |
| 26. | Nalahs | 5.3 | 0.01 |
| | Total | 44472.27 | 100 |

## Change During 1986-1991

During 1986-91 period, major changes have been observed in two land use classes, agricultural and built-up land. Crop area has been decreased considerably, whereas built-up land is increasing, to accommodate the increasing population of the city. New colonies are being developed on the agricultural lands/ open lands.

New plantation areas have been developed in various land use areas, like degraded forest, gullied area and sandy area in 1991. Most of the dense forest areas have been changed to degraded forest and only 112.32 hectares of land has remained unchanged due to droughts or decline in vegetative cover.

A considerably large part of land under transformation has been utilized for residential purpose (2500.76 hectares) and industrial purpose (74.30 hectares). This shows that the Jaipur city has expanded at faster rates, especially in southern and western directions.

*Change During 1986 to 1991 (Aggregate Classes)*

| S.No. | Class Name | Area (in Hec.) | Area (in%) |
|---|---|---|---|
| 1. | High density | 1268.12 | 2.85 |
| 2. | Moderate density | 2372.26 | 5.33 |
| 3. | Low density | 1766.42 | 3.97 |
| 4. | Built-up development | 3161.72 | 7.11 |
| 5. | Crop to settlement | 703.53 | 1.58 |
| 6. | Open area to settlement | 630.89 | 1.42 |
| 7. | Land trans. to settlement | 2500.76 | 5.62 |
| 8. | New settlement | 56.68 | 0.13 |
| 9. | Crop area | 13727.63 | 30.87 |
| 10. | New crop area | 1068.88 | 2.4 |
| 11. | Land under transformation | 8846.67 | 19.89 |
| 12. | Fallow land | 700.19 | 1.57 |
| 13. | New plantation | 295.49 | 0.66 |
| 14. | Park/playground | 51.84 | 0.12 |
| 15. | Degraded forest to open area | 32.2 | 0.07 |
| 16. | Degraded forest | 4318.68 | 9.71 |
| 17. | Dense forest | 112.32 | 0.25 |
| 18. | Industrial area | 808.24 | 1.82 |
| 19. | New industrial area | 87.26 | 0.2 |
| 20. | Mining area | 167.27 | 0.38 |
| 21. | New mining area | 50.8 | 0.11 |
| 22. | Wasteland | 583.03 | 1.31 |
| 23. | New wasteland | 119.35 | 0.27 |
| 24. | Barren rocky | 786.36 | 1.77 |
| 25. | River | 249.35 | 0.56 |
| 26. | Nalahs | 6.34 | 0.01 |
|  | Total | 44472.27 | 100 |

## Change During 1991-2003

During 1991-2003 time period, major change was found within/between two land use classes: agricultural and built-up land. Crop area has decreased consideably, whereas built-up land is increasing very fast, as population of the city is increasing at a fast rate. A major positive transformation occurred in this time period; 4222.77 hectares of land of this class changed to plantation. This is because of afforestation programmes, and this is further reclassified as new plantation category.

*Change During 1991 to 2003 (Aggregate Classes)*

| S.No. | Class Name | Hectares | % Image |
|---|---|---|---|
| 1. | High density | 2367.42 | 5.32 |
| 2. | Moderate density | 3639.92 | 8.18 |
| 3. | Low density | 3808.92 | 8.56 |
| 4. | Built-up development | 61.98 | 0.14 |
| 5. | Crop to settlement | 1750.87 | 3.94 |
| 6. | Open area to settlement | 490.69 | 1.1 |
| 7. | Land trans. to settlement | 2978.84 | 6.7 |
| 8. | New settlement | 599.67 | 1.35 |
| 9. | Crop area | 10340.01 | 23.25 |
| 10. | New crop area | 1185.29 | 2.67 |
| 11. | Land under transformation | 7003.12 | 15.75 |
| 12. | Fallow land | 435.05 | 0.98 |
| 13. | New plantation | 4304.22 | 9.68 |
| 14. | Park/playground | 26.78 | 0.06 |
| 15. | New park | 99.53 | 0.22 |
| 16. | Degraded forest to open area | 354.99 | 0.8 |
| 17. | Degraded forest | 38.13 | 0.09 |
| 18. | Dense forest | 523.64 | 1.18 |
| 19. | Industrial area | 322.79 | 0.73 |
| 20. | Mining area | 65.43 | 0.15 |
| 21. | New mining area | 28.97 | 0.07 |
| 22. | Wasteland | 355.28 | 0.8 |
| 23. | New wasteland | 305.45 | 0.69 |
| 24. | Barren rocky | 767.92 | 1.73 |
| 25. | River | 255.86 | 0.58 |
| 26. | Road network | 2153.32 | 4.84 |
| 27. | Rail network | 143.08 | 0.32 |
| 28. | Nalahs | 65.09 | 0.15 |
| | Total | 44472.27 | 100 |

## Results and Discussion

Jaipur city was bounded by a strong wall for the living of 2.5 lac people. The city has shown a higher rate of expansion after 1973 and has covered 151.1 sq. km. area under the residential category. The major expansion is observed in the western, southern and south-eastern parts and along the national highway Nos. 8, 11 and 12. Towards the south, it has expanded about 20 km from Ajmeri Gate covering Sanganer town and even along the Tonk Road. Towards the western direction it has almost reached Bagru town, which is about 35 km away from the city.

## Main Findings

- Jaipur city has expanded towards south and western directions in an uncontrolled manner, engulfing used productive cropped areas.
- The urban area has cover over the surroundings towns, which are developing as the satellite towns like Sanganer, Bagru, Chomu, Achrol, Kanota, etc.
- Jaipur city is putting heavy pressure on the ecological sensitive areas, due to deforestation and mining in Nahargarh and Jhalana reserve forest areas.
- The Remote Sensing Technology has proved its importance for Urban land use analysis, Change detection and Planning.

## Suggestions

- The planning authorities should prepare the future development in advance to avoid unplanned expansion of the city.
- The planner should use Remote Sensing technology for identification of suitable lands for establishment of new colonies, industrial and institutional areas without harming ecology and productive areas.
- The planners can utilize Remote Sensing tools for route planning, circuitry route (byepass routes), requirement of flyovers, treatment plants and various facilities. It can also

be used for identification of unplanned and slum area, so that they can be taken up on priority of plan.

- The authorities responsible for the supply of various facilities and amenities should prepare a comprehensive database for the entire city, so that they may know the status of the facilities available in different parts/colonies of the city. It would help in identifying the area where the facilities are lacking or needs immediate measures.
- Various types of simulation modelling can be possible by combination of Remote Sensing and GIS techniques to facilitate the planners for effective planning.

*Acknowledgement*

A part of this work is supported by the earlier study done by Dr. M.S. Nathawat (second author). Visually interpreted and classified maps were used in the project for the ready reference of the study area. Authors would like to express their thanks to Mr. B.K. Singh (System Analyst, B.M.B.S.T.C., Jaipur) and Mr. V.S.P. Sinha (Project Scientist, B.M.B.S.T.C., Jaipur) for their association, help and support in the project.

## Select Bibliography

Behra, G. and Nageswara Rao: "Growth Of Bangalore City Since 1900 Based on Maps And Satellite Imagery". In Technical Report, ISRO, November 1985.

Das, Bhaskar and Gupta, A.R.: "A Remote Sensing System For Effective Land Transformation And Management". In Proceedings of National Symp. on Remote Sensing in Land Transformation and Management, pp. 23-30(1987).

Master Development Plan – 2011, Jaipur Region, Part – 1, Technical Report, July, 1995.

Mathur, Indu: "Jaipur: A Social Profile". In Nathawat G.S. et al. (ed.) *Human Ecology: An Indian Perspective*, pp. 134-135(1985)

Nathawat, M.S.: "Monitoring Urban Land Transformation In Jaipur And Its Environs". A Field Study, Jaipur, University of Rajasthan (1992).

Nathawat, G.S.: "The Jaipur City: Its Growth And National Resources". In Proceedings of The International Conference on Jaipur and Its Environs; Jaipur, University of Rajasthan (1979).

Pathan, S.K., Jothimani, P., Sampat Kumar, D., and Pandharkar, S.P. (1989) – Urban Landuse Mapping Of Bombay Metropolitan Region Using Remote Sensing Data – SAC (Ahmedabad). Photanirvachak – *Journal Of Indian Society Of Remote Sensing,* Vol. 17, No. 3, 1989

Prasad V.B. and Krishnan, N., "Landuse/Landcover Study of Urban Environment Using Remote Sensing Information. Jaipur City and Its Environs". For Third Module Project Work, HUSAG, IIRS, Dehradun, (1989).

Subudhi, A.P. and Azad, H.S., "Application of Remote Sensing Technology in Urban Analysis and Planning".

**Map 1**

*Landuse/Landcover Map (1975)*

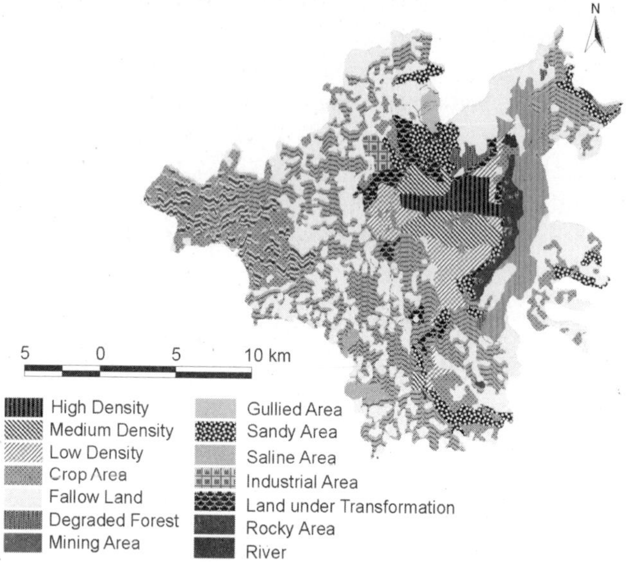


**Map 2**

*Landuse/Landcover Map (1986)*

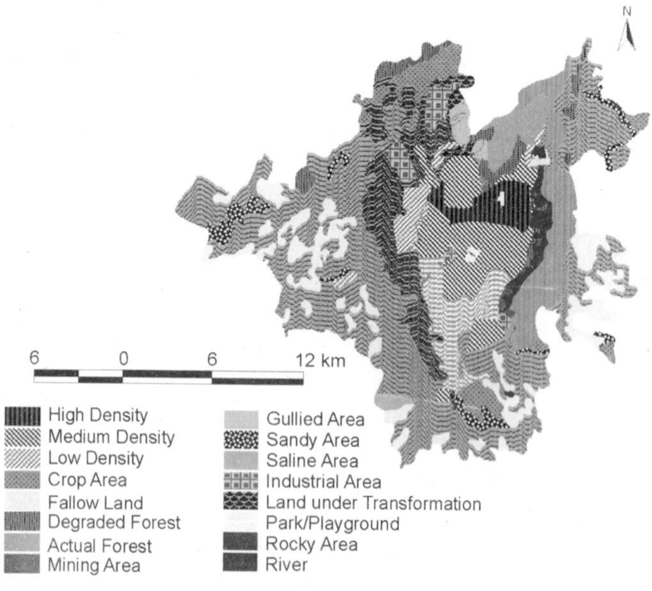

**Map 3**

*Landuse/Landcover Map (2003)*

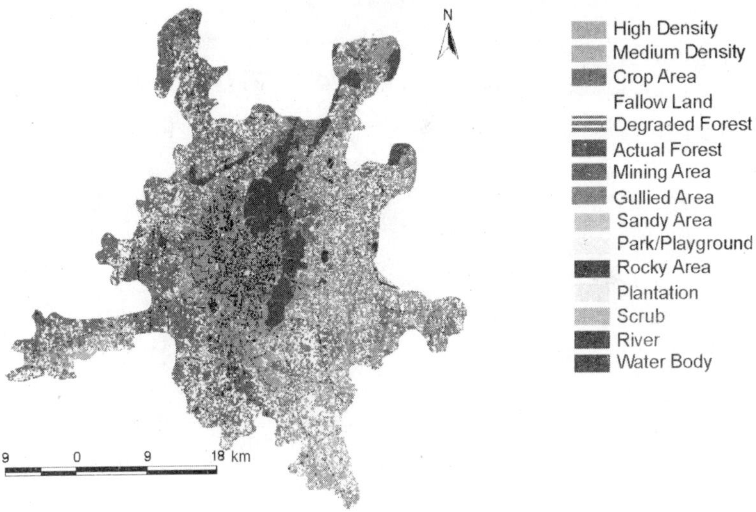


**Map 4**

*Land Transformation During 1975 to 1986*

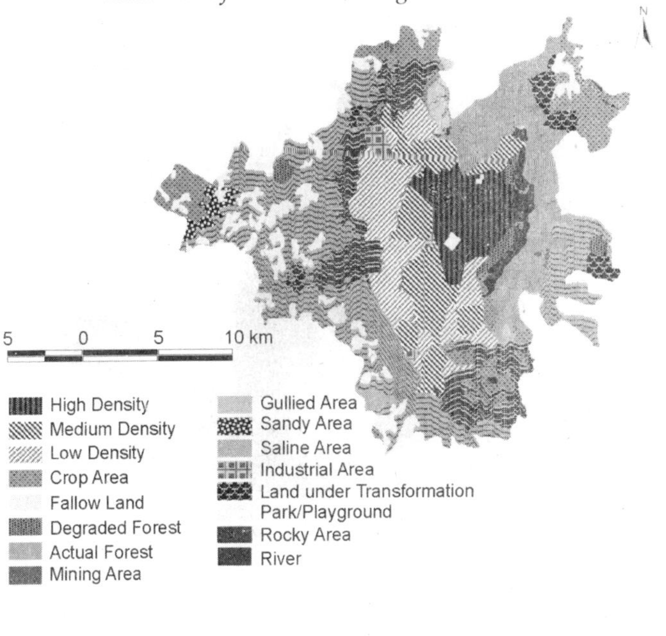

**Map 5**

*Land Transformation During 1986 to 1991*

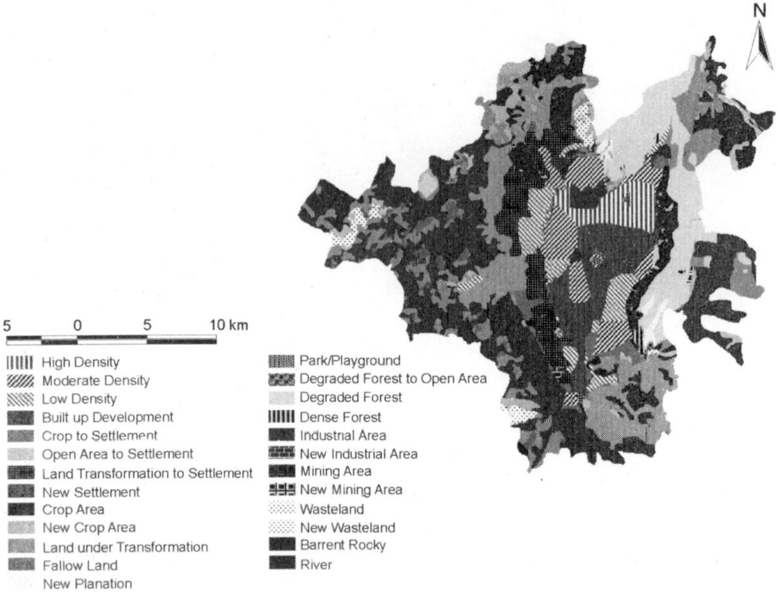


**Map 6**

*Land Transformation During 1991 to 2003*

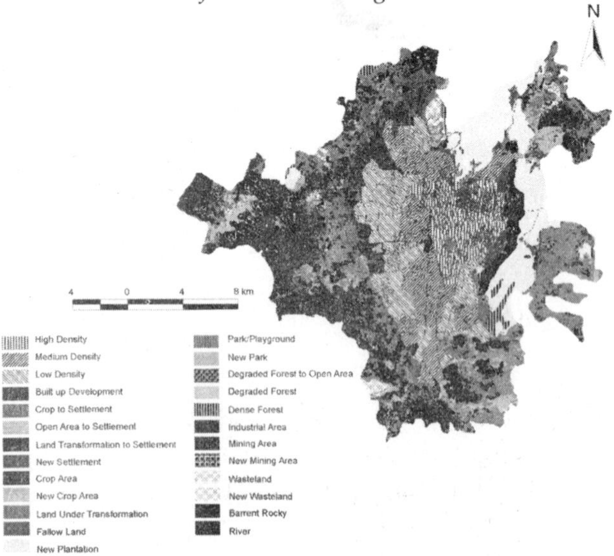

# IV

# Geoinformatics for Natural Resources Mapping

# 29

# Natural Resource Database Management in Margajo Watershed, Upper Barakar Basin, Hazaribagh Using GRAM++ GIS

Ashok Kumar

Large portion of agriculture lands within the Margajo watershed are mono-cropped (mostly rainfed Kharif crop) whereas area receives nearly 1200 mm normal rainfall. The rainfall within the Upper Barakar Basin (UBB) provides sufficient scope for the development of water resources for double cropping. It is also feared that unplanned extraction of presently underutlized water resources in the plateau region may have other long-term consequences as witnessed in the southern part of India. So utlization and management of the resources are required to be looked into simultaneously. This gives opportunity to create database on the land and water system of the watershed and thereafter look into the utilization and management of the available natural resources. In a first hand account of the area, it looks that surface and groundwater development is poor. At present groundwater is being tapped through dug-wells. Even shallow aquifer is underutlized. Therefore, scientific study of water resource is required to be given priority. Besides this, it has other problems such as soil erosion and decrease in the forest cover. These problems have impacts on the Damodar Valley Corporation (DVC) Tilya Dam, as all the drainage system of the Margajo watershed drains into the Dam area. The soils are susceptible to severe water erosion hazard. The soils experience partial waterlogging in early stages of crop growth, followed by seasonal droughtiness during the rest of the crop growth period. Subsoil is graveliness and coarse texture at places, reduce AWC. In the recent years the stakeholders of the available resources within the watershed are giving more stress on the groundwater development. For them, groundwater development is easy to manage. There is fear that the recent trend may deplete the groundwater resources if proper care is not taken from the early stage. The area witnesses high fluctuation of water table. The aquifer is also inhomogeneous and erratic. This makes the large scale and high yield bore-well development difficult. The stake-holders are not ready to take up the bore-well development due to uncertainty about the success.

Various ongoing rural development schemes like construction of dug-well, check dam, ponds, afforestation, etc. under the Integated Watershed Development Programme (IWDP), Drought-Prone Area Development Programme (DPAP) and Afforestation and Wasteland Development Programme are being implemented in the region. The development of these Government sponsored schemes needs scientific database on various natural resources, so that the objective of sustainable development of natural resources may be achieved.

## Study Area

### *Upper Barakar Basin (UBB)*

The entire UBB falls under the Hazaribagh and Chatra districts of Jharkhand. Chatra districts jurisdiction is to a small part of W4 and W1. The details of the divisions of the watersheds are given in the Table 1.

**Table 1**
*Divisions of Upper Barakar Basin*

| *Name of Watershed* | *Watershed Code* | *Area (ha.)* | *Administrative Jurisdictions* | *Number of Villages* |
|---|---|---|---|---|
| Kewta | W7 | 12500 | Ichak and Barhi blocks of Hazaribag | 50 |
| Barakar-I | W6 | 09000 | Ichak and Barhi blocks of Hazaribag | 26 |
| Anjanwa | W5 | 08500 | Ichak, Barhi and Chauparan blocks of Hazaribag | |
| Barakar-II | W4 | 11000 | Barhi block of Hazaribag and Itkhori blocks of Chatra | 45 |
| Nadhadwa | W3 | 10000 | Barhi blocks of Hazaribag | 112 |
| Margajo | W1 | 15000 | Barhi and Chauparan blocks of Hazaribag, Itkhori blocks of Chatra | 115 |

The Margajo watershed (AIS and LUS, 1988) is a fifth order drainage basin, geographically located between latitude 24°15′ – 24°30′N, longitude 85°15′ – 85°30′E. In regional sense, it is a part of lower Hazaribagh plateau and forms the 370-410 (above m.s.l.) metres erosion surface. On the whole, the plain is undulating with some minor ridges interrupting the level nature topography. The area may be termed as buried pediplain. The cover material is formed by coarse alluvium in the immediate valley of streams while rest of the pediplain has a ferruginous soil. The watershed has total areal extent of 15,000 ha. Area on average receives 1,000 mm of rainfall. Land utilization statistics (Table 3) derived from the remotely sensed data indicates that 9.82 per cent area is under rabi crop and 13.30 per cent area is under forest.

## Agro-climate

The climate of the area is characterized by hot summers and cool winters. The area receives an annual rainfall of 1200 to 1600 mm of which 70-80 per cent is received between July and September. The annual rainfall meets about 60 per cent of annual PET demand (1400 to 1500 mm) resulting in an annual water deficit of

**Figure 1**

*Showing Study Area Falling Under Agro-Eco-Region 11*

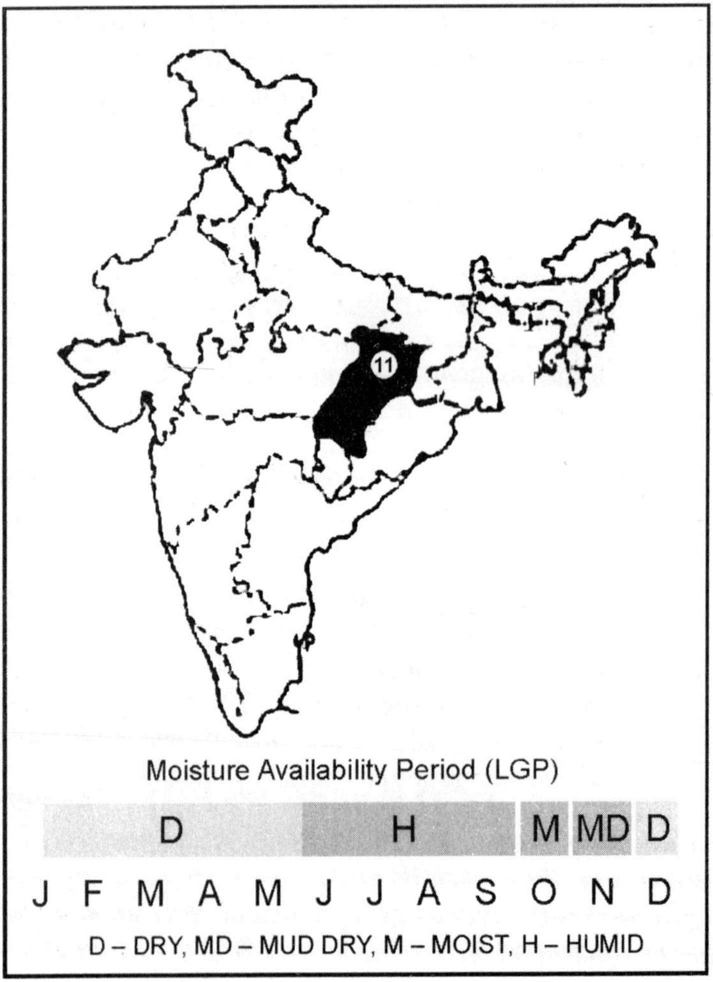

**Figure 2**
*Showing Water Balance Graph in the Region*

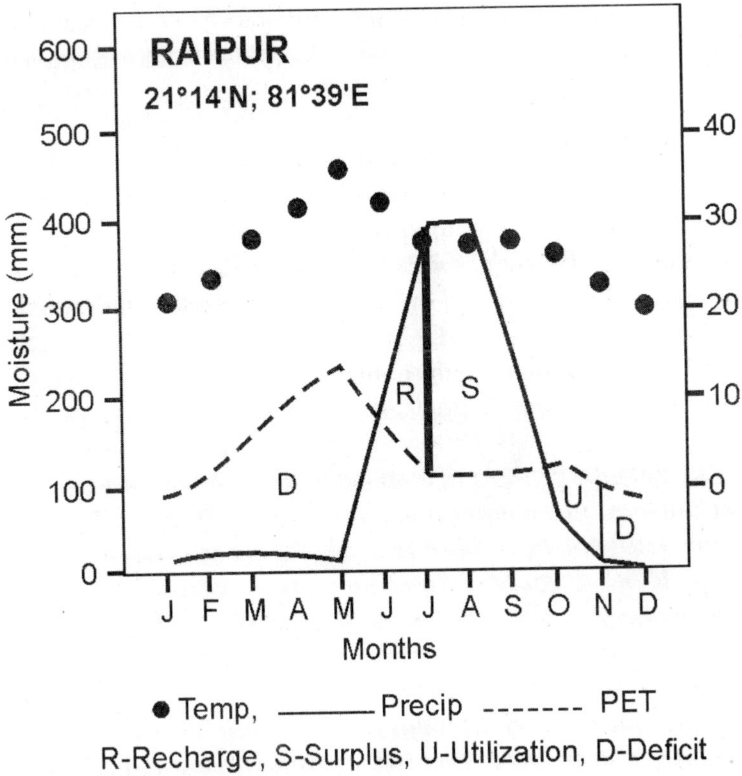


500 to 700 mm. The water balance diagram (Figures 1 and 2) shows that water need (PET) exceeds water supply (precipitation) from October to June next. The soils remain dry from December to May/June suggesting Typic Ustic soil moisture regime. The mean annual soil temperature of more than 22°C qualifies the area for Hyperthermic soil temperature regime. The LGP ranges between 150 and 180 days in a year.

## Objectives

Following were the pointed objectives:

- Evaluate geo-hydrology

- Work out detailed sub-surface aquifer geometry for knowing the aquifer storage and retrieval (ASR)
- Correlation of aquifer geometry of the UBB with the other area particularly with Siwane Sub-basin located on the Upper Hazaribagh Plateau
- Determination of aquifer reserve and its utilization
- Designing practical solutions to meet specified goals
- Optimizing designs for economic efficiency and account for environmental effects (optimization)
- Groundwater recharge sites identification

**Figure 3**
*IRS-LISS II data of Feb. 1999*

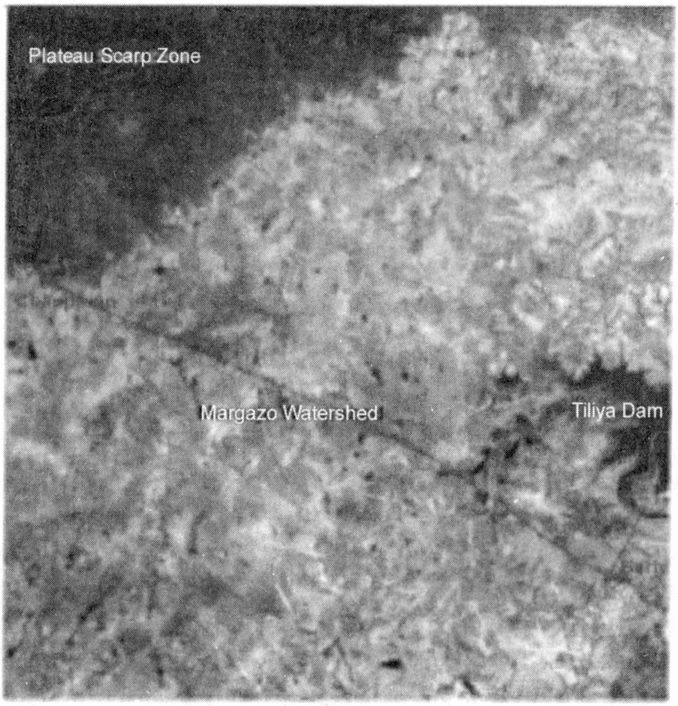

**Table 2**

*Drainage Analysis*

| Watershed name and code | Geographic location | Area (sq.km) | Drainage order-wise streams length (km) | | | | | Cumulative length (km) | Drainage density (km/sq. km.) |
|---|---|---|---|---|---|---|---|---|---|
| | | | 1st | 2nd | 3rd | 4th | 5th | | |
| Margajo (W1) | Northern portion of basin | 150 | 125.01 | 47.76 | 26.86 | 22.79 | 7.73 | 230.17 | 1.53 |

## Natural Resources Mapping

It has been realized that groundwater analysis for its development and management would not be complete if database on other related natural resources is not available. Therefore, primary objective of the study is to generate the spatial and non-spatial natural resource database. Study includes:

- Terrain modelling of the area
- Drainage analysis
- Land use/Land cover analysis
- Surface water analysis
- Groundwater Information System
- Integration of all natural resources inputs and planning for sustainable utilization of natural resources within the watershed

**Figure 4**

*Drainage and Water Body Map of Watershed*

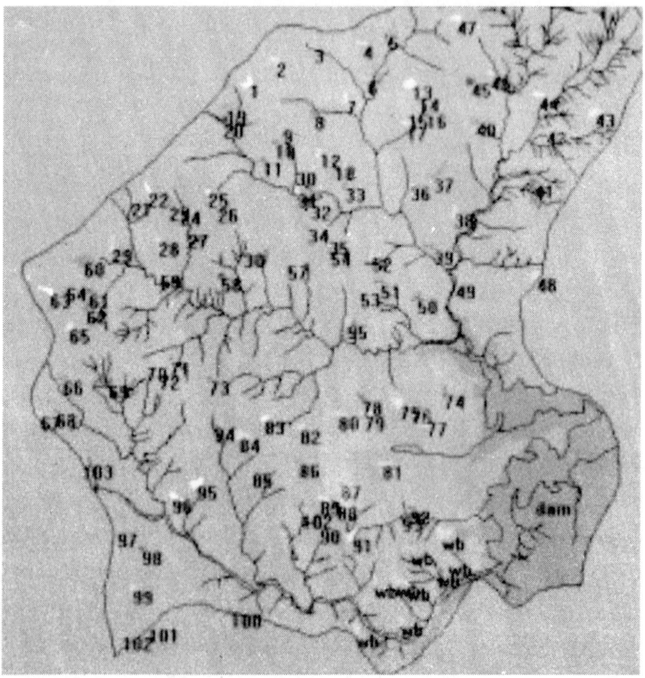

- Preparation of database of natural resources on GRAM (GIS) environment.

## Drainage Analysis

Drainage analysis is an important part for watershed development activity. The drainage channels have been updated from the remotely sensed data. With help of GRAM++ GIS drainage length, cumulative length of each drainage order has been calculated (Table 2). The results indicate that Margajo watershed is a fifth order drainage basin. Drainage density is 1.53 km/sq km. The drainage pattern being dominantly of dendritic nature indicates crystalline basement.

## Terrain Modelling

The topography plays important role in surface and groundwater management. In unconfined aquifer system of weathered granite gneisses, the surface topography governs the upper surface of the aquifer and movement of groundwater. DSTM will also help in identifying the recharge sites. It indicates that mid-eastern portion of watershed is relatively a levelled plain, otherwise the entire watershed is a typical undulating pediplain. Upland, i.e., micro-water divides, are broad and slopes smoothly towards depression channels.

## Land Use/Land Cover analysis

Land use/land cover mapping has been carried out with the help of IRS-1C LISS-III data of February 1999 (Figures 3). In Margajo watershed, kharif crops are grown under rainfed condition. Rabi crops are grown under irrigated condition. Therefore areal extent of rabi crops indirectly indicates the utilization of surface and groundwater. The final land use/land cover map has been inputted into GRAM++ GIS environment where village-wise statistics has been generated and it is an integral part of GIS database. The results indicate that 1473 ha. area under rabi crops within the watershed (Table 3). Most of the rabi crops are spatially located near the village settlements. Its areal extent is comparatively more near the tanks/ponds. The forestland within the watershed is in a degraded condition. Compared to other parts of adjoining watersheds, land with and without scrub is 23 ha. only. This clearly indicates that only 10 per cent of area is under cropping and there is further scope for utlization of agricultural land for rabi season.

**Table 3**

*Land Use/Land Cover Statistics Based on Remote Sensing Data*

| *S.No.* | *Map Legend* | *Class name* | *Area (ha.)* |
|---|---|---|---|
| 1. | 1 | Rabi (wheat etc.) | 1473.62 |
| 2. | 3 | Agriculture plantation | 315.46 |
| 3. | 4 | Close forest | 140.17 |
| 4. | 6 | Scrub forest (mixed) | 261.26 |
| 5. | 7 | Scrub forest (*Lantana Indicia*) | 831.82 |
| 6. | 8 | Scrub forest (coppice) | 517.75 |
| 7. | 10 | Forest blank | 225.23 |
| 8. | 11 | Forest plantation | 180.63 |
| 9. | 13 | Scrub (broken land) | 23.13 |
| 10. | 14 | Scrub (*Lantana Indicia*) | |
| 11. | 15 | Scrub (coppice) | |
| 12. | 16 | Scrub (mixed) | |
| 13. | 18 | Seasonal fallow | 6039.74 |
| 14. | 18 | Permanent fallow | 335.73 |
| 15. | 19 | Water body (including Tiliya dam) | 409.55 |

## Groundwater Information System (GWIS)

Weathered material and fractured basement is a principal aquifer system in the area. Groundwater exploration in this area is mainly a search for depth and nature of weathered and fractured horizon. In groundwater storage and retrieval, surface and basement topography or aquifer geometry also plays an important role. Therefore, for sustainable development and utilization of groundwater in watershed perspective, climatic, physiographic and aquifer geometry has to be taken into consideration. For the analysis of unconfined aquifer basement topography derived with the help of electrical resistivity survey data, geohydrological and physiographic data has been analysed. In the study area, groundwater is being utilized for drinking and irrigation purpose. Groundwater is generally augmented through dugwells which are not capable of sustaining long duration pumping. To understand the aquifer storage and retrieval system of aquifer, 3-D aspects of aquifer has been generated for watershed.

Groundwater occurs under water table condition and aquifer is unconfined/semi-confined in nature. Area has limited groundwater reserves. The yield of dug-wells depends on the thickness of weathered and fractured horizon. It has been observed that dug-well depth normally does not go beyond 11 m below ground level. The pre-monsoon variation of water table is 5 to 6 m in deeply weathered zone. In moderately weathered zone, water table varies between 1 to 3 m. In shallow weathered zone, water table in the month of January is below 4 to 5 m on uplands. Based on the groundwater related data, i.e., water table, litho-log, chemical quality, etc., Ground Water Information

**Figure 5**

*Representation of Fence Diagram Using GWIS*

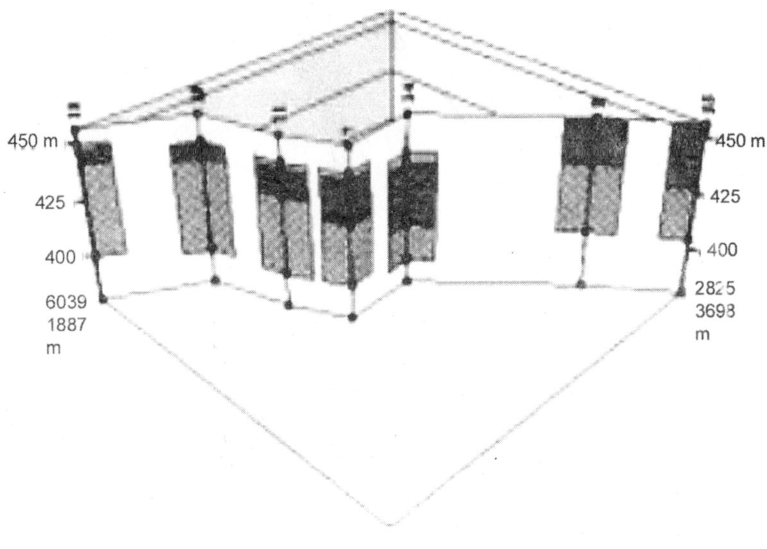

**Figure 6**
*Analysis of Hydrographs of Observation Well of CGWB using GWIS*

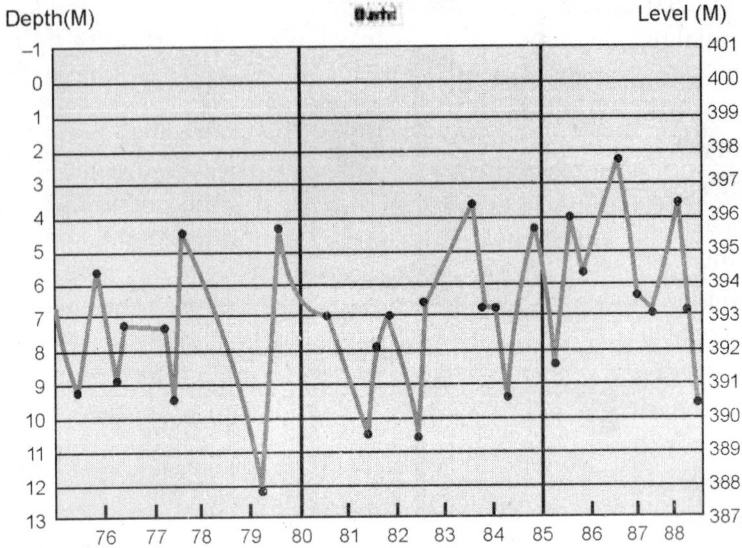


**Figure 7**
*Hydrogeomorphic Zonation of Watershed Based on Remotely Sensed Data*

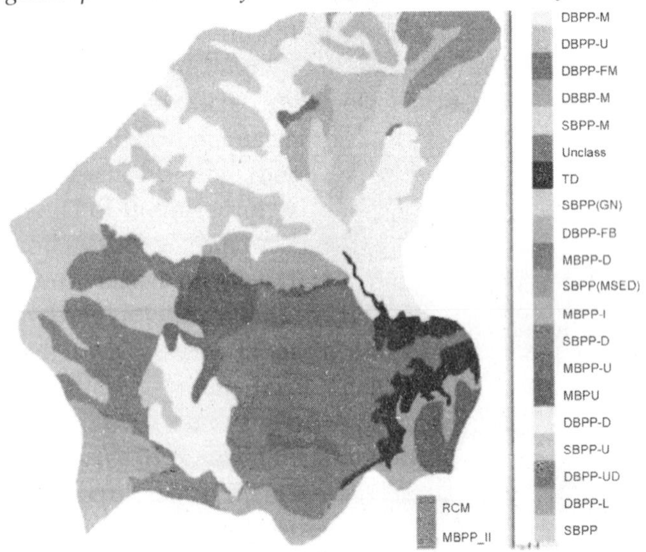

System (GWIS) of the watershed has been developed (Kumar, et al., 2001). The GWIS has facility to present the spatial and aspatial database in graphical form (Figures 5, 6 and 7).

Estimation of existing utlization of groundwater indicates that 84 ha.m water is required for providing drinking water to the watershed population. The remote sensing based mapping indicates that a 1473 ha. of rabi crop (Table 3) is grown in irrigated condition. All the irrigation demands are being met through groundwater. Irrigation of rabi crops utilize 589.44 ha. m of replenished groundwater. Finally 856.84 ha.m of replenished groundwater remains unutlized.

Hydro-geomorphologic mapping has been also carried out with the help of remotely sensed data (Figure 7 and Table 4a). It indicates that 65 per cent of the watershed (Table 4b) is deeply weathered pediplain. Interestingly, upper reaches of watershed divide are more weathered compared to channel depression or topographic low. However hydro-geomorphic mapping fails to identify the two prominent basement depressions mid-north of the watershed through DBTM. Separate class within the same hydro-geomorphic zone has been assigned for its relative geomorphic location, i.e., upland, midland, lowland. This is essential because within the same hydro-geomorphic zone water table and its seasonal variation on upland, midland and lowland are different.

## Digital Basement Topographic Model (DBTM)

Depth of basement has been derived from the interpretation of VES data (Zohdy, 1989). In general, slope of the basement surface is in the same direction as surface slope. Only difference is in steepness of the slope. Basement surface slope is more steep compared to surface. Besides this there are two major enclosed sub-surface depressions in the basement surface has been observed. These depressions can also be used for development of high-yield bore-well and recharge sites identification.

**Table 4a**

*Details of Hydro-geomorphic Zones*

| S. No. | Hydrogeomorphic Zone | Geomorphic Location | Geology | Symbol | Area (hec.) |
|---|---|---|---|---|---|
| 1. | Deeply Weathered Pediplain | Upland | Granite Gneiss | DBPP-U | 3691 |
| 2. | Deeply Weathered Pediplain | Mid-land | Granite Gneiss | DBPP-M | 5761 |
| 3. | Deeply Weathered Pediplain | Low-land | Granite Gneiss | DBPP-L | 355 |
| 4. | Moderately Weathered Pediplain | Upland | Granite Gneiss | MBPP-U | 431 |
| 5. | Moderately Weathered Pediplain | Undifferen-tiated | Granite Gneiss | MBPP-I | 163 |
| 6. | Moderately Weathered Pediplain | Undifferen-tiated | Granite Gneiss | MBPP-II | 33.8 |
| 7. | Shallow Weathered Pediplain | Upland | Granite Gneiss | SBPP-U | 256 |
| 8. | Shallow Weathered Pediplain | Mid-land | Granite Gneiss | SBPP-M | 414 |
| 9. | Shallow Weathered Pediplain | Undifferen-tiated | Granite Gneiss | SBPP | 276 |
| 10. | Shallow Weathered Pediplain | Undifferen-tiated | Mica Schist | SBPP (MSED) | 430 |
| 11. | Deep Weathered Pediplain | Dissected | Granite Gneiss | DBPP-D | 670 |
| 12. | Moderately Weathered Pediplain | Dissected | Granite Gneiss | MBPP-D | 935 |
| 13. | Shallow Weathered Pediplain | Dissected | Granite Gneiss | SBPP-D | 855 |
| 14. | Water Body | | | TD | 528 |
| 15. | Rock Outcrop | | | RCH | 42.6 |

**Table 4b**

*Area Extent of Broad Geomorphic Zone*

| S.No. | Major Groundwater Prospects Zone | Area (hec.) | Percentage |
|---|---|---|---|
| 1. | Total deeply weathered pediplain area including (U, M, L) | 9822 | 65.48% |
| 2. | Total moderately weathered pediplain area including (U, M, L) | 628 | 4.19% |
| 3. | Total shalow weathered pediplain area including (U, M, L) | 1376 | 9.17% |

Geographical area between different depth zones has been calculated (Table 6). The 1.2 per cent area falls within 5-10 m b.g.l. basement depth range. This zone is marginally suitable for dugwell development. The 98.8 per cent area having basement depth greater than 10 m. and this zone is suitable for dugwell development. The 54.8 per cent area having depth of basement greater than 20 m and in this zone dug-cum-borewell is best alternative to tap the maximum possible aquifer thickness. The 29.84 per cent area having depth of basement greater than 25-m and this zone is suitable for borewell development.

Based on the true resistivity value of the aquifer material at depth of 11m, variation of resistivity value at depth of 11m has been also analysed. Area extent within different resistivity range is given in Table 5. This has helped in understanding the aquifer material and its saturation. This has been used to determine the sustainability of the dug-wells.

**Table 5**

*Resistivity Value at Depth of 11 m from Ground Level*

| *S. No.* | *Range (ohm-m)* | *Area (sq. m)* |
|---|---|---|
| 1. | <20 | 3702201.00 |
| 2. | 20-30 | 12169242.40 |
| 3. | 30-40 | 21357233.00 |
| 4. | 40-50 | 38335528.00 |
| 5. | 50-60 | 22130158.00 |
| 6. | 60-70 | 15564769.80 |
| 7. | 70-80 | 13628298.49 |
| 8. | 80-90 | 3862842.10 |
| 9. | 90-100 | 12865997.70 |
| 10. | 100-110 | 2693674.90 |
| 11. | 110-120 | 1975576.30 |
| 12. | 120-130 | 1432550.80 |
| 13. | >130 | 801450.05 |
| | Total | 150519522.54 |

**Table 6**

*Depth of Basement from Ground level*

| *S. No.* | *Range (m)* | *Area (sq. m)* |
|---|---|---|
| 1. | <10m | 1589334.00 |
| 2. | >10-<15m | 7290347.00 |
| 3. | >15-<20m | 30662563.90 |
| 4. | >20-<25 | 49103020.00 |
| 5. | >25-<30 | 35220942.00 |
| 6. | >30 | 14885150.00 |
| 7. | >35-<40m | 4868096.40 |
| 8. | >40-<45m | 3153586.20 |
| 9. | >45m | 3935143.83 |
| Total | | 150708183.33 |

In the entire Margajo watershed, 1800 ha. m replenished utilizable groundwater reserves are available (Kumar et al., 2002). This reserve has potential to provide irrigation to 3600 ha (24 per cent) area of watershed. Besides that, huge amount (4870ha. m) of groundwater reserves lies below the existing dug-well depth (i.e., 10 m b.g.l.) and basement surface. This untapped static groundwater reserve has potential to irrigate 12177 ha. (81%) land of the watershed. But this untapped static groundwater reserve can be utilized if suitable recharge mechanism is developed so that withdrawal amount may be replaced through recharge. Estimation indicates that huge volume of static reserves exists within the watershed (Kumar, et al., 2002). It has been noticed that major portion of the static reserves (2918 ha. m) lies between 15 to 30 m b.g.l. To tap this zone of aquifer, deep dug-well or dug-well cum borewell are the feasible structures. Balance static reserves can be tapped only through large diameter bore-well.

## Decision Support System

Integration of different natural resource for optimal and sustainable utilization within all available natural resources is the

**Table 7**

*Database Field Available in GRAM++ GIS database of Margajo Watershed*

| *Village* | *Area* | *Block* |
|---|---|---|
| District | Residential_House | Population |
| Schedule_caste | Schedule_tribe | Literate |
| T_Worker | T_indcatel | T_Marginal |
| T_Non_Worker | Education_facility_km | Medical_facility_km |
| Drinking_water_km | Post_Telegraph_km | Market_hat_km |
| Communication_km | Road | Nearest_Town_km |
| Electricity | Physiograpgy | Rainfall_mm |
| Geographical_Area_ha | Area_within_watershed_ha | Wheat_Irrigated_acre |
| Tomato_Irrigated_acre | Potato_Irrigated_acre | Onion_Irrigated_acre |
| Cauliflower_Irrigated_acre | Sag_Irrigated_acre | Wheat_unirrigated_acre |
| Gram_unirrigated_acre | Kesari_unirrigated_acre | Arhar_unirrigated_acre |
| Masur_unirrigated_acre | Tisi_unirrigated_acre | Mustard_unirrigated_acre |
| Maize_unirrigated_acre | Peas_unirrigated_acre | Barley_unirrigated_acre |
| Brinja_unirrigated_acre | Total_Agriculture_Land_ha | Total_Rabi_ha |
| Permanent_fallow_ha | Seasonal_fallow_ha | Agriclture_plantation_ha |
| Total_Agrculture_wasteland_ha | Land_without_scrub_ha | Scrub_broken_land_ha |

Contd...

Contd...

| | | |
|---|---|---|
| Scrub_lantana_ha | Scrub_coppice_ha | Scrub_mixed_ha |
| Total_forest_land_ha | Close_Forest_ha | Open1_forest_ha |
| Open2_forest_ha | Watertable_m | Total_groundwater_recharge_ha_m |
| Total_groundwater_draft_ha_m | Balance_groundwater_reserves_ha_m | Groundwater_Irrigation_potential_balance_ha |
| Gwp_mod_ha | Gwp_pmod_ha | Gwp_vp_ha |
| Gwd_Not_Suitable_ha | Gwd_Marginal_Suit_ha | Gwd_less_suit_ha |
| Gwd_Suitable_ha | Depth_Rock_Less_5_m | Depth_Rock_5_10_m |
| Depth_Rock_10_15_m | Depth_Rock_15_20_m | Depth_Rock_20_25_m |
| Depth_Rock_25_30_m | Depth_Rock_30_m | Groundwater_development_feasibility_No_ha |
| Dugwell_feasibility_Marginal_ha | Dugwell_feasibility_Suitable_ha | Dug_cum_borewell_ha |
| Dug_cum_borewell_Borewell_ha | Surface_water_Number | Surface_water_Area_ha |
| Harvested_SW_Volume_ha_m | Total_SW_Volume_avai_ha_m | Bal_Untapped_volume_ha_m |
| Effective_SW_VOL_AVA_ha_m | Irrigation_Potential_ha | Number_Tanks_possible |

present day need. This is the main objective for using Geographic Information System (GIS) technology for natural resource. The use of GIS platform has provided good opportunity for integrating the groundwater geophysics to other natural resources (Kumar, 1997). To integrate the spatial and aspatial information and to facilitate the decision makers of the watershed, GRAM ++ GIS package (DST/CSME, 2001), a low cost user's friendly GIS package has been used for analysis, integration and creation of interactive computerized database for individual resource or in combination or integration of two or more natural resource parameters for achieving the meaningful conclusion/results. It has helped in establishing dynamic link between each natural resource. The GIS database of the watershed contains village-wise information on various

**Figure 8**

*Village Boundary Map, Report can be Generated Village Wise*

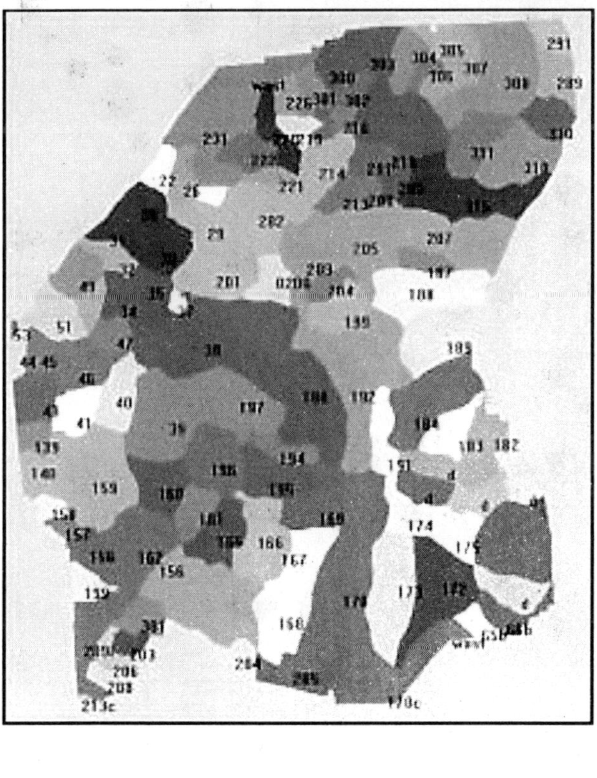

parameters (Table 7). Presentation of database for various themes, spatial and aspatial query (Figures 8, 9 and 10) can be generated which will facilitate the decision makers concerned with the sustainable development of natural resources of the watershed.

The database has been developed village-wise (Figure 8) for vector analysis. All thematic maps can be views separately and two or more themes are overlaid in Raster Analysis module of GRAM++ GIS package. From database it is possible to generate graphical representation of attributes (Figure 9). It is also possible generate class based thematic map generation such as socio-economic data (Figure 10), crops, etc.

**Figure 9**

*Village wise Graphical Representation of Population*

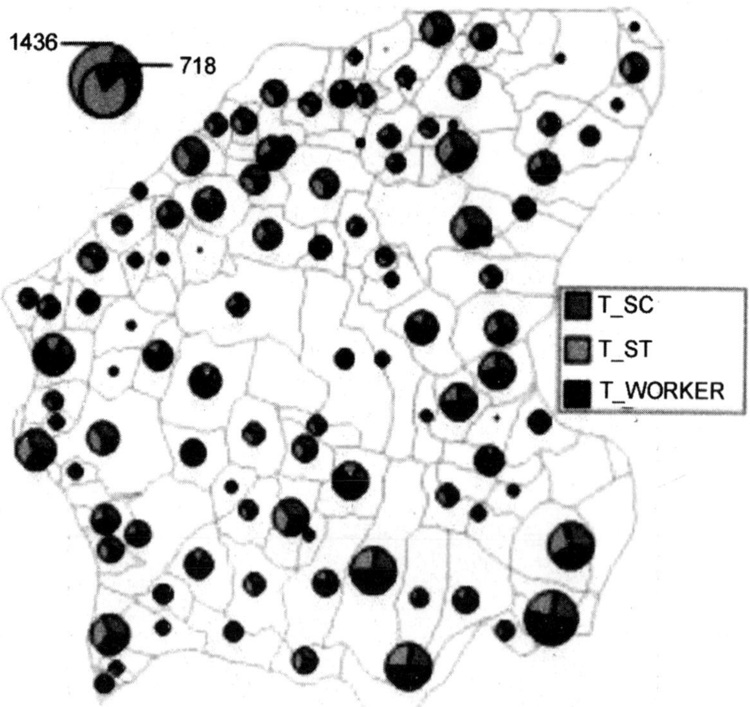

Based on the database in GRAM++ GIS environment non-spatial query is also possible, i.e., Tomato_crop_acreage > 5 hectares. This query will provide list of all villages having tomato crop acreage is greater than 5 hectares as well graphical presentation will be also be displayed for the village fulfilling the criteria set in the query.

**Figure 10**

*Class Based Thematic Map Generation of Socio-Economic Data*

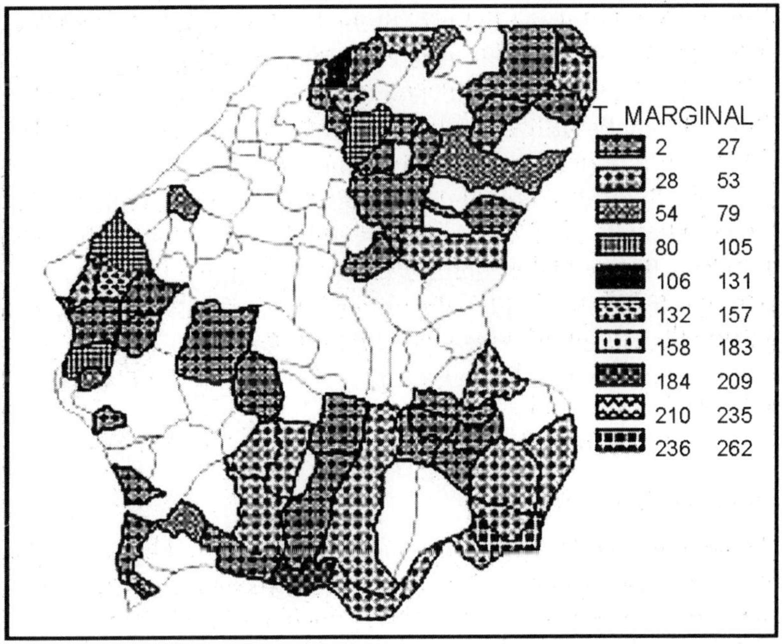


Major emphasis has been given to develop the spatial decision support system of groundwater development. For this purposes, numerical 3D groundwater modelling technique were used and different spatial scenario (Kumar, 2001) of water table has been generated. The numerical modelling result for variation of water table under different rainfall has been inputted into the GRAM++ GIS environment. This has facilitated the rural development authority to find the village going to be affected in summer, well in advance.

## Conclusions

Results indicated that huge storage of static groundwater reserve is available within the watershed. This has the potential to provide irrigation to more than 100 per cent area of watershed. But at present only 27 per cent of the total reserve is replenished. The estimation of total reserve has given further scope for planning groundwater development beyond the replenished limit. It may be utilized only when same volume will be replenished through natural and/or artificial recharge.

The integration of all natural resources on GIS platform has given enough scope for understanding the interrelationship between the available natural resources within the watershed. This has facilitated to plan the utilization of each natural resource optimally so that adverse effect on each natural resource can be minimized. The low-cost and users' friendliness of the GRAM++ GIS package may be integral part to the managers, decision makers of the rural development agencies of the watershed.

### *Acknowledgements*

Author is thankful to Prof. (Dr.) Amitabh Ghosh, Project Director, Bihar Remote Sensing Application Centre, Patna for giving constant encouragement for research work. Author is thankful to NRDMS Division, Deptt of Science and Technology, Govt. of India for providing financial assistance to the project.

## References

AIS and LUS, (1988), *Watershed Atlas of India*, Ministry of Agriculture, Government of India, New Delhi.

BCST/DST, (1999). "Aquifer Geometry Analysis and Natural Resources Management in GIS Environment, Upper Barakar Basin, Koderma, Bihar". Unpublished report of BCST, Patna.

GRAM++ (2001). Geo-Referenced Area Management (GRAM++) GIS package, CSME, IIT, Mumbai and NRDMS Division, Department of Science and Technology, Government of India, New Delhi.

Kumar, Ashok (1997). Natural Resource Database Management for Sustainable Utilisation of Water Resource in Lapasiya Watershed, Hazaribagh, Bihar – A Case Study, Proceeding of workshop on Remote Sensing and GIS Application in Water Resource Engineering, Organised by CBIP, New Delhi, 17-19 September, Bangalore, pp. V(A)23-34.

Kumar, Ashok (2001). Groundwater modelling of unconfined aquifer system of crystalline Area – A case study in Lapasiya Watershed, Hazaribagh, India, International Workshop on GIS technologies for sustainable development at local level, 9-11 October New Delhi, DST-UNDP, pp. 124-127.

Kumar, Ashok and Prasad, L.B. (2002). DSTM and DBTM for Estimation of Static and Dynamic Groundwater Reserve Estimation in Hard Roack – A case study in Margajo Watershed. International Conference on Hydrology and watershed management with a focal theme on water quality and conservation for sustainable development, 18-20th December, CWR, IPGSR, JNTU, Hyderbad.

Kumar, Ashok L.B. Prasad, B.B. Prasad and Nisha Mendiratta (2001). GIS and GWW is tool for creating Ground Water Information System (GWIS) – A case study of Upper Barakar Basin, Bihar, Proceeding of International Conference in Remote Sensing, ICORG 2000, JNTU, Hyderabad, Vol. II, pp. 144-151.

Zohdy A.A.R. (1989). "A new method for automatic interpretation of Schlumberger and Wenner sounding curves", *Geophysics,* Vol. 54, No. 2, pp. 245-253.

# 30

# Land Use/Land Cover Mapping of Itanagar Circle, Papumpare District, Arunachal Pradesh, Using Remote Sensing Techniques

S. Deb and A.D. Patgiri

Itanagar, the capital town of Arunachal Pradesh, is a very fast growing town. According to 1991 census, the total population of Itanagar circle was 18,402 which went up to 39,173 in 2001. To manage with the increasing population, the town is growing in a haphazard manner without any proper and systematic planning. Hence, there is a need to study for the sustainable development of the town. The first step in this regard is to map and monitor the land use/land cover of the area, which will provide necessary inputs to the decision makers/planners to make a plan for a systematic growth of the town. Information on land use/ land cover permits a better understanding of the land utilization aspects on cropping pattern, forests, wastelands and surface water bodies which is vital area for developmental planning (Rout, et al., 2000). Remote Sensing through satellite imagery is found to be an effective technique for mapping and monitoring of land use/land cover of an area. The spatial distribution pattern and location aspects of different categories can be studied more precisely by using Remote Sensing technique.

The present study area covers the Itanagar circle of Papum-pare district of Arunachal Pradesh. The circle falls in the geographical coordinate of 27°00′ N to 27°11′ N and 93°30′ E to 93°44′ E covering a total area of 176.8 sq.km. The objective of the study was to map and analyse the land use/land cover pattern of the study area.

## Methodology

IRS-1D LISS-III digital data of Feb,2003 was used for delineating different categories of land use/land cover using ERDAS 8.6 software. A base map layer was prepared first from SoI

**Figure 1**

*Land Use/Land Cover Map of the Study Area*

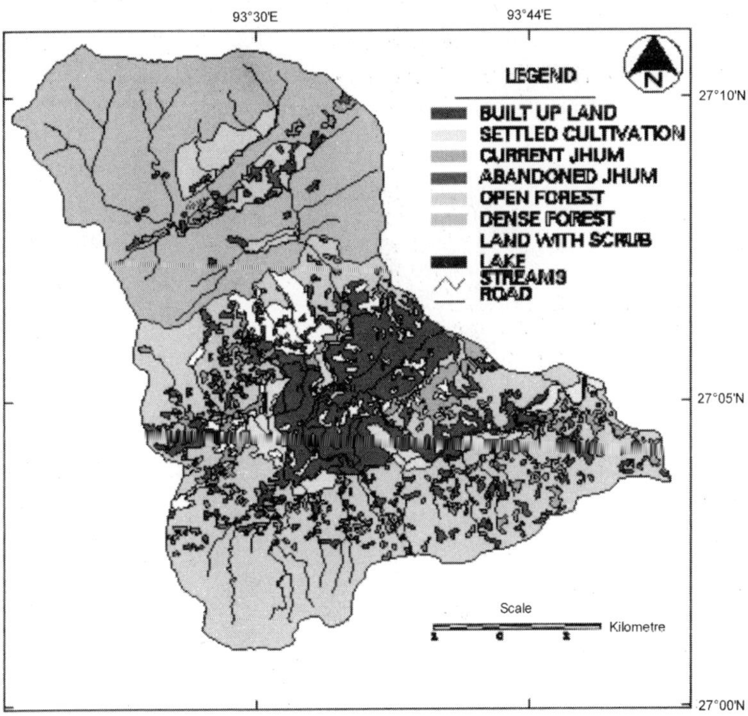

toposheet incorporating major roads, settlement and major drainages of the study area. This base map layer was then put over the satellite image and a layer of land use/land cover was generated through interpretation based on image characters like tone/colour, texture, pattern, shape, size, location, association. This map along with the satellite data was then carried to the field for verification. Then the map was finalized and statistics was generated.

## Results and Discussion

According to the map generated through Remote Sensing (Figure1), the different land use/land cover categories covering the study area includes built-up land, areas under settled and shifting cultivation, open and dense forest, land with scrub and a lake. The detail area statistics are shown in the Table 1 and Figure 2.

**Figure 2**

*Percentage-wise Area Distribution of Different Land Use/Land Cover Categories*

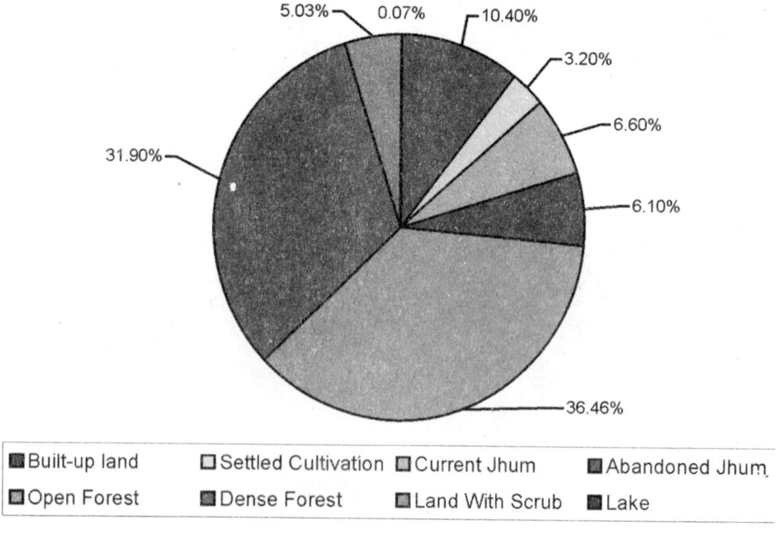

**Table 1**

| *S.No.* | *Category* | *Area (sq. km)* |
|---|---|---|
| 1. | Built-up land | 18.465 |
| 2. | Settled cultivation | 5.715 |
| 3. | Shifting cultivation (current) | 11.805 |
| 4. | Shifting cultivation (abandoned) | 10.845 |
| 5. | Open forest | 64.470 |
| 6. | Dense forest | 56.475 |
| 7. | Land with scrub | 8.900 |
| 8. | Lake | 0.125 |

*Built-up land:* Built-up land includes areas under Itanagar capital complex and some village clusters namely, Niori, Bebi, Bubub, Doria, Chulupu, located outside the capital complex. This is a major category of land use in the study area occupying about 10 per cent of the total area of the circle (Figure 2).

*Settled Cultivation:* About 571 ha. of land is found to be under settled cultivation in Itanagar circle. These areas are mostly plain areas.

*Shifting Cultivation:* Shifting cultivation or Jhuming is an age old practice of primitive agriculture practised by tribal which consists of slash and burning of forests by tribal who cultivate these areas till soil productivity declines and then move over to new sites (Venkataratnam, et al., 2002). 1180 ha. of land is currently under shifting cultivation and 1084 ha. of land is found to be abandoned sites of shifting cultivation in the study area.

*Forest resources:* The forest resources covering the study area falls under the tropical evergreen broad-leaf forest type. Forest areas are divided into two categories, dense and open, according to the density of the crown cover. These wo categories together occupy about 70 per cent of the study area.

*Land with Scrub:* These areas fall under the wasteland category. These occupy 5 per cent of the total area of the circle.

*Lake:* Ganga Lake, a tourist attraction, is located in the study area.

*Drainage:* Chingke, Senkhi, Budhibeta, Dokhoso, Chimpu and Pachin are the main streams flowing through the study area. From the north, Chingke nalah comes downward to meet Senkhi and from the east, Dokhoso and Budhibeta meet Chimpu to join with Senkhi forming Pachin nalah which flows eastward to meet Dikrong river.

## Conclusion

(i) The present study highlights the fact that Remote Sensing technique is found to be useful for delineating land use/ land cover categories of Itanagar Circle of Arunachal Pradesh.

(ii) As Itanagar is having vast forest resources, emphasis should be given to forest conservation.

(iii) The deforested wastelands, found in the area, should be controlled.

(iv) Integration of land use/land cover with other thematic information will be useful in formulating strategy for sustainable development of the area.

### *Acknowledgements*

The authors acknowledge the support given by Dr.G.Ch.Chennaiah, Director, State Remote Sensing Application Centre, Itanagar and Dr. Swapna Acharjee Deb, Principal Investigator, DST Fast Track Project, State Remote Sensing Application Centre, Itanagar.

## References

Rout, D.K., Parida, P.K. and Behera, G. 2000. Land use/Land cover Mapping of Delhi and its Surrounding area using Digital Image Processing Technique. *NNRMS Bulletin*-24 Jan, 2000, DOS, GoI, Bangalore, pp. 38-40.

Sahu, K.C., Pradhan, S., Misra, B.N. and Sudhakar, S. 2002. Landuse/Landcover Mapping of Vishakapatnam District, Andhra

Pradesh, India using Remote Sensing and GIS Proceedings of the ISPRS Commission VII Symposium, December 3-6, Hyderabad, pp. 581-583.

Vekataraman, L. and Das, S.N. 2002. Assessment and Management of Land Degradation in India—Status and Issues. Proceedings of the ISPRS Commission VII Symposium, December 3-6, Hyderabad, pp. 685-694.

# 31

# Land Use/Land Cover Assessment and Change Detection Analysis in a Part of Outer Garhwal Himalaya Using Multi-Temporal Satellite Images

Md. Surabuddin Mondal, A.C Pandey, Minakshi Kumar and R.D. Garg

More than two thousand years ago, the Greek sage Heraclitus claimed that "it is impossible to step into the same river twice". About the 1600 years ago, the Chinese Taoist, Gehong, noted that "the seas change into mulberry fields, while the mulberry fields change into seas". During Heraclitus and Gehong's time, nature was probably still the major factor resulting in land cover change. Today, however, mankind is the most important force in altering the land surface. Through human activities to provide food, building materials, building sites, firewood and clothing, man has been continuously altering land cover.

Remote sensing provides synoptic coverage and temporal data. It helps in monitoring and mapping land use/land cover and assessment of changes. The repetitive coverage provided by satellite data help in monitoring temporal changes on land use/land cover.

## Study Area

The study area falls within the geo-coordinates 30°02′ to 30°12′ North latitudes and 78°25′ to 78°31′ East longitudes covering an area of approximately 15.53 sq. km (Figures 1 and 2). The area encompasses parts of Dehradun, Tehri Garhwal and Puri Garhwal districts, covering Rishikesh and surrounding regions. It is located in the foothill zone of Outer Garhwal Himalaya characterized by denudational hills, piedmont zone and flood plain topography. The elevation ranges between 300 to 500 metres above mean sea level.

The climate in the region is sub-temperate to sub-tropical with mild winter (December to March) and hot summer (May to

**Figure 1**

*IRS-1C LISS 111 Image of Study Area 11th February 1997*

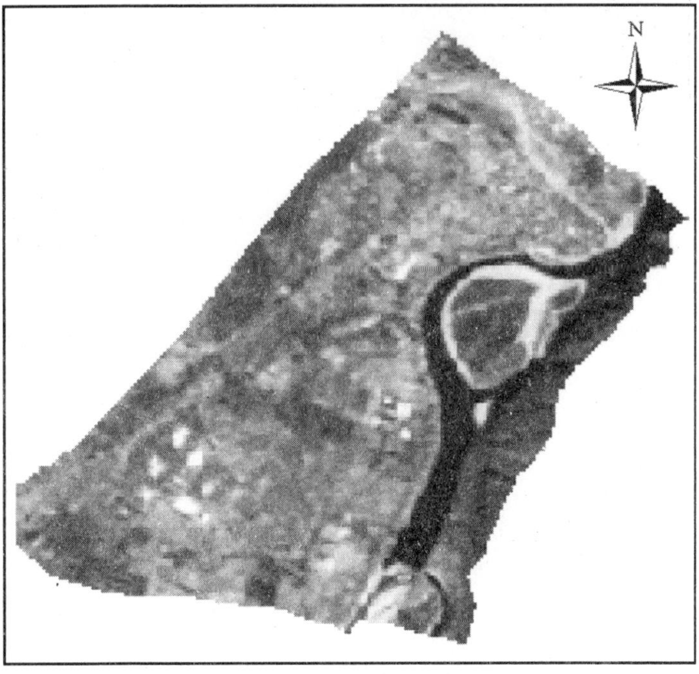

**Figure 2**

*IRS-1C LISS 111 Image of Study Area 6th January 2004*

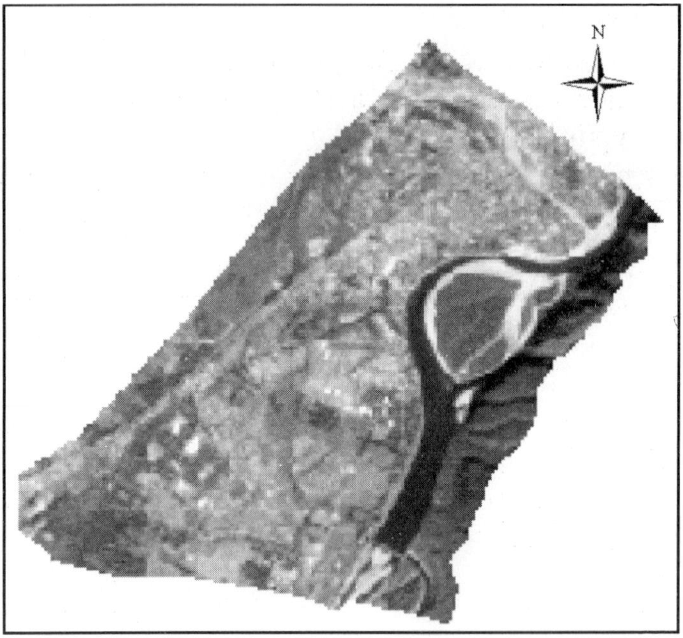


June). The mean winter temperature is 11.2°C whereas mean summer temperature of 25.7°C. Usually monsoon starts in late June and ends in the middle of September. The mean annual rainfall ranges between 1000-1800 mm.

## Database

The area is located in the Survey of India (SoI) topographical sheets 53 J/4 on 1:50,000 scale. The SoI guide map of Rishikesh area on 1:20,000 scale was also used. IRS – 1C LISS-III sensor data of 11th February 1997 and IRS –P6 LISS III sensor data of 6th January 2004 with limited ground truth has been used for the preparation of land use/land cover map.

*Data Used*

| *S.No.* | *Data* | *Year* |
|---|---|---|
| 1. | IRS 1-C LISS-III digital satellite data | 11th Feb., 1997 |
| 2. | IRS 1-C LISS-III digital satellite data | 6th Jan., 2004 |
| 3. | Topographic maps – 53J/4, 53J/8 | 1964-65 |
| 4. | Guide map of Rishikesh area | 1985-86 |
| 5. | Ground data collection | 2004 |

## Methodology

The present study was carried out for preparation of land use/land cover maps for the year 1997 and 2004. To land use and land cover changes analysis at Level–I of LU-LC classification (NRSA, 1995) was performed for the period of 1997 and 2004 due to availability of satellite data of these periods.

During the analysis multi-date images were used. One of the image was geo-referenced through topographical map (rectified toposheet) through image registration and other image was geo-referenced through image to image registration. The RMS error obtained was less than 0.5 which is acceptable for analysis.

To extract information from the remotely sensed digital data, multispectral classification techniques are most often used. The supervised classification technique was used during LU-LC classification. The training sets at various locations covering various categories of land use/land cover were demarcated using image signature and a prior knowledge of the study area and field investigation. Using supervised classification approach, six classes of LU-LC were delineated on the digital satellite data for the years 1997 and 2004.

The rectified and classified images of two dates were compared on a pixel by pixel basis using change detection matrix by using a model in ERDAS 8.7 software. The output in the form of a change image map was generated consisting of brightness values from 1 to 60. Then specific from–to classes were selected for analysis. A select number of possible off-diagonal from–to

**Figure 3**
*Flow Chart of Methodology*

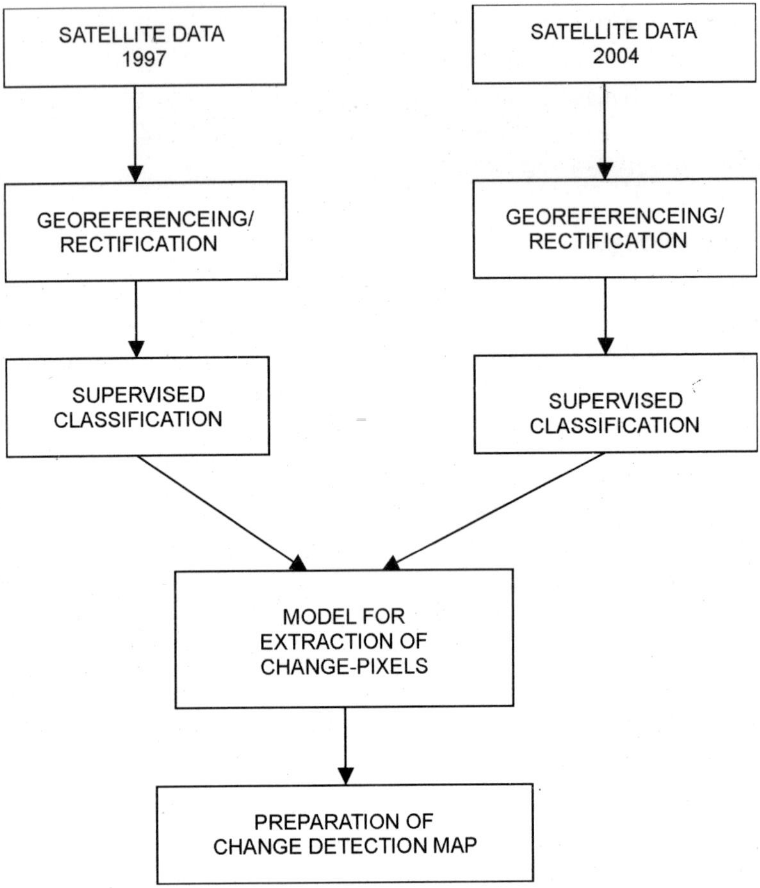


land cover change classes were summarized in the change matrix (Figure 3).

## Results and Discussion

### *Land Use/Land Cover Mapping*

The land use/land cover distribution the year 1997 and 2004 was assessed for about 15.53 sq. km area. The statistics of various classes mapped in the area are given in the Tables 1 and 2 and presented in the Figures 4 and 5 respectively. Six categories of land use/land cover were identified and mapped in the study area for the years 1997 and 2004 for change detection analysis.

**Table 1**

*Distribution of Land Use/Land Cover (2004)*

| *S.No.* | *Land Use/Land Cover* | *Area (sq. m)* | *Area (sq. km)* | *Area%* |
|---|---|---|---|---|
| 1. | Water Bodies | 905.14 | 0.91 | 5.86 |
| 2. | River Channel | 494.82 | 0.49 | 3.16 |
| 3. | Forest | 5035.97 | 5.04 | 32.45 |
| 4. | Agricultural Land | 1482.79 | 1.48 | 9.53 |
| 5. | Vacant Land | 1192.86 | 1.20 | 7.73 |
| 6. | Built-up Area | 6413.28 | 6.41 | 41.27 |
| Total | | 15524.86 | 15.53 | 100 |

**Table 2**

*Distribution of Land Use/Land Cover (1997)*

| *S.No.* | *Land Use/Land Cover* | *Area (sq. m)* | *Area (sq. km)* | *Area%* |
|---|---|---|---|---|
| 1. | Water Bodies | 895.20 | 0.90 | 5.80 |
| 2. | River Channel | 811.81 | 0.81 | 5.2 |
| 3. | Forest | 7093.65 | 7.10 | 45.72 |
| 4. | Agricultural Land | 751.06 | 0.75 | 4.83 |
| 5. | Vacant Land | 2233.85 | 2.23 | 14.36 |
| 6. | Built-up Area | 3739.29 | 3.74 | 24.07 |
| Total | | 15524.86 | 15.53 | 100 |

## *Land Use/Land Cover Changes*

The built-up area in 1997 was about 24.07 per cent (3739.29 sq. m) and 41.27 per cent (6413.28 sq.m) in 2004. The increment is about 17.2 per cent during the period 1997 to 2004. Average annual addition of built-up area was around 381 sq. m. The increase in the built-up area caused partial decrease in the areal spread of forests, river channel and vacant land areas as the later were transformed into to settlements.

**Figure 4**
*Land Use/Land Cover (1997)*

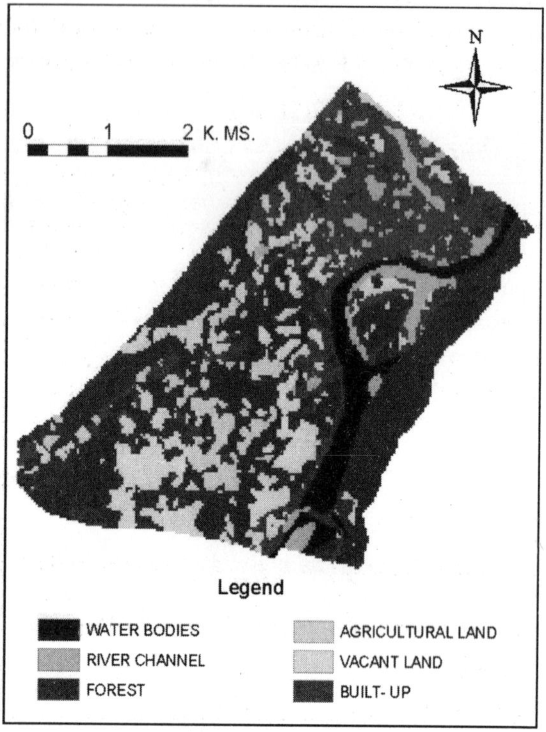


The agricultural land is increased by about 731.73 sq. m. (4.7%) during 1997 to 2004. The agricultural areas increased mainly due to encroachment in forest areas and vacant lands.

It has been noticed that majority of land transformation was from forest to others areas like settlement and agricultural land. The forest land is reduced by about 3104.75 sq.m. (13.27%) from 1997 to 2004. The transformation of forest to settlement is about 1698 sq. m. and forest to agricultural land is about 1026.08 sq. m. during 1997 to 2004.

The other major changes was from vaçant land to others land use/land cover categories like forest, settlement, etc. The changes from were vacant land to settlement is about 703.57 sq. m. whereas from vacant land to forest is about 629.01 sq. m. The

**Figure 5**

*Land Use/Land Cover (2004)*

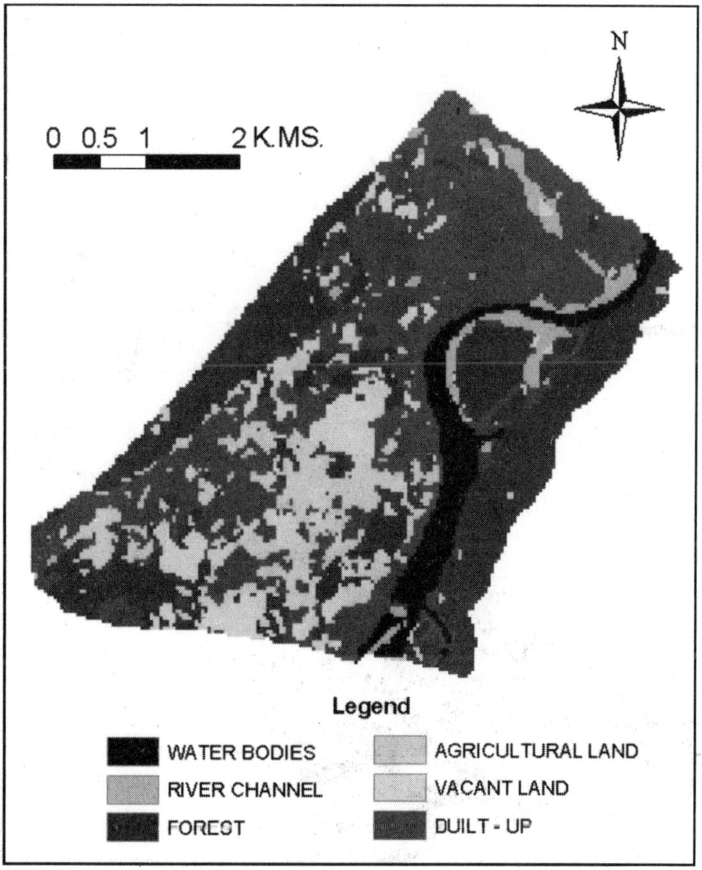


transformation of vacant land to forest primarily resulted due to forest plantation in areas of vacant land.

Another remarkable change was observed in conversion of areas around river banks to built-up causing spread of built-up by about 390.99 sq.m. The banks of non-perennial Chandrabhaga river were utilized for developing settlements due to its proximity to other developed settlements. The urban sprawl trend show development of settlements towards the area in the vicinity of river banks (Tables 3, 4 and 5 and Figure 6).

**Figure 6**

*Land Use/Land Cover from 1997 to 2004*

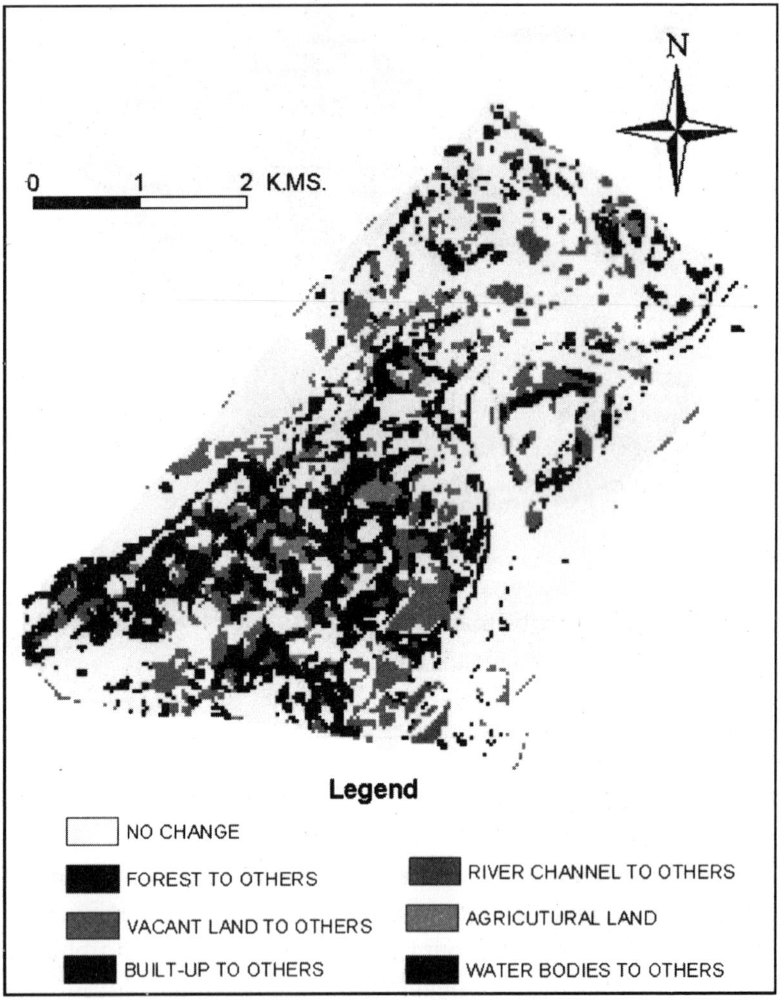

**Table 3**

*Land Use/Land Cover Change (From 1997 to 2004)*

| *S. No.* | *Land Use/Land Cover* | *Change (sq. m.)* | *Change (%)* |
|---|---|---|---|
| 1. | Water Bodies | 9.94 | 0.06 |
| 2. | River Channel | –316.99 | 2.04 |
| 3. | Forest | –2057.68 | –13.27 |
| 4. | Agricultural Land | 731.73 | 4.7 |
| 5. | Vacant Land | –1040.99 | –6.63 |
| 6. | Built-up Area | 2673.99 | 17.2 |

**Table 4**

*Total Changes of Land Use/Land Cover From 1997 to 2004*

| *S. No.* | *Land Use/Land Cover (From 1997)* | *To Others (To 2004)* | *Change (sq. m.)* |
|---|---|---|---|
| 1. | Water Bodies | Others | 60.19 |
| 2. | River Channel | Others | 457.82 |
| 3. | Forest | Others | 3104.75 |
| 4. | Agricultural Land | Others | 381.61 |
| 5. | Vacant Land | Others | 1534.71 |
| 6. | Built-up Area | Others | 532.91 |

**Table 5**

*Land Use/Land Cover Change (From 1997 to 2004)*

| *From (1997)* | *To (2004)* | *Change (sq.m.)* |
|---|---|---|
| Water Bodies | River Channel | 22.64 |
| | Forest | 13.25 |
| | Agricultural Land | 0 |
| | Vacant Land | 0 |
| | Built-up Area | 24.30 |
| River Channel | Water Bodies | 33.69 |
| | Forest | 16.57 |
| | Agricultural Land | 1.11 |
| | Vacant Land | 15.46 |
| | Built-up Area | 390.99 |

Contd...

Contd...

| | | |
|---|---|---|
| Forest | Water Bodies | 41.42 |
| | River Channel | 2.76 |
| | Agricultural Land | 1026.08 |
| | Vacant Land | 336.32 |
| | Built-up Area | 1698.17 |
| Agricultural Land | Water Bodies | 0 |
| | River Channel | 0 |
| | Forest | 10.49 |
| | Vacant Land | 134.20 |
| | Built-up Area | 236.92 |
| Vacant Land | Water Bodies | 0 |
| | River Channel | 8.84 |
| | Forest | 629.01 |
| | Agricultural Land | 193.29 |
| | Built-up Area | 703.57 |
| Built-up Area | Water Bodies | 91.12 |
| | River Channel | 140.27 |
| | Forest | 179.48 |
| | Agricultural Land | 25.40 |
| | Vacant Land | 96.64 |

## Conclusion

Multi-temporal satellite images are ideal source for land use/land cover change detection studies. Level 1 land use/land cover map were generated by applying supervised classification techniques on digital satellite data. Two classified maps compared on a pixel by pixel basis using a change detection matrix could effectively present final change detection map. This is the most widely and easily understandable technique for preparing change detection of multi-date satellite images.

*Acknowledgements*

The authors are extremely grateful to Dr. V.K. Dadhwal, DEAN, and Prof. S.K. Govil, Head, PRS division, Indian Institute of Remote Sensing, Dehradun, for their guidance, suggestions and encouragement. Md. Surabuddin Mondal is specially grateful to Dr. M.S. Nathawat, Prof.

and Head, Dept. of Remote Sensing, Birla Institute of Technology, Mesra, Ranchi, who always motivated him to right direction.

## Select References

Elman, R. Aleskerov and Minakshi Kumar, 2004, "Mapping and monitoring of land use/land cover and lake boundary using Remote Sensing and GIS techniques – a case study of Udaipur city, Rajasthan", Thesis Report, CSSTEAP, IIRS, Dehradun.

Jahne, B., 1991. *Digital Image Processing – Concept, Algorithms and Scientific Application*, S. Pringer – Verlay, New York.

Jenson, John R., 1996. *Introductory Digital Image Processing – A Remote Sensing Perspective*, second edition, Prentice Hall, New Jersey.

Maithni, S. and B.S. Sokhi, 2002, "Modeling land transformation using remote sensing and GIS: a case study, Haridwar and surrounding areas", *ITPI Journal* Vol. 2. No. 2 (181).

Ross, S. Lunetta and Christopher D. Elvidge, "Remote sensing change detection – Environmental Monitoring Methods and Applications".

# 32

# Water Resources Development in Hard Rock Terrain: A Geoinformatics Approach

M.S. Nathawat, V.S. Rathore, A.K. Sinha and Praveen Kumar

In India people who are living in hard rock terrain, face drinking water problem particularly in the months of March to June because of inadequate surface water and groundwater development. The present study entails for delineation of potential groundwater areas as well as suitable sites for water harvesting so that surface water storage capacity can be enhanced in the Amlai district, Shahdol (M.P.). This study was carried out by applying Satellite Remote Sensing technique which enables to narrow down the target areas, followed by geophysical exploration (Electrical Resistivity Survey) technique in order to know the approximate location of occurrence and extension of aquifer. In the present study the satellite data of IRS-1C/1D LISS III was used to delineate the potential groundwater zone in hard rock terrain of Barakar Sandstone. Then the geophysical survey (Resistivity Soundings) at selected locations was done. Apart from this the attempt was also made to find out the suitable locations for surface water harvesting and estimate the approximate storage capacity of the suggested structures.

The present study has the following broad objectives:

1. To find out the sufficient groundwater regime in and around the Amlai village.
2. To provide alternate source of water to be used in lean period by selecting sites for the surface water storage.

## Study Area

The study area includes the hard rock terrain of Barakar Sandstone in Shahdol district of Madhya Pradesh and falls in the Survey of India topographic sheet No. 64 E/12 having latitudes 23° 0′ to 23° 15′ and longitudes 81° 30′ to 81° 45′. Physiography of the area is characterized by hilly to undulating plain. Some circum-denudation hills/hillocks and plateau are existing in the

**Figure 1**

*IRS 1C/1D Satellite Image with Location of the Study Area*

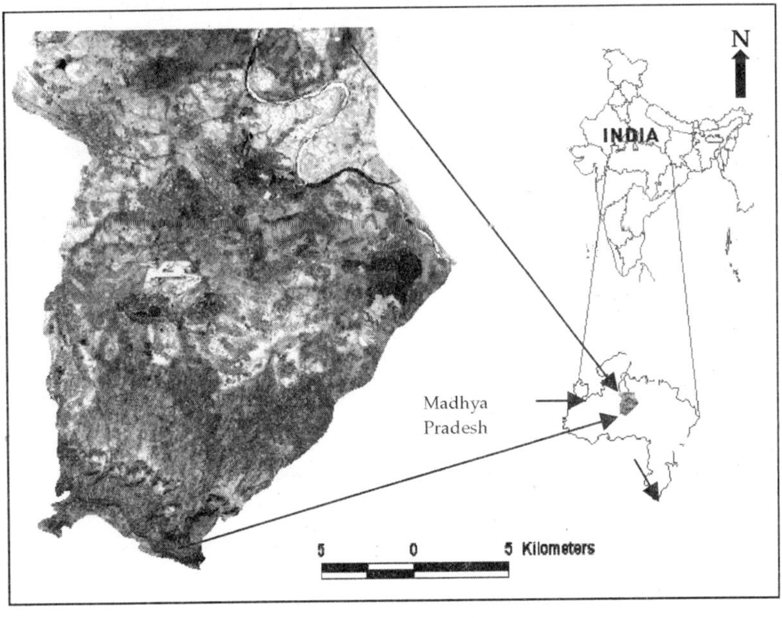

**Figure 2**

*Base Map of the Study Area Near Village Amlai, District Shahdol (M.P.).*

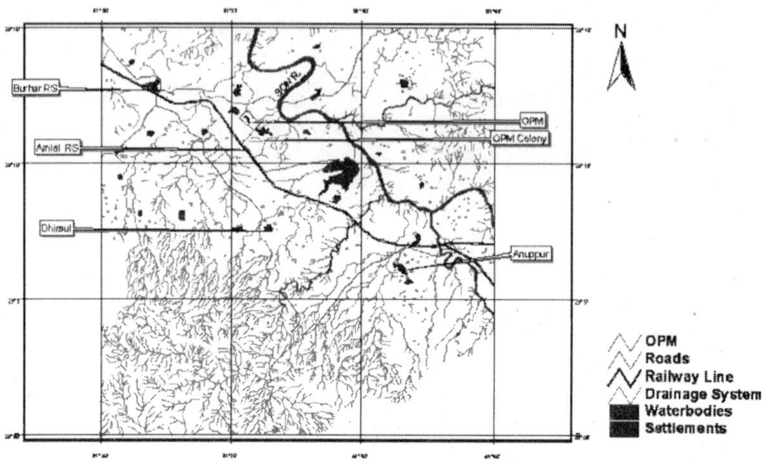


south-west part of the study area and undulating terrain with small ridges occupying the centre and north-western part. The area is mainly drained by the river Son and its tributaries, showing dendritic drainage pattern.

The Barakar sandstone of Gondwana formation occupies the central and northern part of the study area having thickness of 200-300 m which is dominantly felspathic and intermittently workable coal seams. The Lameta beds composed of silicified limestone rest unconformably over the Gondwana beds and are exposed along the fringes of trappean hills in the southern part of the study area.

## Methodology

The satellite technology has capability to provide synoptic view of large area on a repetitive basis as well as gives better understanding of natural resources. It can be successfully used for resource survey and mapping. This latest remote sensing technique has been used in this study to narrow down the target area for groundwater exploration and for selection of water harvesting structure site in identified watershed area.

In this study, digital satellite data of IRS 1C/1D LISS-III of two seasons October 2002 and February 2003 and geo-coded paper prints of IRS 1C/1D LISS-III were used. The digital satellite data was geometrically corrected and processed in the lab. Subsequently, the visual as well as digital interpretation of satellite data was done to prepare various thematic maps such as structural, land use/land cover of the two seasons, landform, geology, digital terrain model and watershed map. Later a limited field verification study was conducted to check and improve the interpretation excellence.

With the help of Survey of India topographic sheet No. 64 E/12 and satellite image the Base map as shown in the Figure 2 has been prepared to show the transport network, settlement, drainage and water bodies. Similarly, drainage map and contour map of the study area were also prepared. The maps thus prepared were scanned and geo-referenced to convert them into vector layers. The vectors of drainage and contours also had been used in the preparation of watershed map as well as digital elevation model map.

Finally, the groundwater prospectus map of the study area has been prepared by integrating all the information of above mentioned themes. Then the targeted area was explored through the Geophysical Survey (Electrical Resistivity Method) at 51 locations to confirm the depth and thickness of aquifer for recommendation of drilling sites. Further a watershed area was also selected to suggest suitable sites for water harvesting structures and calculated for catchment yield which have been discussed separately under the heading of watershed and surface water harvesting structure.

## Results and Discussion

Based on the image characteristics, physiography, digital elevation model, nature of rocks/sediments, geology, etc., we

**Figure 3**

*Landform Map of the Study Area Near Village Amlai, District Shahdol (M.P.).*

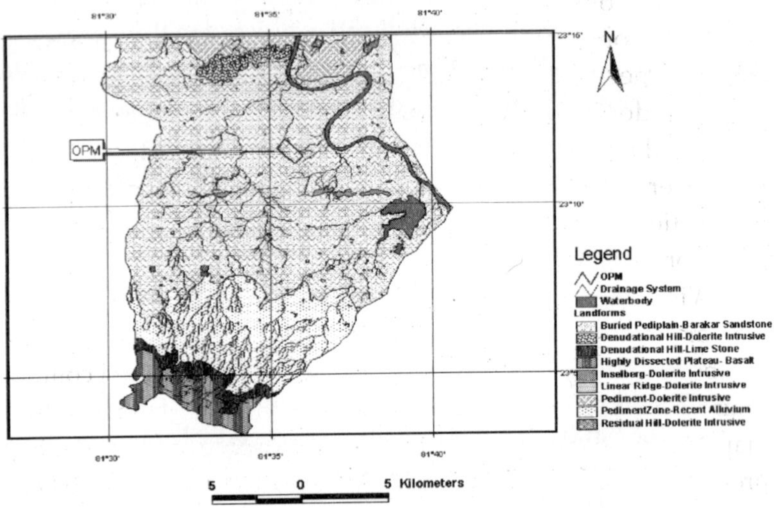


have come out with the landform, structure, land use/land cover and lithology maps.

Information on landforms is an important input for identification of potential zones of groundwater occurrence, as the different landforms of different origin have variable capability of holding groundwater. Thus, a landform map (Figure 3) showing different landform categories in different lithology such as Denudational hills (Dolerite intrusive), Denudational hills (Limestone), Residual hills (Dolerite intrusive), Inselberg (Dolerite intrusive), Linear ridge (Dolerite intrusive), pediment (Dolerite intrusive), pediment zone (Recent alluvium) Plateau highly dissected (Basalt) and Buried pediplain (Barakar sandstone) has been prepared which was further used for preparation of groundwater prospectus map.

The different types of primary and secondary geological structures can be interpreted from satellite imageries, as these

**Figure 4**

*Lineament Map of the Study Area Near Village Amlai District Shahdol (M.P.), Showing Criss-Cross Lineament Pattern*

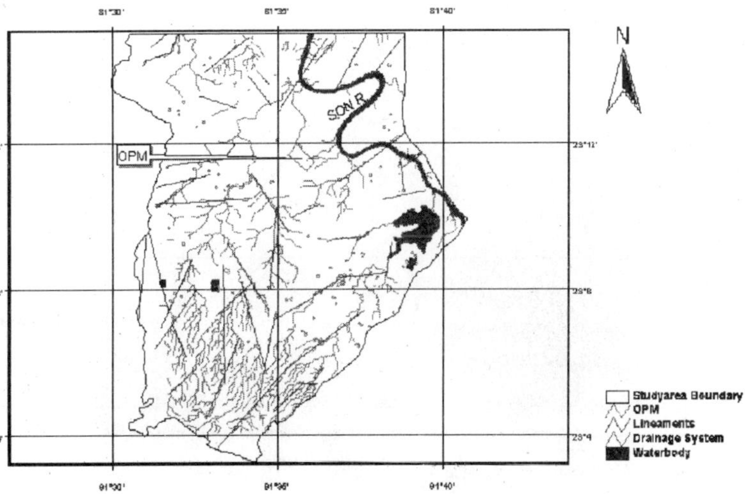


appear as linear features on the satellite imageries which are termed as lineaments. The lineaments control the occurrence and movement of groundwater in hard rock terrain and their significance in groundwater exploration has been proved beyond doubt because they represent the secondary porosity. They occur in parallel sets in different directions indicating different tectonic events. Thus, a map with the help of satellite imagery has been prepared to analyse the structures of the study area. It is observed that the area has criss-cross pattern of lineaments of varying length ranging from 1km to 10 km approx. which is shown in the Figure 4.

## Groundwater Prospects

The characteristics of landforms of various origins, structures, lithology, infiltration characteristics, and water table behaviour coupled with satellite data interpretation enabled us to prepare a

**Figure 5**

*Groundwater Prospects Map of the Study Area, Near Village Amlai, District Shahdol, (M.P.).*

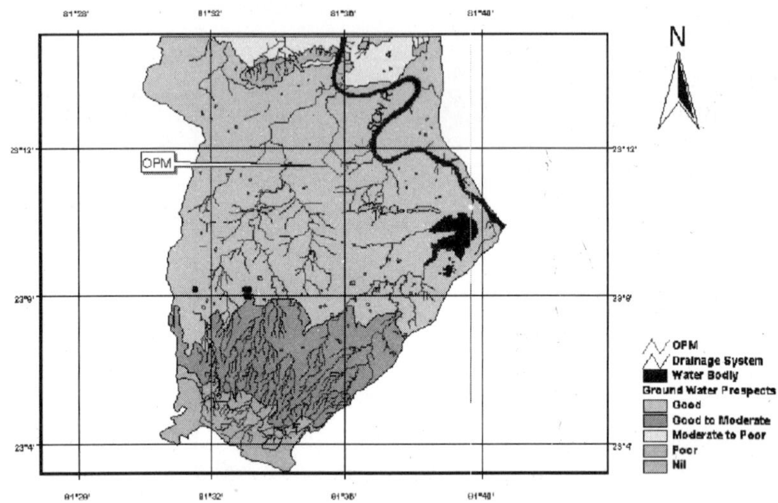


groundwater prospectus map (Figure 5). The perusal of the map reveals five types of aquifer conditions existing in the different landforms that are nil (groundwater discharge is zero or negligible), poor (groundwater discharge is 10-50 LPM), moderate to poor (groundwater discharge is 50-100 LPM), good to moderate (groundwater discharge is 100-200 LPM) and good (groundwater discharge is 200-400 LPM).

## Watershed and Surface Water Harvesting Structure

Watershed is rainwater shed-off area of a stream in which rainwater moves from upper reaches to lower levels. In other words it is an area encompassed by surface water divide of a stream. It is a synonym of catchment or basin of a river. Attempt should be made to control the rain water at the source through watershed management. In this way, the stored rain water will serve dual purpose, as surface water storage which can be taken directly to use, else it improves the groundwater that can be exploited through tube-wells in the down stream.

It has been observed that the groundwater yield in this area is moderate; hence the heavy discharge is not possible to fulfil the requirement of surrounding people throughout the year from tube-wells or bore-wells alone, until there are collector wells or infiltration wells in or along the river Son. Therefore, the study for conservation of surface water has also been carried out for a watershed area located in the study area and sites for construction of water harvesting structure has been suggested which is shown in the Figure 6.

### *Flood Discharge for Proposed Structure No.1*

A weir is proposed at a suitable location as shown in the Figure 5. In the absence of stream flow data and the hydrological data for the catchment, the flood discharge is estimated using Dicken's formula,

$$Q = C\,A^{3/4}$$

Assuming $C = 19$, $Q = 19\ (49.13)^{3/4} = 352.6\ m^3/s$

Assuming the difference between the highest flood level (HFL) and the full tank level (FTL) as 1 m and the coefficient of discharge of the weir as 0.6,

$$Q = (2/3)\ C_d\ (2g)^{1/2}\ L\ H^{3/2}$$

$$352.6 = (2/3) \times 0.6 \times (2 \times 9.81)^{1/2} \times L \times 1^{3/2}$$

**L = 200 m**

Thus, the length of weir may be around 200 m, if the afflux is 1 m. The detailed design of the weir and embankment may be done after obtaining the required field data and other necessary information.

### *Catchment Yield for Proposed Structure No.2*

The catchment area of the concerned stream is calculated as 119.00 km$^2$. Again, assuming the average annual rainfall over this catchment as 1000 mm.

Dependable rainfall = $0.75 \times 1000 = 750$ mm = 0.75 m
Yield of the catchment = CAP
Assuming C as 0.1,

$$\text{Yield} = 0.1 \times 119.00 \times 10^6 \times 0.75 = 8.92 \text{ Mm}^3$$

Assuming evaporation and seepage loss from the tank as 20 per cent,

$$\text{Water available} = 0.8\ '\ 8.92 = 7.14 \text{ Mm}^3$$

With a consumption of about 0.031 $\text{Mm}^3$ per day or 8 MGD (US gallon), the stored water in the tank is likely to be sufficient for seven and a half months.

### *Flood Discharge for Proposed Structure No.2*

A weir is proposed at a suitable location as shown in Figure 4. In the absence of stream flow data and the hydrological data for the catchment, the flood discharge is estimated using Dicken's formula, $Q = C\,A^{3/4}$.

Assuming $C = 19$, $Q = 19 \times (119.00)^{3/4} = 684.6 \text{ m}^3/\text{s}$.

Assuming the difference between the highest flood level (HFL) and the full tank level (FTL) as 1 m and the coefficient of discharge of the weir as 0.6,

$$Q = (2/3)\, C_d\, (2g)^{1/2}\, L\, H^{3/2}$$

$$684.4 = (2/3) \times 0.6 \times (2 \times 9.81)^{1/2} \times L \times 1^{3/2}$$

**L = 391 m**

Thus, the length of weir may be around 391 m, if the afflux is 1 m. The detailed design of the weir and embankment may be done after obtaining the required field data and other necessary information.

## Conclusion

Groundwater prospects map and the results of geo-electric survey indicate that the area is having multi-aquifer system. Both confined and unconfined types of aquifer exist in the area. The first aquifer occurs between 5-35 m depth and a second aquifer of 10-15 m thickness exists between 40-60m depth. The third aquifer has been noticed at one location which is likely to exist between 120 m-170 m depth. The cumulative yield (discharge) of all aquifer system is ranging from 10,000-20,000 lph and it can sustain throughout the year, provided a continuous recharge is there from surface water bodies which can be done by constructive water harvesting structures at selected sites. As the area is underground and open cast coal mining operations which is disturbing the hydraulic gradient, it results in drying up of some aquifers and flooding of some of the abandoned underground mines. Thus, authors cautioned drilling operations without the knowledge of the existing underground mines.

Based on the study following are the recommendations for water resource development in the study area:

I. Looking to the conditions and constraints of the study area it is suggested that a study is to be conducted for construction of infiltration wells/infiltration galleries along or inside the river bed at two/three suitable locations, which may give 4/5 MGD of water even in the lean period.

II. In this proposed study the subsurface flow and alluvium thickness in the riverbed itself is to be measured during the month of December to March so that the subsurface flow can be calculated in the lean period for the construction of infiltration wells or infiltration galleries in or along the riverbed (Infiltration well is a large diameter well, which receives the subsurface water radially from the riverbed through the perforated screen and these have been the successful cases at many places to meet out the requirement).

III. Apart from this, the surface water harvesting structures are suggested at two locations to meet out the balance requirement. This will not only store the run-off of catchment but also recharge the groundwater that later can be exploited in the downstream by constructing tube-wells/bore-wells.

IV. The estimated yield of water harvesting structure No.1 is 2.9 $Mm^3$ which has been suggested for upper catchment and yield of water harvesting structure No.2 is calculated as 7.14

$Mm^3$ for lower catchment. Hence, it is suggested that one weir may be constructed at location No. 2. Water stored in this reservoir thus, may be used for fulfilling the requirement of the residents.

## Select References

Agarwal, Chaitanya S. and Garg, P.K. (2000), *Remote Sensing in Natural Resources Monitoring and Management*, First Ed., Wheeler Publishing, New Delhi.

Nathawat, M.S., Srivastava, A. and Rathore, V.S., Watershed Management in Chhota Gamharia Macro Watershed Using Remote Sensing and GIS, Published in proceeding of All India Seminar on Water Resources Planning in Jharkhand State, held on July 6-7, 2002.

Pandey, A.C., Nathawat, M.S., Srivastava, A. and Rathore, V.S., Water Resource Mapping and Management Using Remote Sensing Technique, Published in Proceeding of National Conference on Water Resources Development and Management, held on February 9-10, 2002.

Pandey, Shiv, N., (2001), *Principles and Applications of Photogrammetry*, New Age International (P) Ltd. Publishers, New Delhi.

Raghunath, H.M., (2000), *Hydrology Principles, Analysis and Design*, New Age International (P) Limited Publishers, New Delhi.

Srivastava, P. and Bhattacharya, A., (2000), Delineation of Ground Water Potential Zones in a Hard rock Terrain of Bargarh district, Orissa using IRS Data, *J. Indian Soc. Remote Sensing*, Vol. 28, No. 2 and 3, pp. 130-140.

Thornbury, W.D., (1969), *Principles of Geomorphology*, 2nd Ed., John Wiley and Sons, New York.

Todd, D.K. (1980), *Ground Water Hydrology*, John Wiley and Sons, New York.

# 33

# Water Management System Using GIS

Sushil Kumar Singh and A.K. Singh

Life was born in water. In Sanskrit language the other name of water is 'Jiban' which means 'life'. Water has been considered in all religions, from the antique time till today, as the purifier of body and soul. In fact, water and health is not only the inseparable bond to maintain the life system but also the health, the development and the environmental quality of our settlement centres. Access to a basic quantity of fresh water is therefore a fundamental social right and a prerequisite to all social developments. Although water is the primary factor of all development activities, it is also the carrier of many diseases and waste materials, and it has become very dangerous to public health when there is a lack of adequate purification and treatment facilities, particularly in the developing countries. According to WHO's statistics, the number of deaths caused by waterborne diseases and polluted water is larger than all other cumulated causes of death. About 40, 000 people die every day in these countries from unsafe water consumption of which the majority are children, which is equivalent to 100 jumbo jet crash every day! Water and health are interdependent and constitute the primary link of the life and growth system of the human settlements. Whenever there is water shortage or quality degradation, it affects immediately the public health safety. During any disaster event, such as war, earthquake, floods and draughts, safe water is the primary requirement before any other aids to avoid epidemics and loss of further human lives. Or, paradoxically, the demand for safe drinking water surpasses hardly more than 1 to 2 per cent of the total water production and this amount does not vary much between the developed and the developing countries. Water is the most essential thing of our life. No life can exist without water, since water is as essential for life as air is. It has been estimated that two-thirds of human body is constituted of water. Water is essential not only for survival of human beings but also for animals, plants and all other living beings. People waste water frequently because they do not give much importance to water. Due to the wastage of water, water scarcity has become a major problem today. Therefore, in order to ensure the availability of sufficient quantity of water, it has become almost imperative to plan and build water scheme, which may provide water to the different sections of community in accordance with their demands and requirements, i.e., management of water has become essential. So we need to design a system for water management.

## Methodology

To design a system for water management we have taken various topics into consideration:

1. Population forecasting, in which various methods of forecasting are considered and also factors affecting population growth being taken into consideration.
2. Water demand, in which various types of water demands like domestic demand, commercial demand, institutional demand and average per capita demand being taken into consideration.
3. Water resources, in which various forms of water resources and types of water resource like groundwater and rainwater being taken into consideration.

4. Flow in pipes, in which types of flow, head loss through pipes and losses of energy in pipes being taken into consideration.
5. Flow through open channel, in which classification of flow in open channel, rate of discharge through open channel and design of most economical section of open channel being taken into consideration.
6. Hydrology, in which various methods of estimating rainfall, methods of estimating run-off and unit hydrograph being taken into consideration.
7. Pumps for lifting water, in which various types of pumps, factors affecting the selection of particular type of pump, pump discharge and location of pumping station being taken into consideration.
8. Distribution reservoir, in which types of distribution reservoir, storage capacity of reservoir and location of reservoir being taken into consideration.
9. Wastewater treatment, in which constituent of wastewater, types of treatments like primary, secondary and tertiary and BOD and COD of wastewater being taken into consideration
10. Rainwater harvesting, in which various methods like rainwater harvesting using roof and courtyard, rainwater harvesting using natural slope, rainwater harvesting on bare rocks and rainwater harvesting using roads as catchments being taken into consideration.
11 Water reuse, in which various applications of water reuse keeping in mind pubic health, environmental issues and water reclamation technologies being taken into consideration.

In the previous models, study is limited up to a certain extent because in some models, study is done for rainfall, run-off and in some other models study is done for hydrologic parameter, groundwater parameter, so there is a need for extensive study

## Study Area

Educational Institution Campus.

## Data Collection

Different data has been collected for the study. Population data has been collected for Campus. In Campus there are nearly 5,000 people living within the campus, and serving these people there are 18 wells for the desirable water demand. Well-discharge reading has been taken at different wells. Water demand and wastewater data has been collected. Rainfall data and roof area of different buildings has been collected. Roof area of different staff quarters, different boys hostels, faculty division1, faculty division2, and faculty division3 has been measured. Different points of elevation has been taken by Theodolite surveying—Also use of GIS for generating 2-D Plots.

## Generation of Forecast Model

Program for Newton's Divided difference formula:

```
INTEGER N, XP
REAL FP, SUM, PI, X, F, A, D
DIMENSION X(10), F(10), A(10), D(10, 10)
OPEN (unit=12, file = 'file name.inp')
READ(12, *) N
```

*Plots Generation Using GIS*

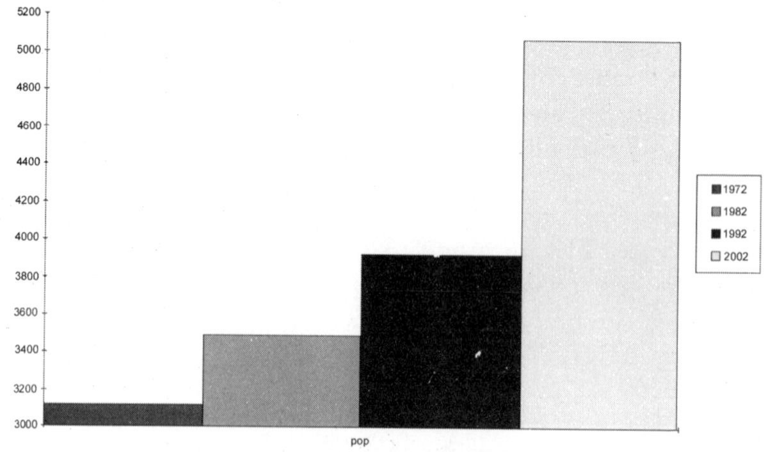

*Decade-wise Population Variation Plot*

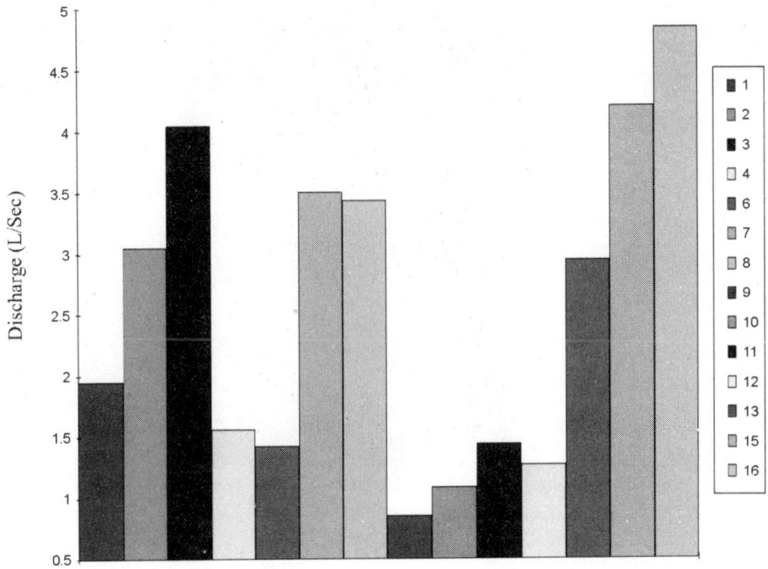


*Decade-wise Water Demand Variation*

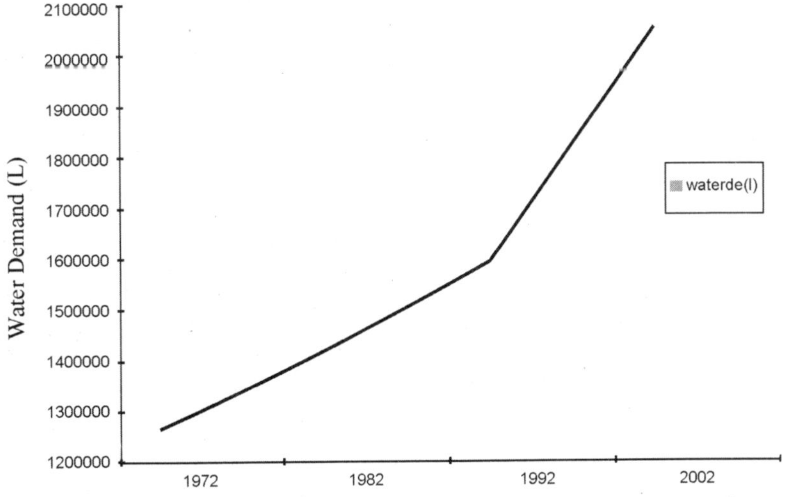

*Run-off Variation for Different Buildings in 2002*

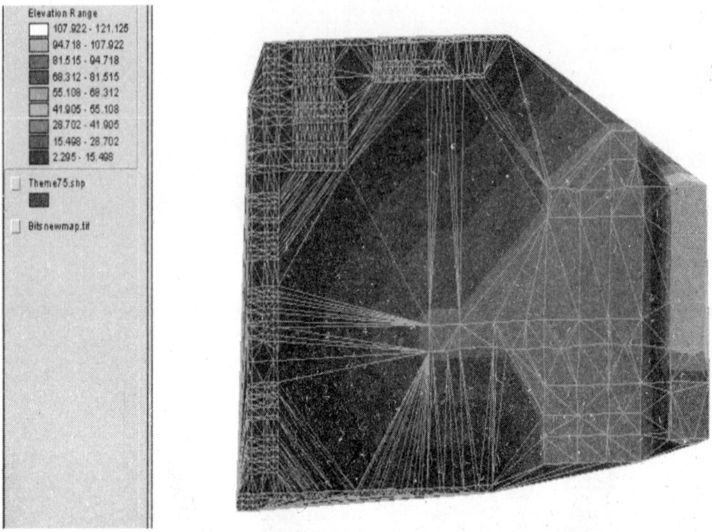


*Decade-wise Run-off Variation Plot*

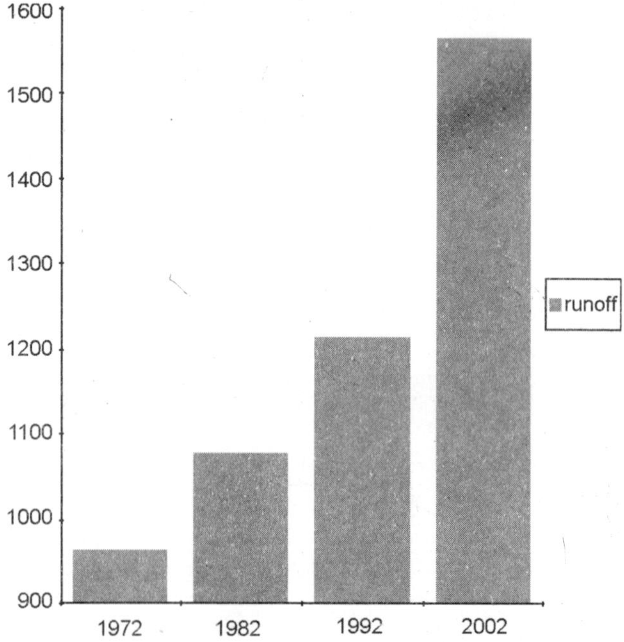

*Decade-wise Wastewater Variation*

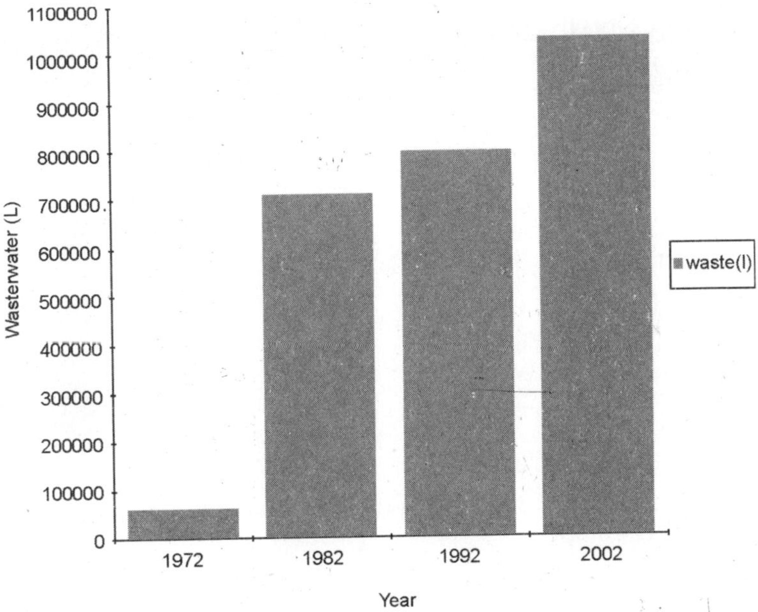


*Elevation Variation Plot*

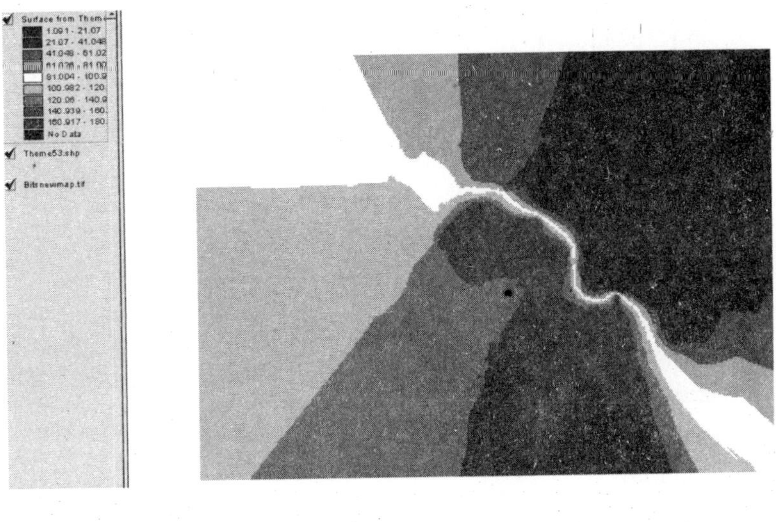

```
DO 10 I=1, N
READ(12, *) X(I), F(I)
10 CONTINUE
DO 20 I=1, N
D(I, 1)=F(I)
20 CONTINUE
DO 40 J=2, N
DO 30 I=1, N-J+1
D(I, J)=(D(I+1, J-1)-D(I, J-1))/(X(I+J-1)-X(I))
30 CONTINUE
40 CONTINUE
DO 50 J=1, N
A(J)=D(1, J)
50 CONTINUE
WRITE(*, *) 'Input XP where interpolation is required'
READ(*, *) XP
SUM = A(1)
DO 70 I=2, N
PI=1.0
DO 60 J=1, I-1
PI= PI*(XP-X(J))
60 CONTINUE
SUM=SUM+ A(I)*PI
70 CONTINUE
FP = SUM
WRITE(*, *)
WRITE(*, *) 'NEWTON INTERPOLATION'
WRITE(*, *)
WRITE(*, *) 'INTERPOLATED FUNCTION VALUE'
WRITE(*, *) 'AT X= ', XP, 'IS', FP
WRITE(*, *)
STOP
END
```

## Result and Discussion

Generated forecast model predicts about various things like population, water demand, wastewater, run-off, etc., on the basis of previous available data. We have previous data of population, water demand, wastewater and run-off. So we have predicted population water demand, wastewater, and run-off. On the basis of wastewater and run-off we have designed wastewater and rainwater treatment plant. The location of wastewater treatment plant will be decided by generated DEM using GIS. The location of rainwater treatment plant will be decided by generated TIN using GIS. Results obtained using forecast model are given below:

Year: 2007
Population =6103
Water demand=2480515 l litre per day
Waste water =1240258 litre per day
Runoff=1888.67m^3/hr
Design of wastewater treatment plant:
Net area of screen openings required =. 0 258m^2
Design of grit chamber:
L=7.75m
B=0.17m
D=0.5m
Design of sedimentation tank:
L=24m
B=3.1m
D=2m
Design of filter:
Volume of filter media (V) =179m^3
Area of filter media (A) =225m^2
Design of rainwater treatment plant:
Design of sedimentation tank:
L=72m
B=14m
D=6m
Design of filter media:

Area of filter media (A) =377.6m^2
Year: 2012
Population =7556
Water demand =3071852 litre per day
Wastewater =1535926 litre per day
Runoff =2338.11m^3/hr
Design of wastewater treatment:
Net area of screen openings required =. 04m^2
Design of grit chamber:
L=7.75m
B=0.256m
D=0.5m
Design of sedimentation tank:
L=24m
B=3.9m
D=2m
Design of filter:
Volume of filter media (V) =216m^3
Area of filter media (A) =272m^2
Design of rainwater treatment plant:
Design of sedimentation tank:
L=72m
B=19m
D=10m
Design of filter media:
Area of filter media (A) =467.76m^2

## Conclusion

Datas have been collected for Campus for generating different 2-D plots. These plots have been generated using GIS. Now these plots have been used for generating forecast model. The forecast model predicts about population, water demand, wastewater, and run-off on the basis of previous data. On the basis of forecast data wastewater and rainwater treatment plant have been designed. These forecast data will be useful for future generation.

## References

Belainch, G., Perelta, R.C. and Trevor, C. Hughes (1999) "Simulation/ Optimization Modeling for Water Resources Management". Water Resources Planning and Management, *ASCE*, Vol. 125 (3).

Biggs, D. and Williams R. (2001) "A Case Study of Integrated Water Resource Management Windhoek, Namibia". Workshop on Demand Management Practice, Policy, Data and Technologies, International Hydrological Programme Symposium.

Casanova, L.G.C. (2001) "Sustainable Management of Municipal Wastewater and Stormwater: Unep-Ietcs Information and Capacity Building Resources". Workshop on Technological Outlook for the Future, International Hydrological Programme Symposium.

Castlan, E. (2001) "Water Management in the Mexico City Metropolitan Area: The Hard Way to Learn". Workshop on Integrated Urban Water Management for the Future, International Hydrological Programme Symposium.

Chakrarbarti, P.G.D. (2001), "Roof Top Rain Water Harvesting an Alternative Technology for Fresh Water Augmentation in Chronically Deficient Urban Agglomerates of India". Workshop on Technological Outlook for the Future, International Hydrological Programme Symposium.

Cunningham, A.B. and Amend, J.R. (1984) "Water Management Using Interactive Simulation". Water Resources Planning and Management, *ASCE*, Vol. 110(3).

Feng, G. and Feng, Y. (2001) "Heightening Urban Water Supply Capacity and Reliability Reconstructing Integrated Regional Water Resources Systems". Workshop on Demand Management Practice, Policy, Data and Technologies, International Hydrological Programme Symposium.

Fisher, P.G. and Chiang, T.T. (1980) "Practical Aspects of Town Water Planning". Water Resources Planning and Management, *ASCE*, Vol. 106 (WR1).

Garbrecht, J., Ogden, F.L., DeBarry, P.A. and Maidment, D.R. (2001) "GIS and Distributed Watershed Models". Hydrologic Engineering, *ASCE*, Vol.6 (6).

Geldof, G.D. (2001) "Society and Water Quality: Self Organised Critical Water Systems". Workshop on Demand Management Practice, Policy, Data and Technologies, International Hydrological Programme Symposium.

Hamdy, A. (2001) "Wastewater as a Source: Perspectives and Challenges". Workshop on Recycling and Reuse at different scale, International Hydrological Programme Symposium.

Hole, S.W. and Rhoads, P.B. (1982) "Water Resources and Growth Management in South Florida". Water Resources Planning and Management, *ASCE*, Vol. 108, (WR3).

Juanico, Friedler and Dalia, K. (2001) "Wastewater Reuse-Integration of Urban and Rural Water Resources Management". Workshop on recycling and reuse at different scale, International Hydrological Programme Symposium.

Kaplan, E. (1984) "Water Conservation, Recycling, and Reuse: U.S. Northeast". Water Resources Planning and Management, *ASCE*, Vol. 110 (4).

Kyessi, A.G. (2001) "Community Based Urban Water Management Under Scarcity in Dar Es Salaam Tanzania". Workshop on Demand Management Practice, Policy, Data and Technologies, International Hydrological Programme Symposium.

Larsen, T.A., Rauch, W. and Gujer, W. (2001) "Waste Design Paves the Way for Sustainable Urban Wastewater Management". Workshop on Technological outlook for the future, International Hydrological Programme Symposium.

Lazarova, V. (2001) "Recycled Water: Technical-Economic Challenges for its Integration As a Sustainable Alternative Resource". Workshop on recycling and reuse at different scale, International Hydrological Programme Symposium.

Leipnik, M. R., Kemp, K.K. and Loiciga, H.A. (1993) "Implementation of GIS for Water Resources Planning and Management". Water Resources Planning and Management, *ASCE*, Vol. 119 (2).

Mckinney, D.C., Maidment, D.R., and Tanriverdi, M. (1993) "Expert GIS for Texas Water Planning". Water Resources Planning and Management, *ASCE*, Vol. 119 (2).

Meyer, S.P., Salem T.H. and Labadie, J.W. (1993) "Geographic Information Systems in Urban Storm-water Management". Water Resources Planning and Management, *ASCE*, Vol. 119 (2).

Nie, L. and Schillang, W. (2001) "Development of Sustainable Water Management in Beijing, China". Workshop on Integrated Urban Water Management for the future, International Hydrological Programme Symposium.

Patry, G.G. and Marino, M.A. (1984) "Two Stage Urban Runoff Forecast Model". Water Resources Planning and Management, *ASCE*, Vol. 110 (4).

Ross, M.A., and Tara, P.D. "Integrated Hydrologic Modeling with Geographic Information Systems". Water Resources Planning and Management, *ASCE*, Vol.119 (2).

Sally, L.R., Scott, C., Hassan, M.U., Ensink, J., Matsuno, Y. and Hoek, W.V.D. (2001) "Rethinking Urban Wastewater Re-Use: Opportunities and Challenges". Workshop on Water and Health, International Hydrological Programme Symposium.

Shea, C. Grayman, W., Darden, D., Males, R. and Sushinsky, P. (1993) "Integrated GIS and Hydrologic Modeling for Countywide Drainage Study". Water Resources Planning and Management, *ASCE*, Vol. 119 (2).

Tshkay, A.A. and Morozova, Y.S. (2001) "GIS Hydro Manager for Urban Areas Water Control". Workshop on Demand Management Practice, Policy, Data and Technologies, International Hydrological Programme Symposium.

Tsou, M.S. and Whittemore, D.O. (2001) "User Interface for Ground-Water Modeling: Arc View Extension". Hydrologic Engineering, *ASCE*, Vol.6 (3).

Yemi, Samsons A. (2001) "Partnership for Sustainable Urban Water Utilization and Management in South-West Nigeria: Issues and Prospects". Workshop on Private Participation in the Provision of Urban Water Services, International Hydrological Programme Symposium.

# 34

# Characterization Soil Slope of Drought-Prone Residual Chota Nagpur Plateau

P.S. Arun, R. Jana and M.S. Nathawat

Complex interactions in space of time between, parent material, topography, climate, and organisms and different human management practices cause soil properties to vary spatially. Information on soil variability has been very limited in the past because of the limitation of technology. Terrain attributes (e.g., elevation, slope aspect, curvature) may aid spatial estimation of soil properties because relief has great influence on soil formation (Jenny, 1941). During the past decade, advances in computer technology, geographical information systems (GIS), global positioning systems (GPS), remote sensing and application of geo-statistics, etc., have greatly increased our ability to identify, record, analyse, and manage soil spatial variability. Advance techniques like precision agriculture (PA) or site-specific farming are gaining popularity. A better understanding of the differences of soil spatial variability is necessary to apply management strategies in farming. Soil erosion is one of the most serious environmental problems faced today and threatens agriculture and the natural environment. A lot of effort is put into the study of the erosion characteristics and the possible measures to reduce it. Number of models are being developed to estimate the erosion, where the two parameters soil type and the slope form important inputs that have profound effect on the output. Hence, accurate information regarding soil and slope of the area has to be obtained, so as to get outputs simulated by the models as close as possible to the actual values. Soil and slope also form the major contributor to watershed related studies like run-off and watershed management (Biswas, et al., 2002). Detailed maps of soil properties are hence needed for soil use and management decisions. But these are usually created with traditional survey methods or by interpolation of grid soil sampling data. Often, these maps do not meet desired quality (Mueller, et al., 2001). Scientists are trying to improve the estimation of soil properties and enhance map quality with specific methods ranging in complexity from simple linear regression to geo-statistical techniques. Detailed Digital Elevation Data of the area is a necessity to make it easier to find out the spatial variation in the terrain. But digital elevation data that is obtained from topographic sheets are not always sufficient enough. In such cases it becomes necessary for improving quality of DEM (Digital Elevation Model) obtained from available coarse data.

DEM is a surface generated using interpolation of elevation data obtained from the topographic sheets in general using algorithms. But DEM obtained from the contour data with 20 m obtained from the topographic sheets using usual softwares is not sufficient to describe the variation in slope especially in an undulated residual terrain. Unless the DEM is improved the slope derived from it will also not be good enough. This study is focused on improving over the accuracy of the software derived DEMs. The accuracy of the DEMs derived from these softwares when applied to residual area which usually have less slope variation are low and have major problem of pits and flat areas in them. The algorithms given by Martz and Garbrecht (1992) are best suitable in such area. It was pointed out that the methods used by Jenson and Domingue (1988) and Martz and Garbrecht (1992) tend to produce parallel flows which are unrealistic.

Suitable measures have been suggested to improve these short-comings of parallel flows by Turcotte, et al. (2001) delineation by taking drainage as well as other environmental factors into consideration by Vogt, et al. (2003). The present study deals with analysis of the improvements in soil slope combinations obtained from this improved DEM.

## Study Area and Data Acquisition

The present study area is Gandheshwari watershed in Bankura district of West-Bengal region. The extents of study area lies in between latitudes 23°15′ N and 23°30′N and longitude 86°45′ E and 87° 15′ E (Figure 1). The total area of the watershed is 35, 386 hectares. This area is a sub-watershed of upper Dwarakeshwar watershed. The whole study area is in the Drought-Prone Area Programme (DPAP) under Indian governance. The area faces climatic variability with summer being very hot during the months of March-May. Primary cause for moisture stress in this area is because of the uneven rainfall, as well as the soil

**Figure 1**
*Location Map of the Study Area*

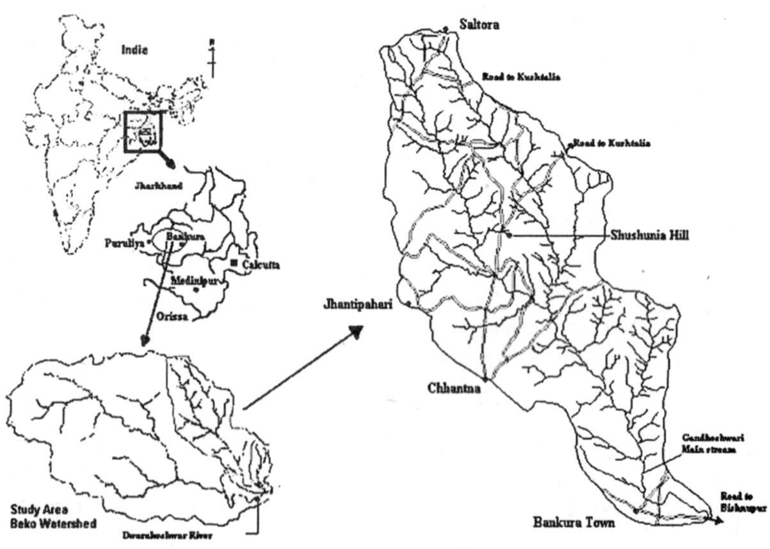

characteristics, it in turn affects the agriculture of this area. Agricultural drought is the major problem in the study area. The principle crop of this region is paddy which suffers maximum scarcity of water.

A spatial approach is adopted in this study where, the satellite data is a LISS-III image with a resolution of 23.5 metres of the month April (1998). Maps like soil maps, geological maps and other land base data were obtained to support the spatial study. Land based information are collected from NRDMS Centre, Bankura and National Bureau of Soil Survey (NBSS) and Land Use Planning (LUP), Nagpur and from field surveys which mainly consisted of Survey of India toposheets, Satellite image, Geophysical maps and other land based information.

## Analysis

In the present study a Digital Elevation Model (DEM) is obtained for the watershed area using contour data obtained from topographic sheets. The DEM obtained from software with a 20 metre contour interval did not reflect the conditions in the field. However, an improved DEM was generated by applying flat area removal algorithm (Martz and Garbrecht, 1992). The slope thus obtained from the improved DEM took care of minute slope categories. The improvement of the DEM was validated with the TIN obtained from these DEM showing more number of triangulation than that obtained from software derived DEM. Soil maps obtained from the NBSS and LUP were combined with this improved slope map so as to bring up information regarding the slope variation within each of the various soil types. Further these soil maps are combined with slope map derived from improved DEM to form the soil and slope combination map which will be used for correlation studies of soil, slope and land use/land cover. Moreover, the study brings about improvement in slope categories especially in the lower ranges varying from 0-3 per cent slope. The interfaces in the present study are a developed programme in C language, ERDAS 8.6, Arcview 3.1 and EASI PACE software.

## Improvement of DEM

The general topography of the study area is undulated with a major elevation difference in Sushunia hill, but these undulations are not reflected in topographic sheets. Elevation data obtained from the available topographic sheets of 1:50, 000 scales are of 20m contour interval which is used as an input to prepare the digital elevation model. The general methods used in interpolation of contour data is by using TIN (Triangular Irregular Networks) for the intermediate points between two contours. Each point in one contour is connected to nearest two points surrounding it to form a triangle. The intermediate points are calculated based on the Delaunay triangles and the vertex points. However this interpolation technique has errors when there are closed contours, which result in flat areas. Hence DEM generated

**Figure 2**

*DEM Derived from Software (ERDAS 8.6)*

by the software like ERDAS 8.6 is less informative for delineation of watershed accurately (Figure 2). The DEM is improved (Figure 3) by applying flat area removal algorithms, a numeric technique proposed by Martz and Garbrecht (1992). It induces a gradual elevation difference considering surrounding eight pixels of a centre pixel. The contour data that were obtained from the topographic sheets were digitized using ERDAS 8.6 vector module. The digitized contours were rasterized having contour height as the attribute. The algorithm is developed in such a way that it retains the characteristics of the actual input. The algorithm is captured in C-programming environment.

**Figure 3**

*DEM Derived from the Algorithm*

## Validation of Improved DEM using Triangular Irregular Network (TIN)

The DEM generated from the same contour layers using ERDAS 8.6 and those generated by present algorithm are used for the comparison. The DEM generated by software give many flat areas and also loose original contour information. Using ERDAS 8.6 the Triangular Irregular network (TIN) was generated using both the software derived DEM and the improved DEM from the programme. The program generated DEM showed a better result in comparison. The figures are self explanatory to show the difference in the amount of interpolation that has been generated in programme generated DEM. The Figure 4 and Figure 5 shows a part of the study area where the software derived DEM shows

**Figure 4**
*TIN Generated by ERDAS Derived DEM*

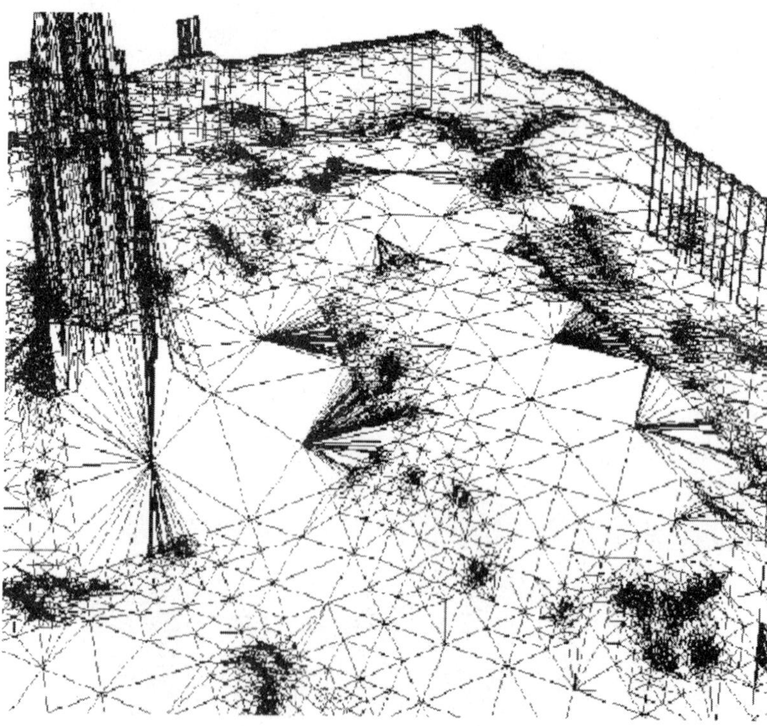

**Figure 5**

*TIN Derived Using Program for Inducing Forced Elevation*

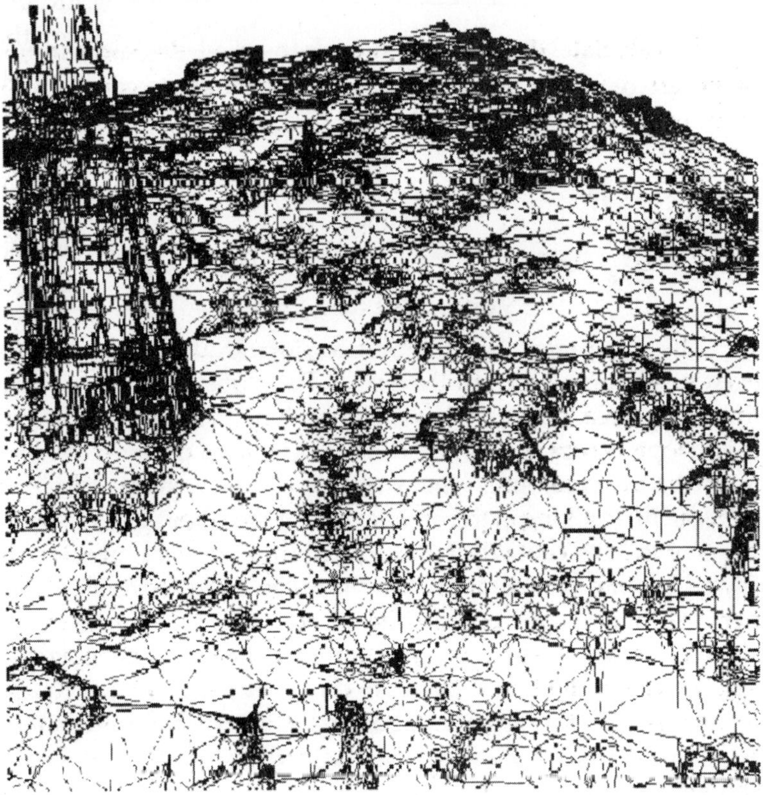

flat areas where as the DEM from program shows variation in slope. Moreover, it is seen that the forced elevation do not exceed the contour interval and algorithm maintains the original contour elevation data.

## Slope Map from the Improved DEM.

The slope map is derived from the improved digital elevation model by applying a program so that the minute slope difference can be taken into consideration. The slope map is obtained by calculating the slope of each pixel as shown below.

| A | B | C |
|---|---|---|
| D | E | F |
| G | H | I |

To calculate the slope of the centre pixel the surrounding pixels are read and the following steps are done to calculate the slope of the centre pixel E. where,

$Dx1 = C\text{-}A$ $Dy1 = A\text{-}G$

$Dx2 = F\text{-}D$ $Dy2 = B\text{-}H$

$Dx3 = I\text{-}G$ $Dy3 = C\text{-}I$

$dx = (Dx1+Dx2+Dx3)/3^*x$ ; $dy = (Dy1+Dy2+Dy3)/3^*y$

$$\text{Slope} = \text{Sqrt}\ (dx^2 + dy^2)/2$$

The slope map the variation of slope in the study area for each pixel obtained from the DEM. The slope maps shows that the study area does not have major slope categories falling in high slope except for some of the central area of the watershed near the residual hill. The watershed has a gentle slope in general. But looking at the areal extent of the slope category of 1.5 per cent to 3 per cent which occurred only at the contour intervals it was feasible to recombine the slope categories of range 1.5—3 per cent and >3 per cent in the same category. Moreover, the slope class of 0.9—1.5 per cent showed a marked areal coverage and hence was retained. This is because the software was unable to incorporate the slope categories below 0.9 per cent values. Thus, at last the slope map only had three major categories as shown in Table 1.

**Table 1**

*Recoded Values of Soil and Recoded Values of Slope*

| *Sl. No* | *Soil types as per map* | *Recoded values* | *Sr. No* | *Slope range (%)* | *Recoded value* |
|---|---|---|---|---|---|
| 1. | 21 | 1 | 1 | 0-0.9 | 3 |
| 2. | 18, 8, 9 | 2 | 2 | 0.9-1.5 | 2 |
| 3. | 22 | 3 | 3 | > 1.5 | 1 |
| 4. | 1 | 4 | | | |
| 5. | 3, 4 | 5 | | | |
| 6. | 17 | 6 | | | |

## Integration in GIS

The soil information is obtained from the NBSS and LUP maps. The map is geo-referenced and converted to a vector layer in GIS environment. The soil types in the map were classified on the basis of grain size, its depth at that location and also taking association of physiographic units into consideration. The corresponding slope of the physiographic unit also becomes an important factor for the soil type classification. The soil categories in the map also considered soils in different range of slopes but those slope categories could not be visualized in the map. In all,

**Figure 6**
*Land Use/Land Cover Map*

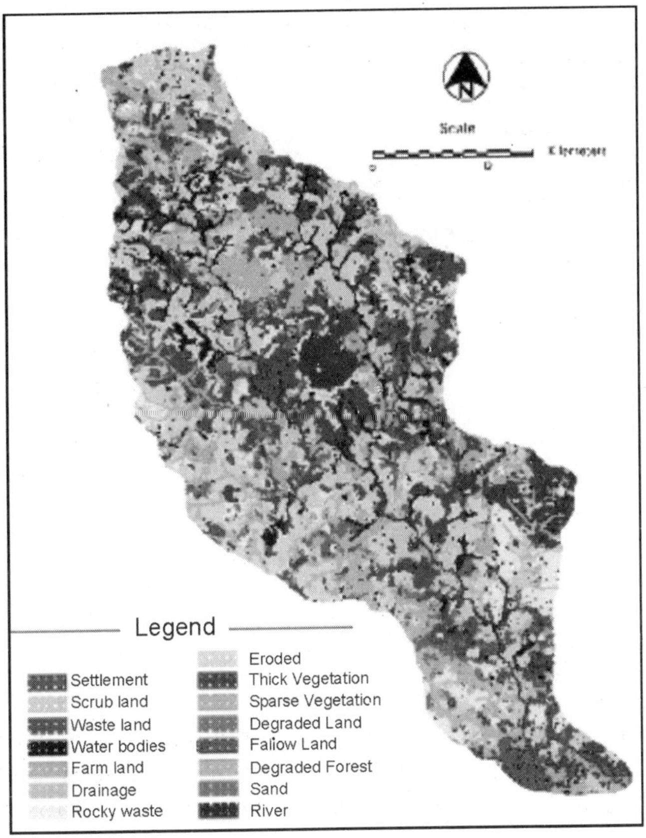

six categories of soil were found to be present in the watershed area. The categories of soils which are classified in the map and the respective classes they were grouped into are shown in the Table 1. As land use/land cover factor is influenced by soil and slope condition an analysis on the variation of land use/land cover is carried out.

The land use/land cover pattern (Figure 6) is generated from the IRS LISS-III satellite image by visual interpretation. The whole watershed area is divided into different categories of land use/ land cover as shown in Table 2. The soil layer and slope layer thus obtained in vector data format were rasterized for further GIS analysis.

**Table 2**

*Recoded Values of Land Use/Land Cover*

| *S.No.* | *Land use/Land cover Class* | *Recoded Values* |
|---|---|---|
| 1. | Settlement Area | 1 |
| 2. | Fallow land | 4 |
| 3. | Rocky waste | 1 |
| 4. | Thick vegetation | 2 |
| 5. | Degraded land | 3 |
| 6. | Sand | 1 |
| 7. | Scrub | 3 |
| 8. | Eroded land | 1 |
| 9. | Sparse vegetation | 4 |
| 10. | Degraded forest | 2 |
| 11. | Waste land | 4 |
| 12. | Farm land | 5 |
| 13. | Water-bodies | 6 |
| 14. | Drainage | 4 |
| 15. | River | 1 |

The two layers of soil and slope of the watershed reveal their corresponding area distribution over the watershed. Soil layer consists of six categories of soil and the slope categories obtained from the DEM were reclassified into three major categories which dominated the study area. The GIS layers were integrated to get all the possibilities of combination of each parameter in individual layers. Thus, in all, there were 3(slope) × 6(soil) i.e., 18 categories of soil and slope combination to form a combined map of slope and soil map (Figure 7) using GIS. The initially obtained land use/land cover of fifteen classes was reclassified into six classes and recoded so as to suit for agricultural purpose. The land use ranked 6, was most favourable while other land use were categorized in decreasing order of suitability.

**Figure 7**

*Soil and Slope Combination Map*

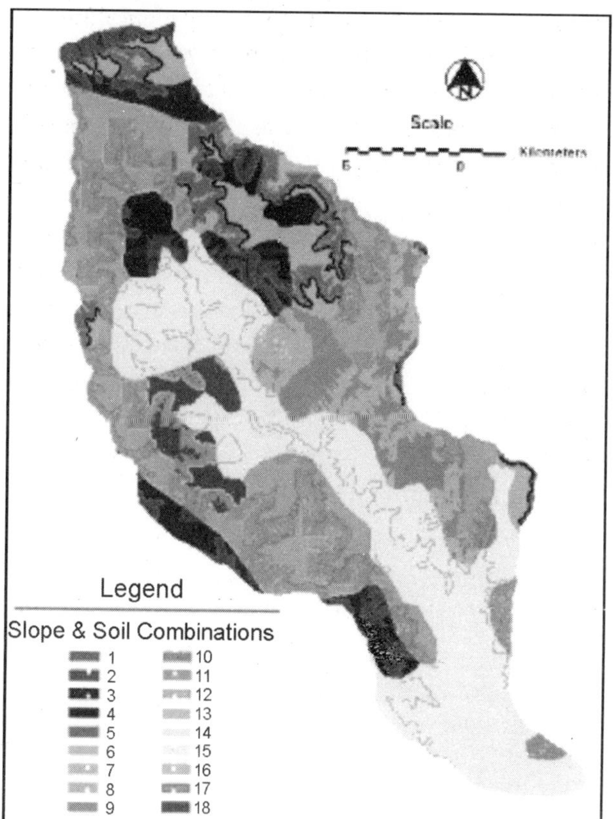

The land use/land cover of the watershed area was further overlaid with the soil and slope combination to obtain 18 × 6 total 108 combinations. The statistics of the data regarding the type of land use/land cover occurring in particular type of soil and slope combination were studied from the output. It was made sure that all the layers were in the same coordinate system and with the same origin.

## Results and Discussion

The slope obtained from the improved DEM had an additional slope class ranging from 0.9 per cent to 1.5 per cent, which could not be retrieved by the software derived DEM. The overlay of the soil layer and slope layer using GIS resulted in the combination of soil and slope categories giving all possible combination of soil and slope. Moreover, the distribution of the type of soils in the slopes ranging in these categories could be visualized in the GIS environment. The visualization of the slope classes derived from improved DEM within each soil class proved to be additional information which is not present in the soil maps, especially without extensive field surveys. While the information provided in the maps are the result of extensive field survey, taking into consideration the physiographic units present in the area. Some of the higher slopes values that were associated along with certain physiographic units could not be interpreted even by the contour data obtained from the topographic sheets, thus causing some minor differences in output of improved DEM when compared to situation given in maps taken as ground truth. The interpolation for abrupt slopes >3 per cent in physiographic units could not be incorporated due to the large contour interval data. But on the other hand the improved DEM gave better results in the rest of the study area where the software derived DEM showed no slope at all.

As per soil map classification each soil class had different slope categories within itself. This variation is not visualised in the maps, but the soil and slope combination map not only gives information as attributes but also represents them pictorially in detail in the map (Figure 8).

The study of the land use/land cover brought about some interesting results. The results show that majority of the area is dominated by land use/land cover category, viz., Lan 1 and Lan 4 (Table 3). Settlement area, eroded land, rocky waste sand and river which form the land use/land cover class 1 dominated the area associated with the flood plains and alluvial fields and form a major part in the slopes less than 1.5 per cent. This was expected as Bankura town the main settlement areas were located in the alluvial fills and flood plains. Thick vegetation and degraded forest are mainly found in the low slope (slope < 0.9%) regions in

**Figure 8**

*Variation within One Soil Types Caused Due to Slope Classes, Representation of Figure 7*

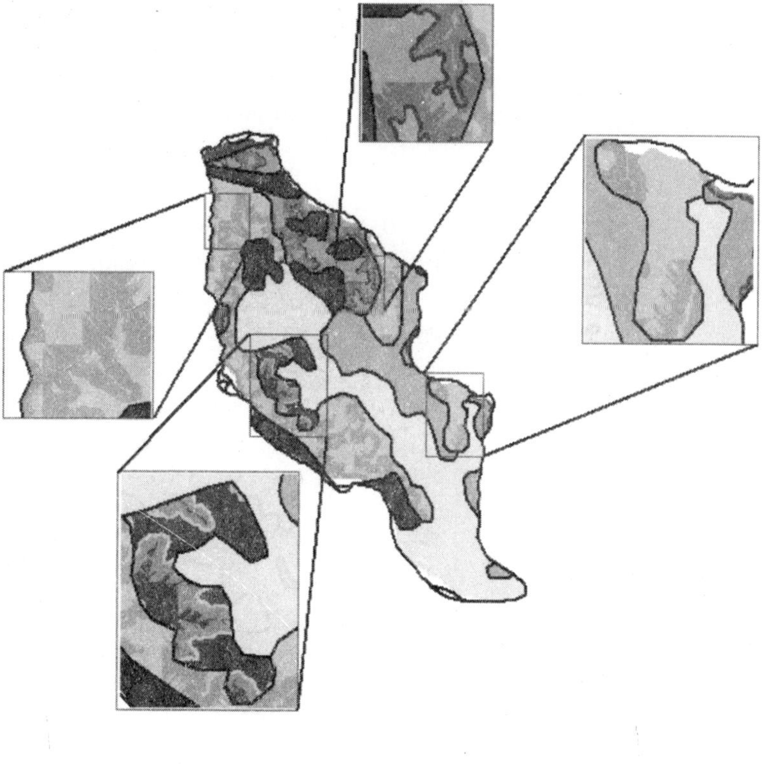

**Table 3**

*Landuse/Landcover Distribution for Corresponding Soil and slope Combination*

| Soil class | Soil types | Slope class Range | Percentage Area Covered by Different Landuse/Landcover Classes* (%Landuse classes per slope class/%slope classes per landuse) | | | | | | | Physiographic Unit Associated with these areas |
|---|---|---|---|---|---|---|---|---|---|---|
| | | | *Lan1 (%) | Lan2 (%) | Lan3 (%) | Lan4 (%) | Lan5 (%) | Lan6 (%) | Total Lan classes under a slope class (ha) | |
| 1 | 21(shallow gravelly sandy loam) | >1.5% | 30.95/1.34 | 12.18/1.56 | 15.54/1.29 | 35.67/1.28 | 3.57/2.40 | 2.08/1.01 | 476.32 | Residual hillocks mounds and pediments |
| | | 0.9%-1.5% | 26.14/3.52 | 15.67/6.23 | 20.50/5.3 | 32.68/3.66 | 2.52/5.26 | 2.50/3.78 | 1482.28 | |
| | | 0-0.9% | 28.51/6.31 | 8.94/5.84 | 21.32/9.06 | 38.11/7.01 | 1.55/5.32 | 1.58/3.92 | 2436.04 | |
| 2 | 18, 8, 9 (moderately to very deep gravelly sandy loam) | >1.5% | 18.98/0.89 | 11.56/1.6 | 24.92/2.24 | 37.88/1.48 | 3.74/2.72 | 2.92/1.54 | 515.88 | Upper undulating alluvial plain and Low dissected plateau interspersed with valley |
| | | | 27.43/4.06 | 3.77/1.65 | 25.89/7.36 | 34.27/4.22 | 3.99/9.18 | 4.65/7.74 | 1629.24 | |
| | | | 25.08/4.99 | 10.08/5.92 | 21.24/8.12 | 39.14/6.48 | 1.86/5.75 | 2.61/5.84 | 2191.24 | |
| 3 | 22 (moderatly deep gravelly sandyloam) | >1.5% | 15.83/1.09 | 50.02/10.17 | 4.13/0.55 | 27.28/1.56 | 0.98/1.05 | 1.76/1.36 | 757.84 | Residual hillocks mounds and pediments |
| | | | 29.70/3.25 | 15.73/5.08 | 7.75/1.63 | 39.54/3.59 | 3.12/5.29 | 4.18/5.13 | 1203 | |
| | | | 19.51/3.42 | 13.39/6.94 | 10.84/3.65 | 52.29/7.63 | 1.16/3.17 | 2.82/5.55 | 1931.84 | |

Contd...

Contd...

| | | | | | | | | | | |
|---|---|---|---|---|---|---|---|---|---|---|
| 4 | 1 (very deep sandy) | >1.5% | 33.34/3.91 | 8.65/3.00 | 15.26/3.44 | 38.85/3.79 | 1.45/2.64 | 2.45/3.23 | 1292.4 | Flood plain and alluvial fills |
| | | | 35.67/11.81 | 8.93/8.74 | 14.37/9.14 | 34.83/9.59 | 3.31/17.01 | 2.89/10.76 | 3645.76 | |
| | | | 30.90/14.71 | 9.73/13.68 | 16.14/14.76 | 38.69/15.31 | 1.63/12.08 | 2.91/15.57 | 5239.52 | |
| 5 | 3, 4 (very deep loam to clay) | >1.5% | 38.03/4.15 | 7.89/2.54 | 14.24/2.98 | 36.38/3.30 | 1.34/2.26 | 2.12/2.60 | 1200.4 | Flood plain and alluvial fills |
| | | | 40.65/12.78 | 8.29/7.69 | 13.19/7.96 | 33.77/8.82 | 1.31/6.39 | 2.80/9.89 | 3459.88 | |
| | | | 32.85/19.21 | 9.84/17.00 | 16.49/18.52 | 36.26/17.63 | 1.54/13.95 | 3.03/19.89 | 6438 | |
| 6 | 17 (very deep clay and loam to clay) | >1.5% | 34.36/0.49 | 6.73/0.28 | 14.04/0.38 | 42.82/0.51 | 0.00/0.00 | 2.05/0.33 | 156.36 | Low dissected plateau interspersed with valley |
| | | | 48.30/1.82 | 5.33/0.59 | 12.54/0.91 | 30.67/0.96 | 0.87/0.51 | 2.29/0.97 | 414.92 | |
| | | | 26.83/2.25 | 6.00/1.49 | 16.79/2.70 | 45.58/3.18 | 3.85/5.02 | 0.95/0.90 | 476.32 | |
| Total slope classes in one Lan class(ha) | | | 11007.96 | 3726.72 | 5731.44 | 13239.68 | 708.52 | 979.6 | 35393.92 | |

* Landuse/landcover classes: Lan1-Settlement area, Rocky waste, sand, eroded land, river; Lan2-Thick Vegetation and degraded forest, Lan3-Degraded land and scrub; Lan4 -Fallow land, Sparse vegetation Waste land and Drainage; Lan5-Farm Land, Lan6-Water bodies.

the flood plains and alluvial fills in association with soil types ranging from deep loam to clay and very deep sandy soil. But thick vegetation is also a major part of residual hill, viz., Sushunia hill, which lies in the slope range greater that 1.5 per cent. This slope class 1(>1.5%) occupies 10.17 per cent of total area in land use class 2 while 17.00 per cent being highest value belonging to the slope class 3 (0-0.9%). Degraded land and scrub land are moderately found in flat areas of residual hillocks mounds, pediments, upper undulating alluvial plain and low dissected plateau interspersed with valley. Other than the alluvial deposits the fallow land are mainly found in the flat areas near the residual hills and the low dissected plateau. The farmland is found mainly in the fluvial deposits of alluvial fills in the gently sloping (0.9%-1.5%) deep sandy and clayey soils, and secondly in between (0-0.9%) slope category 3. The highlight of the results is the indication of farmland to be associated along with all the physiographic units. Farmlands are even moderately found associated with residual hills and mounds and upper undulating alluvial plain and low dissected plateau interspersed with valley, especially at lower slope areas (slope <1.5%). This indicates farming practice in those areas may be largely dependent on the availability of water. This possibility is well supported by the moderate amount of percentage area covered by water bodies (Lan 5) in the similar kind of soil and slope combination area.

Overall, the area is largely dominated by alluvial fills and flood plains and the land use/land cover dominating the study area is landuse/landcover class 4 and class 1.

## Conclusion

The DEM generated using the flat area removal algorithm gives a better slope variation, especially in gently sloping residual areas with slope below 3 per cent in comparison to the usual software derived DEM wherein, the slope in those areas are shown as 0 per cent indicating flat area. The improved DEM can be further used in delineation of watershed and its parameterization. The slope variation in the range of 0-3 per cent has greatly increased which can be realized by using the improved DEM. The studies related to slope and soil mainly are dependent on the accuracy of DEM which can be improved by solving the problems in the interpolation in the areas when there are abrupt physiographic landforms suddenly change the slope keeping the elevation difference within the contour interval. The soil and slope combination adds to the amount of information in maps by incorporating the slope classes within a particular soil. The study of soil slope combination with land use/land cover map brought about general categorized distribution of the land use classes in the different soil and slope combination classes. The study revealed that the watershed mainly dominated by low slope areas and most of them falling with the alluvial fills and flood plains, while Sushunia Hill came under high sloped areas covered with the thick vegetation. The study also brought about redundant information like the farming practice in water scarce area to be predominantly associated with water bodies for the irrigation requirement. The soil slope combination gives all the combination of slope and soil present in the area and this spatial information can be readily used for any future analysis.

## References

Biswas, S., Sudhakar, S. and Desai, V.R. (2002). "Remote Sensing and Geographic Information System Based Approach for Watershed Conservation". *Journal of Surveying Engineering*. 128, (3), 108-124.

Jenny, H. (1941). *Factors of Soil Formation*. McGraw-Hill, New York.

Jenson, S.K. (1988). "Automated Derivation of Hydrologic Basin Characteristics from Digital Elevation Models". *Photogrammetric Engineering and Remote Sensing*. 54, (11), 1593-1600.

Martz, L.W. and Garbrecht, J. (1992). "Numerical Definition of Drainage Network and the Sub-Catchment Areas from Digital Elevation Models". *Computers and Geosciences,* 18, (6), 747-761.

Mueller, T.G., Pierce, F.J., Schabenberger, O. and Warncke D.D. (2001). "Map Quality for Site-Specific Fertility Management." *Soil Science Society of America Journal*. 65: 1547–1558.

Turcotte, R. Fortin, J.P. Rousseau, A.N. Massicotte, S. and Villeneuve, J.P. (2001). "Determination of the Drainage Structure of a Watershed

Using a Digital Elevation Model and a Digital River and Lake Network". *Journal of Hydrology,* 240, (3-4), 225-242.

Vogt, J.V., Colombo, R. and Bertolo, F. (2003). "Deriving Drainage Networks and Catchment Boundaries: A New Methodology Combining Digital Elevation Data and Environmental Characteristics". *Geomorphology,* 53, (3-4), 281-298.

# 35

# Topographic Analysis of a Hydrocarbon Bearing Folded Sedimentary Terrain Around Mashimpur, Assam, Using Remote Sensing and GIS

P. Phukon, S. Phukan and A.A. Laskar

Topographic expression that pertains to the disposition of land in three dimension, is the most distinctive feature of a terrain that reflects the geological factors, viz., lithology, structure as well as climatic and hydrological factors. Remote sensing data together with GIS is found to be a valuable tool in topographic analysis as a component of terrain study. For the purpose of topographic analysis of the present study area, which is part of the Barak valley in south Assam around Mashimpur (92°38′41″–92°50′42″ E : 24°35′52″–24°56′39″N), a 30 m resolution DEM was generated from elevation contours and spot height layers in the scale of 1:50000, using algorithm of Geomatica 9.1. The DEM derivatives viz., shaded relief, DEM profile, slope and aspects together with remote sensing data (RESOURCESAT-1, L3) are found useful in this study. Automatic drainage extraction from DEM (O'Callaghan and Mark, 1984) using various threshold is attempted for the Thakiri Nadi catchment that flows along the axial part of the Mashimpur anticline, to see the applicability of the technique in a folded sedimentary terrain. A flow accumulation layer is generated using 'DWCON' algorithm and drainage at different threshold values (150, 175, 200, and 225) are derived using the 'DRAIN' algorithm in Geomatica 9.1, after performing the preprocessing. The drainage lines were then vectorized for morphometric analysis. The geological features, viz., structure and lithology have been worked out through a combination of shaded relief, geo-referenced satellite data, published collateral maps and field mapping. These features are correlated with the topographic characteristics of the area.

## Geology and Geomorphology

The study area is part of the Tripura-Cachar-Mizoram frontal fold belt within the Greater Assam-Arakan tectonosedimentary basin. About 10 km of Tertiary Neogene clastics form the bulk of the sedimentary sequence, viz., the Barail Group (Oligocene), Surma and Tipam Group (Miocene-Pliocene), and the Dupitila (Plio-Pleistocene), with an alluvial cover of variable thickness. The rocks are highly deformed giving rise to about seven lines of folds, viz., Adamtila-Patharia, Lonpad-Chorgola, Chatachura-Kanchanpur-Chandipur-Badarpur, Hilara-Narayanchara-Shabazpur, Bhairabi-Mashimpur, Rengte-Pathimara, Teidukhan-Tukbai and Bhuban-Digli. The narrow, box type and faulted and thrusted asymmetrical anticlines are separated by broad synclines (Seshavataram, et al., 1998). The present study area comprises the doubly plunging Mashimpur anticline having N-S axial trend that swings to NNE-SSW towards the northern part. A combination of shaded relief (Figure 1A) derived from DEM, collateral published geological maps and satellite image interpretation followed by selective ground verification shows that the structure exposes Miocene Tipam Sandstone along its flanks while the finer sand-shale sequence of Bhuban Group form the core part (Figure 2). The trend lines as well as lithological

**Figure 1 A**

*Shaded Relief from DEM*

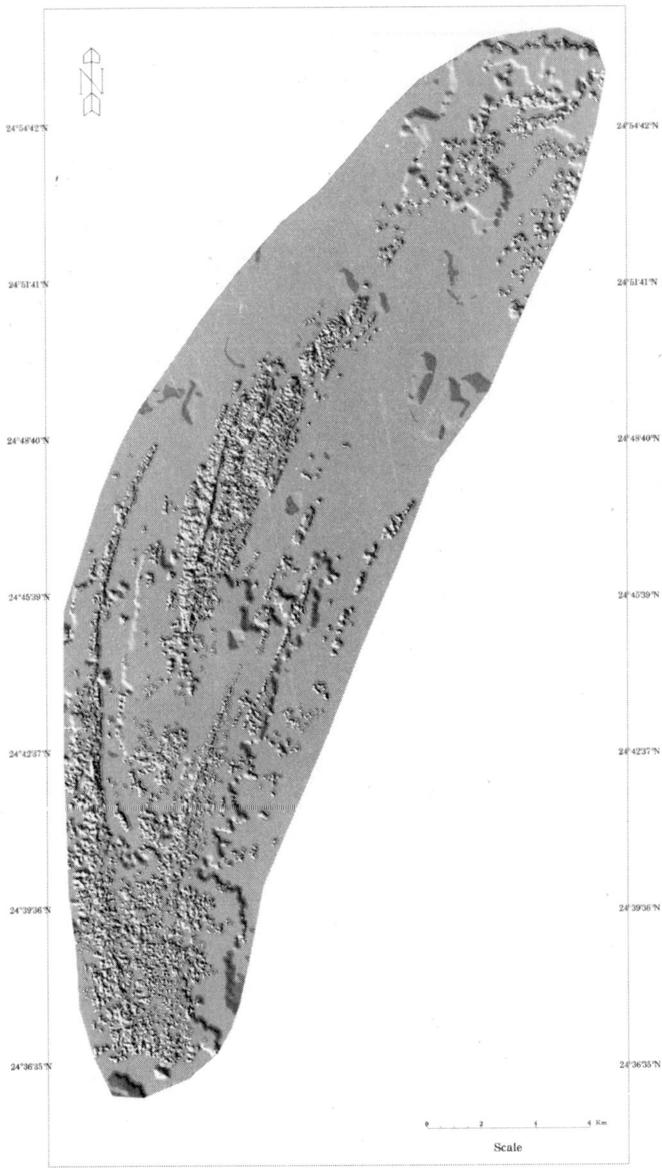

**Figure 1 B**
*Lineament Map*

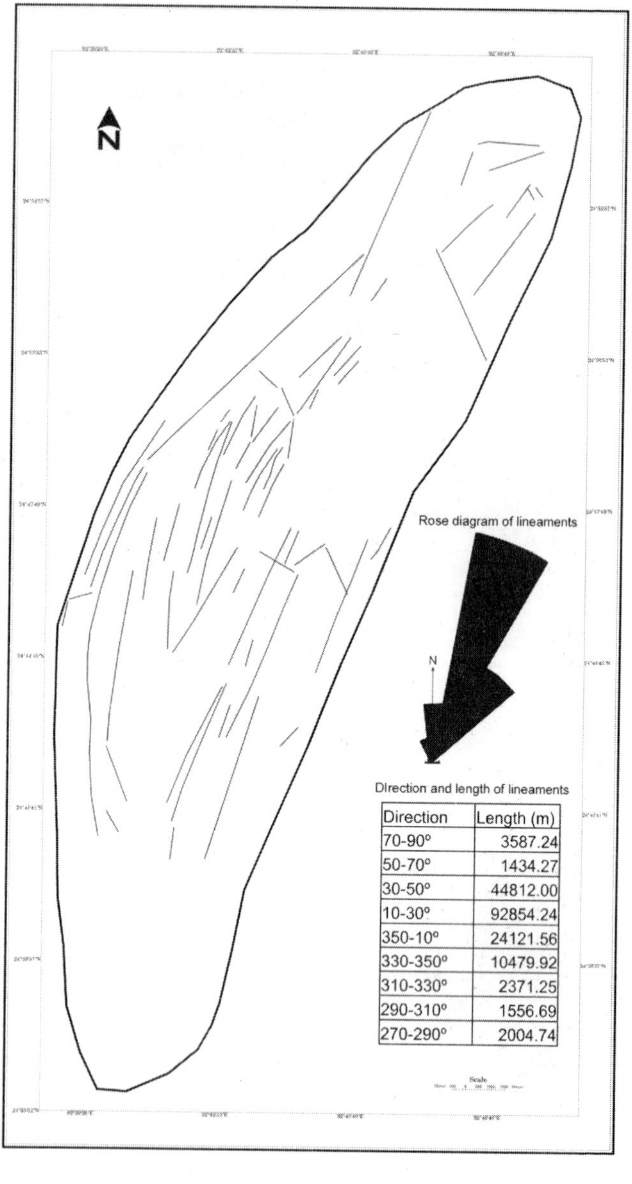


Direction and length of lineaments

| Direction | Length (m) |
|---|---|
| 70-90° | 3587.24 |
| 50-70° | 1434.27 |
| 30-50° | 44812.00 |
| 10-30° | 92854.24 |
| 350-10° | 24121.56 |
| 330-350° | 10479.92 |
| 310-330° | 2371.25 |
| 290-310° | 1556.69 |
| 270-290° | 2004.74 |

**Figure 2**
*Geological Map of Mashimpur Anticline*

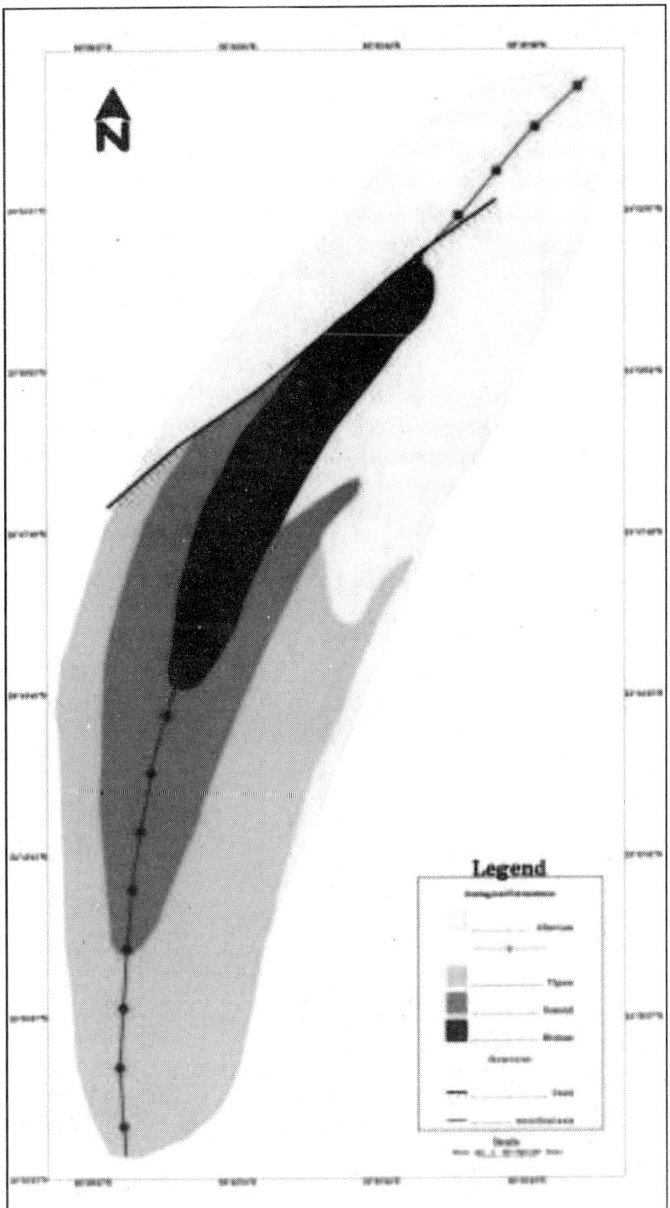

contacts and a few longitudinal fractures are well discernible from the shaded relief. Manually extracted lineaments (Figure 1B) within the structures show a dominant NNE-SSW trend and having an overall lineament density of about 0.522 $km^{-1}$.

The structural and lithological set up of the region is well reflected in the topography of the area. It represents a typical ridge and valley province with anticlines forming the low lying hills interspersed with synclinal broad valleys. Differential erosion of the variable sand-shale lithology of the anticlines, has resulted in linear cuesta ridges along the anticlinal limbs that are dissected mostly by fluvial erosion. A number of linear, litho unit bound static water bodies of syntectonic origin is another unique geomorphological feature in the area. Tectonic control of the river regime is quite apparent from the major tributaries, viz., Madhura, Jatinga, Sonai, Rukni, Jiri, Chiri, Ghagra, Longai and Katakhal flowing through the synclinal low, parallel to the axial trend of the anticlines and synclines. Presence of misfit streams, e.g., the lower reach of Ghagra, deflection of rivers around the

**Figure 3**

*A Fly! View of the Northern Part of the Structure*

anticlinal noses besides well developed trellis pattern of drainage within the structures and some tributaries flowing along the anticlinal axis, e.g., Thakiri nadi along Mashimpur anticline strongly suggest structural and tectonic control of the drainage.

Generalized view of the northern part of the terrain can be seen in the Fly! Model generated by draping the satellite image over DEM in Geomatica (Figure 3).

## Topographic Analysis

DEM profiles represent the topographic expression along a defined orientation which can be effectively correlated with the lithology and structure of a region. They serve as a rapid visual method for identifying the type of topography that exists in an area. Along eight predetermined azimuth, one along the axial trend and seven across the trend, profiles were derived from the DEM to represent the topographic expression of the study area (Figure 4). The corrugated topography of the anticlinal structure is well reflected in all the profiles. While the central part representing core of the anticline is invariably seen as a topographic high, a notable feature is the presence of valleys flanking this high. Field investigations show that these valleys are developed due to differential erosion along the largely argillaceous Bokabil rocks. Profiles 7 and 8 indicate the tapering northern nose of the plunging anticline. Lithological control on the topographic expression of the area is evidenced by a correlation of the profiles and underlying lithology. While the Tipam and the Bhuban rocks form the low lying ridges, Bokabil Group of rocks invariably form the valleys within the anticlinal structure.

Slope is the steepness or gradient of a unit of terrain, usually measured as an angle in degrees or as a percentage. Aspect is the direction in which a terrain unit faces, usually expressed in degrees from the north (Burrough, 1986; Heywood, et al., 2003). The area mostly comprises almost flat terrain with average elevation of less than 30m (msl). Eight slope classes derived from the DEM shows that as much as 77 per cent of the total area is having a slope within 2° while those with slope angle >35°

**Figure 4**

*Elevation Classes from DEM and DEM Profiles*

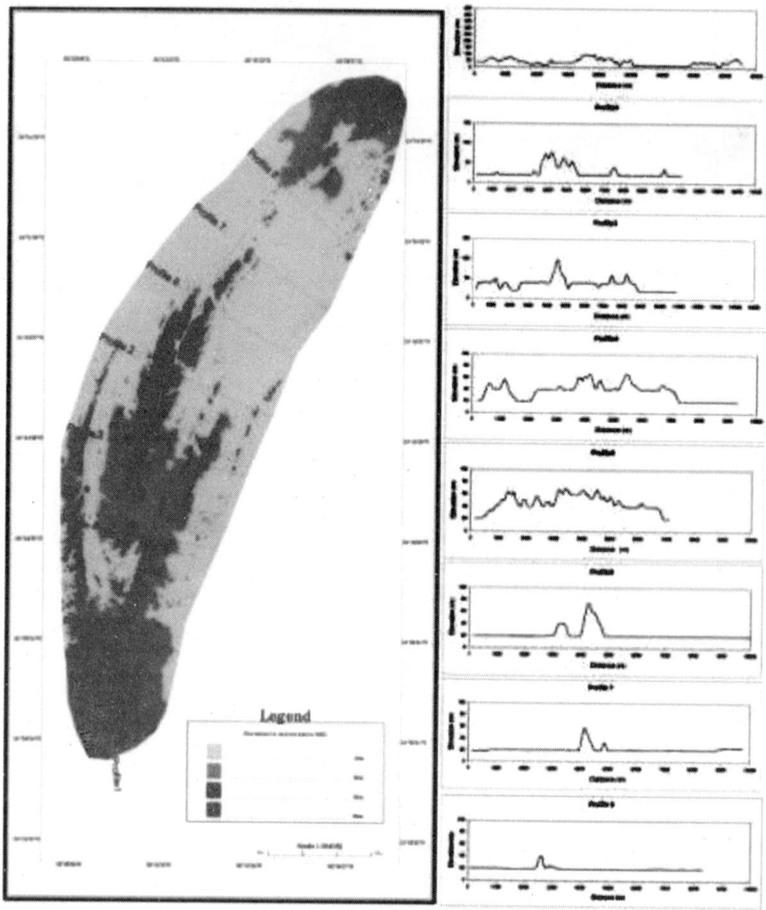

constitute less than 1 per cent. The aspect classes show a dominant trend towards NNE within 0-45° azimuth.

## Drainage Analysis

The Thakiri Nadi catchment occupying the central part of the anticline is delineated from SoI topomap in the scale of 1:50000 and the morphometric parameters, viz., drainage density, frequency and shape factors are deduced to compare with the

**Figure 5**

*Slope and Aspects Distribution*

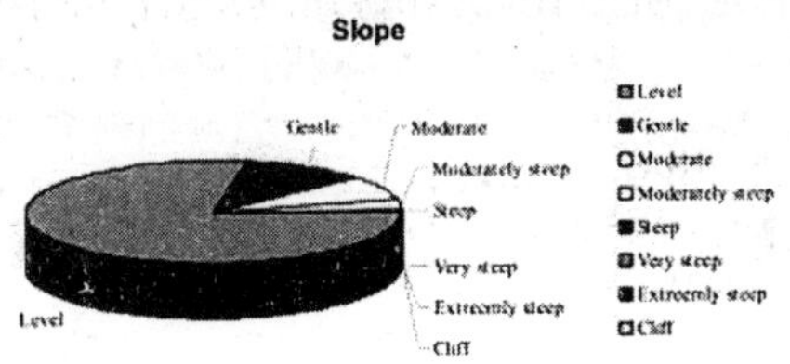


| Slope Classes | | Area (Sq. km) |
|---|---|---|
| 0-2° | Level | 273.285 |
| 2-5° | Gentle | 38.071 |
| 5-10° | Moderate | 27.214 |
| 10-20° | Moderately steep | 11.557 |
| 20-35° | Steep | 1.297 |
| 35-45° | Very steep | 0.140 |
| 45-60° | Extreemly steep | 0.064 |
| > 60° | Cliff | 0.007 |

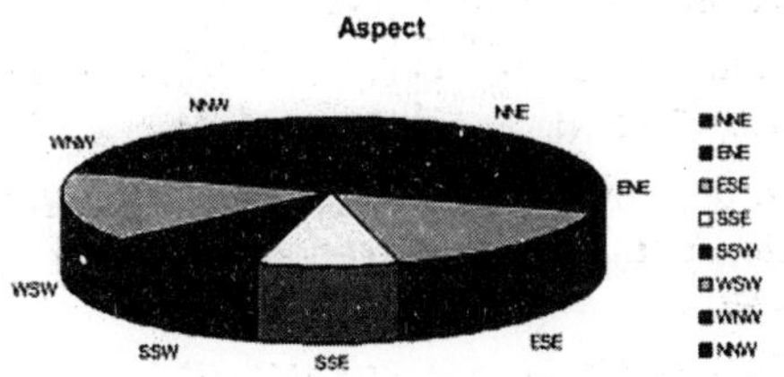


| Azimuth | | Area (Sq. km) |
|---|---|---|
| 0-45 | NNE | 68.303 |
| 45-90 | ENE | 36.114 |
| 90-135 | ESE | 56.749 |
| 135-180 | SSE | 30.708 |
| 180-225 | SSW | 34.590 |
| 225-270 | WSW | 54.308 |
| 270-315 | WNW | 30.828 |
| 315 | NNW | 40.781 |

values for the drainage extracted from DEM for various thresholds (150, 175, 200, and 225). From the bivariate plots of the morphometric parameters it can be inferred that there is strong threshold area influence while extracting drainage from DEM. Repeated tests shows that the drainage derived with 225 threshold is the nearest reflection of the ground condition with drainage density of 4.9 which is close to the drainage density of 4.6 found for the drainage extracted directly from 1:50000 topomap. The standard deviation in the morphometric parameters can also plausible due to generalizations involved in both the cases. As such this simulated drainage basin can be reliably used for further computation of hydrological parameters.

## Conclusion

The present study brings out the usefulness of remote sensing and GIS technique in topographic analysis of a folded sedimentary terrain. It shows lithological and structural control on the ridge and valley topography in and around the Mashimpur anticlinal structure. The river regime as well as the static water bodies also reflect strong litho-tectonic control. The threshold area influence on automatic drainage exraction from DEM is established and a useful threshold of 225 is worked out as the optimum for the study area.

## References

Burrough, P.A., 1986, *Principles of Geographical Information Systems for Land Resources Assessment,* Oxford University Press, p. 194.

Heywood, I., Cornelius, Sarah and Carver, S., 2003, *An Introduction to Geographical Information Systems,* 2nd ed., Pearson Ed., p. 295.

O'Callaghan, J.F. and Mark, D.M., 1984, The extraction of drainage network from digital elevation data, *Computer Vision, Graphics and Image Processing,* 28, pp. 323-344

Seshavataram, B.T.V., Nandi, P.K. and Choudhury, S., 1998, Petroleum Geology of Cachar Fold Belt, in, Proc. Vol. Regional Seminar on Development of Geological Research in North Eeast India, Gauhati University, pp. 327-348.

# 36

# Soil Informatics and Degraded Land Management

S. Chattopadhyay and O.N. Pandey

Land is one of the most important natural resources and nearly 99 per cent of human food supply comes from the land. But the growing human and animal population is making increasing demand on natural resources leading to their unsustainable exploitation and subsequently creating degraded lands. The land resources are tremendously degraded due to a number of reasons, viz., massive destruction of forest, heavy grazing, tremendous increase in various developmental activities like mining, brick kilns, quarrying, urbanization, construction of roads and other modes of communications, reservoirs, etc. About 50 per cent of the total geographical area of India (329.00 m.ha) is under various degradation hazards and about 2.1 million hectare of land is getting degraded and deforested yearly (Anon, 1986). Degraded lands are ecologically most unstable and having poor soil health. A top 15 – 25.5 cm soil, is the real store house for plant growth. Nature takes about 400-500 years to form 1 cm of top soil. It, therefore, means that nearly 2000-2500 years would be required in the formation of even a thin 5 cm soil layer (Gupta, 2002). Thus, a loss of even 5 cm soil layer through erosion cannot be compensated by the process of soil genesis easily and quickly.

But man may lose it due to catastrophic soil erosion. We must, therefore, save our soils against erosion because the soil once lost is difficult and expensive to replace. In order to manage degraded lands, soil conservation measures have to be implemented to reduce soil losses and building up the overall productivity of soil by means of suitable engineering and biological methods of soil conservation.

The extent of land degradation, however, varies from state to state, region to region depending upon and topographical, geological formations, climatological elements and soil characteristics. So, soil related informations which comes under the preview if Soil Informatics, are necessary for effective management of degraded lands or wastelands.

## Status of Degraded Land (Wasteland) in India

The dictionary meaning of wasteland is the land which is unfit for use, uncultivable, ruined, devastated, useless, barren or without vegetation. Bhumla and Khare (1984) have defined the wastelands as the lands which are ecologically unstable, whose top soil has nearly completely lost and which have developed toxicity in the root zones for growth of most plants, both annual crops and trees. ICAR (Anon, 1987) proposed that wastelands are lands which due to degradation are not being utilized to their full potential and classified wasteland from the point of view of establishing vegetative cover (Table 1).

Prasad (1988) classified wastelands on two broad categories, viz., Ecological Wastelands and Developmental Wastelands. Wastelands such as degraded forests, gullied and ravinous lands, naked hill slopes, saline-alkaline soils, desert, sand dunes, shifting cultivation, etc., are under the category of Ecological Wastelands whereas Developmental Wastelands include mine soils and overburdens, waterlogged areas formed by seepage of canal irrigation, foreshores of irrigation reservoirs, industrial wastelands and land strip cleared for electric transmission lines.

So, from the above view on wastelands it can be said that wasteland is the land which is degraded and is presently lying

**Table 1**
*Classification of Wastelands*

| *Level-1 Kinds of Wastelands* | *Level-II: Based on Causative Agents* | | | | | | |
|---|---|---|---|---|---|---|---|
| | 1. | Water | 2. Wind | 3. | Man | 4. | Other |
| A. Utilizable | (i) | Sheet erosion | Sand dunes, Sand bar and coastal sand | (i) | Shifting cultivation | (i) | Shallow soil |
| | (ii) | Rill erosion | | (ii) | Mine spoils | (ii) | Land slides |
| | (iii) | Gullied land | | (iii) | Industrial Wastelands | | |
| | (iv) | Ravinous land | | (iv) | Land affected by roads, kilns and borrow pits | | |
| | (v) | Salinity and sodicity | | (v) | Strip land | | |
| | | (a) Saline soil | | | | | |
| | | (b) Alkali soil | | | | | |
| | (vi) | Diara land | | | | | |
| | (vii) | Marshy land | | | | | |
| | (viii) | Bouldery land | | | | | |
| B. Unutilizable Wastelands | (i) | Bare and rock land | | | | | |

*Source;* (Anon, 1987)

unutilized due to different constraints except as current fallows. So the term wasteland itself is not proper since, each and every piece of land in the universe has been used/exploited by use and hence the term waste is improper to use. So, it may be called as degraded land.

In India, out of 329 million hectares (m.ha) of geographical area, 220.61 m.ha. are degraded (Anon,1994) producing less than 20 per cent of the potential capacity. The present status of land degradation in India is as follows.

| *S.No.* | *Types of Degradation* | *Area in m. ha.* |
|---|---|---|
| 1. | Area subjected to wind and water erosion | 162.00 |
| 2. | Water logged area | 11.60 |
| 3. | Alkali soil | 4.50 |
| 4. | Saline soils including coastal sandy soils | 5.50 |
| 5. | Acid soils | 25.00 |
| 6. | Ravine and gullies | 3.97 |
| 7. | Shifting cultivation | 4.91 |
| 8. | Riverine and torrents | 2.73 |
| | Total Problematic area (degraded lands) | 220.61 |

*Source:* Anon. 1994.

## Soil Informatics and its Use in Degraded Land Management

Soil Informatics is a computerized database system containing a wide range of soil related information obtained from two important techniques, viz., Remote Sensing (RS) and Geographical Information System (GIS). Soil Informatics database consists of two categories of data (i) soil survey data and (ii) ancillary data.

Spatial data map; profile data; data related to morphology, physical, chemical, biological characteristics and soil taxonomy are put in the first category, i.e., soil survey data. Ancillary data include data related to land use, geomorphology, water, relief features, slope, weather and socio-economic condition of the people at microlevel (village level).

Soil Informatics provides the following soil related informations:

(i) Kind and extent of soil degradation in all the degraded pockets.
(ii) Potentiality and limitations of different kinds of soil on productivity.
(iii) Productivity status of the degraded land.
(iv) Spatial distribution and delineation of the extent and type of recoverable and non-recoverable degraded lands.
(v) Indian soil database.

Using Soil Imformatics we can map and monitor degraded lands rapidly. Different reclamation measures for different degraded lands can be suggested based on the soil survey as well as ancillary data of Soil Informatics. Thus Soil Informatics serves as a basis for transfer of technology and it can be the best option today to survey soil in detail and transfer informations for judicious management of degraded lands of Indian subcontinent accelerating both for agricultural as well as forestry crop production and environmental amelioration.

## References

Anon, (1996) "FAO Study Report", Food and Agriculture Organisation, Rome.

Anon, (1987) *Technologies for Wasteland Development,* ICAR Publication.

Anon, (1994) *National Commission on Agriculture Report.* Ministry of Agriculture, Government of India, New Delhi.

Bhumla, D.R., and A. Khare (1984), *Estimates of Wastelands in India.* SPWD, New Delhi.

Gupta, R.D. (2002) "Save the Soil, Save the Nation". *Environ and People* 8(12): 20-25.

Prasad, R. (1988) *Technology of Wasteland Development,* Associate Publishing Company, New Delhi. p. 356.

# 37

# Application of Geoinformatics in Quantifying Water Supply Status and its Futuristic Demand in and around Ranchi City

Praveen Kumar, A.K. Sinha, M.S. Nathawat and A.K. Singh

---

The remote sensing and GIS together have opened the new avenues in studying water resources and analysis for the rapidly growing population. It also can be used in understanding of urban water distribution planning and futuristic water demand aspects, providing an important addition to conventional methods of urban planning in the context of water distribution.

Ranchi being the capital of the newly formed Jharkhand state has tremendous growth of population, resultant experiencing scarcity of water has been chosen for the study. For overall development, the author was attempted to concentrate in providing basic amenities such as sufficient quantity of water. These are few important areas where attention is to be paid by the policy makers. Its importance is not less than any economic project, employment, increased production and raising the national income.

Due to unique topography and lithological characteristic, most of the rainfall goes in the form of run-off, hence the groundwater is also limited. The terrain condition can be envisaged by overlay of surface feature over TIN highlighted the probable direction of movement of surface water for planning of the water harvesting in appropriate locations. The TIN with surface features map is shown in Map 4.

In such a scenario surface water resource becomes very important and plays vital role. The main source of water around Ranchi is river Subarnarekha and few reservoirs namely Hatiya, Kanke and Getalsud.

In this paper data base was created and integrated with demographic data to formulate the futuristic need of water demand in the form of a plan. Such a plan is useful for water resources management and sustainable development. A futuristic water demand model and influence zone of the pump house in Ranchi urban area have been prepared using Geoinformatics technology which certainly will help the decision maker in proper planning and implementing the developmental plan for the growing population of the Ranchi Muncipal Area.

## Study Area

Ranchi has been a hill station with beautiful green bush valley, a midst water falls, hills, forest and rivers. The study area is a part of Chota Nagpur Plateau. The area is bounded by latitude 23° 14′ 42″ N to 23° 30′ 05″ N and longitude 85° 10′ 47″ E to 85° 31′ 02″ E and falls in SoI topographic sheets Nos. 73E/3, 73E/7 and 73E/11. The entire study area has an aerial extent of 562.77 km$^2$. The location of the study area is as shown in Map 1.

The climate is humid and the maximum temperature varies between 37°C to 44°C and the minimum temperature varies between 3°C to 14°C. Most of the rain occurs from southwest monsoon. The average rainfall of area is 1300 mm. (District Census Book).

**Map 1**

*Location Map: Part of Ranchi Dist. Jharkhand, India*

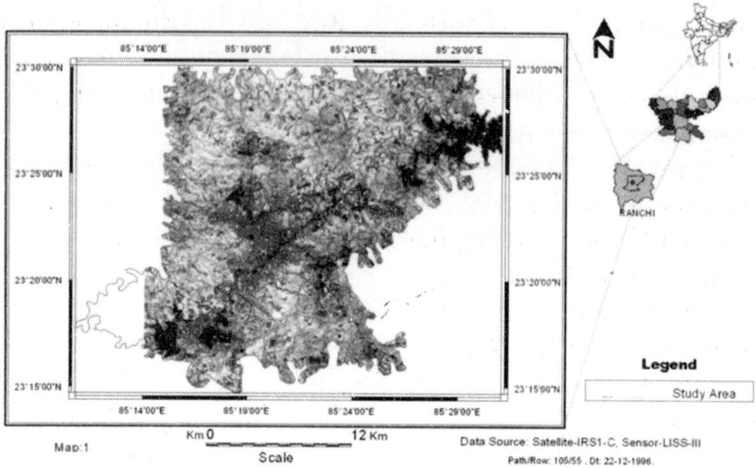


**Map 2**

*Drainage, Road and Settlement Distribution is the Study Area*

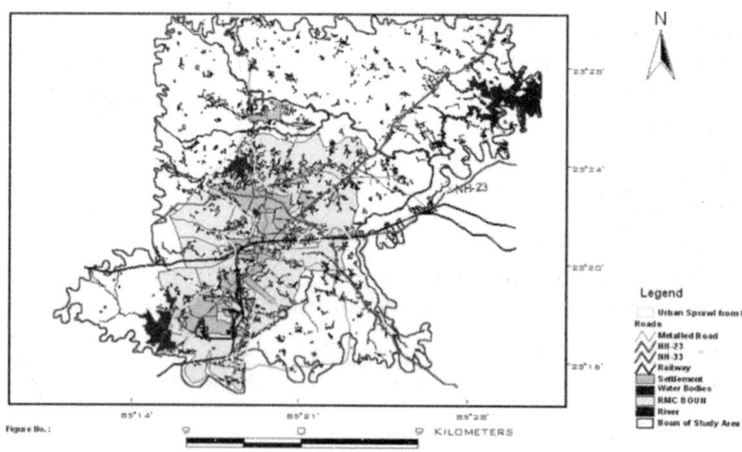

## Methodology

In the present study, IRS 1C LISS-III satellite image was merged with the PAN data to prepare the different maps such as drainage, urban sprawl (Map 2), RMC (Ranchi Municipal Corporation) ward, pump house influence, population density (Map 3) were prepared. With the help of various thematic details and secondary data a Pump House Influence Zone Map was prepared (Map 5). In this study, emphasis was given to predict the future water demand for the growing population in RMC area using statistical analysis based on 2002 data. An analysis of the futuristic water demand for RMC ward area has been calculated by considering the present consumption of water.

In the GIS environment spatial and non-spatial data were analysed to derive the spatial patterns. The spatial pattern highlights the poor distribution of water supply within Ranchi Municipal Corporation (RMC) ward area.

The water supply area was identified and presented in Map 5, which shows the total influence zone of the three pump houses in and around the Ranchi city.

**Map 3**

*Population Density and Pump House Influence in RMC*

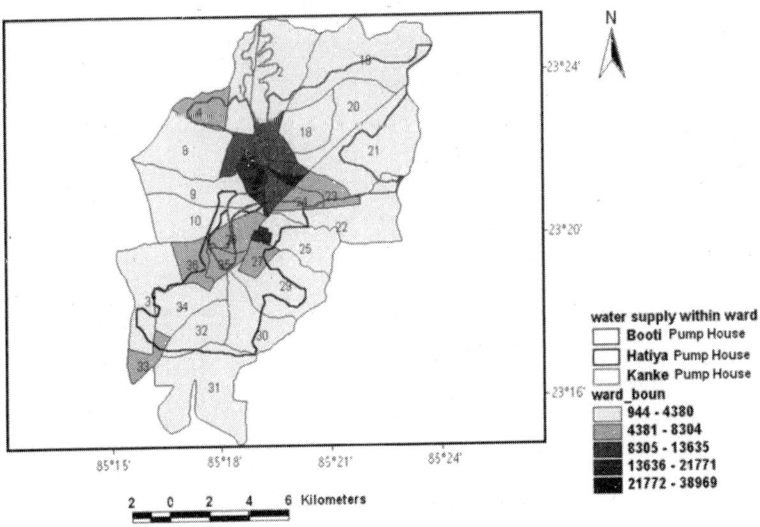

**Map 4**

*Digital Elevation Model (DEM) of the Study Area*

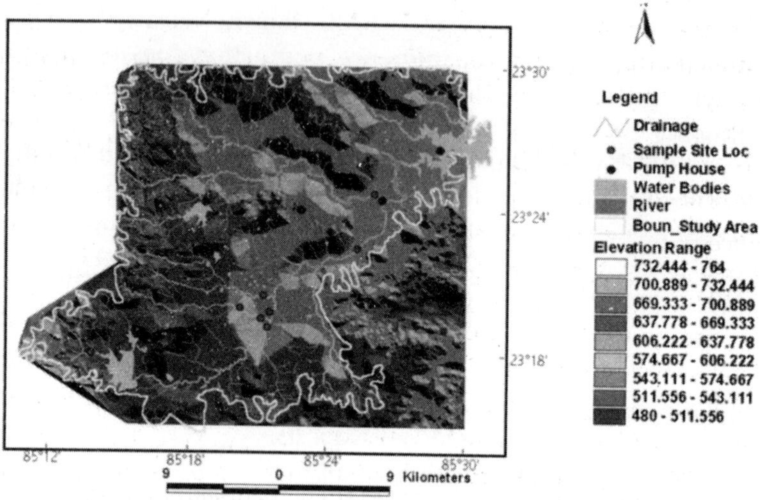


**Map 5**

*Pump House Influence Zone Map of Study Area*

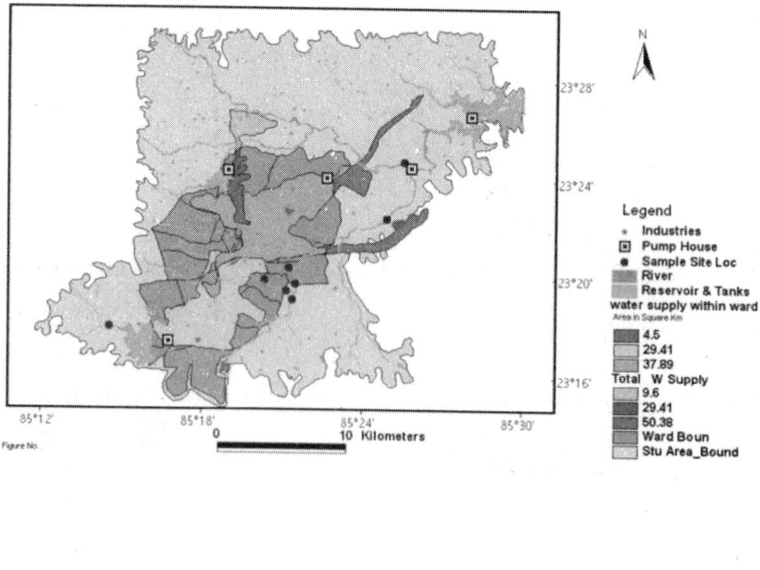

## Results and Discussion

### *Relation of Quantity of Water and Population*

For population estimation, the population data of Ranchi Municipal Ward was taken for calculation of the futuristic need of water demand for various uses. Water consumption estimation is based on population projections.

### *Forecasting of Population*

Since population is always a relevant factor in estimating future water use. It is necessary to predict, in some manner, what the future population will be for next fifty years of the total Ranchi Municipal Ward.

The population data of Ranchi Municipal Ward since 1901 to 1991 was taken for the analysis. The population density of RMC is as shown in the Map 3.

The projected population was calculated using geometrical increase method – Table 1.

This method was again tested on the trend of Ranchi district population expansion.

**Table 1**

*Population Forecasting of Ranchi Municipal Ward*

| *S.No.* | *Year* | *Expected Population* |
|---|---|---|
| 1. | 2001 | 8, 63180 |
| 2. | 2021 | 12, 43238 |
| 3. | 2031 | 25, 79051 |
| 4. | 2041 | 37, 14607 |
| 5. | 2051 | 53, 50148 |

### *Population Growth Modelling*

Population of Ranchi Municipal ward as per the curve fitting method is as shown in Figure No 1.

Figure No: 1 Population Forecasting for Ranchi Municipal Ward Area.

**Figure 1**

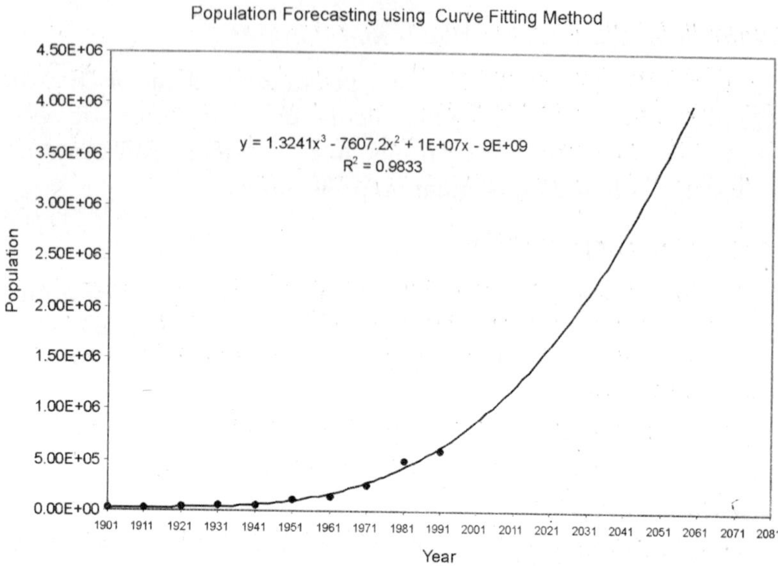


The urban sprawl and population density of Ranchi urban area are as shown in Map 2 and Map 3.

### *Present Status of Water Supply*

The water is mainly supplied by three pump houses namely, Hatiya Pump House, Kanke Pump House and Booti Pump House. The distribution of water has been mapped Figures 3 and 4.

**Table 2**

*Water Supply Within Ranchi Municipal Ward (RMW)*

| *S.No.* | *Pump House (PH)* | *Water Source Reservoir* | *Average Water Supply in MGD** |
|---|---|---|---|
| 1. | Booti | Gatalsud | 19.50 |
| 2. | Kanke | Kanke | 4.32 |
| 3. | Hatiya | Hatiya | 8.00 |
| | Total | | 31.83 |

* Million Gallon per day

**Figure 2**
*Water Demand Forecasting Model*

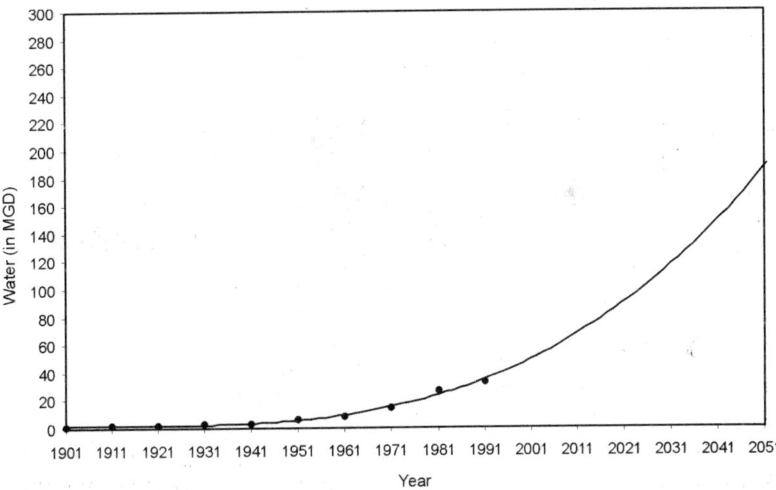


**Figure 3**
*Water Supply Coverage within Ranchi Municipal Area*

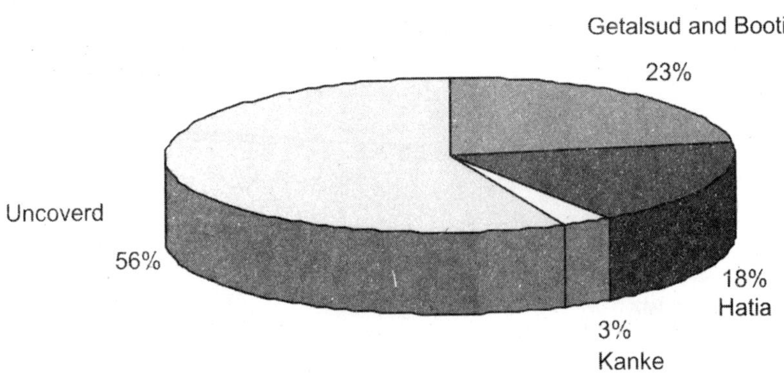

**Figure 4**

*Water Supply Coverage from Different Pump Houses*

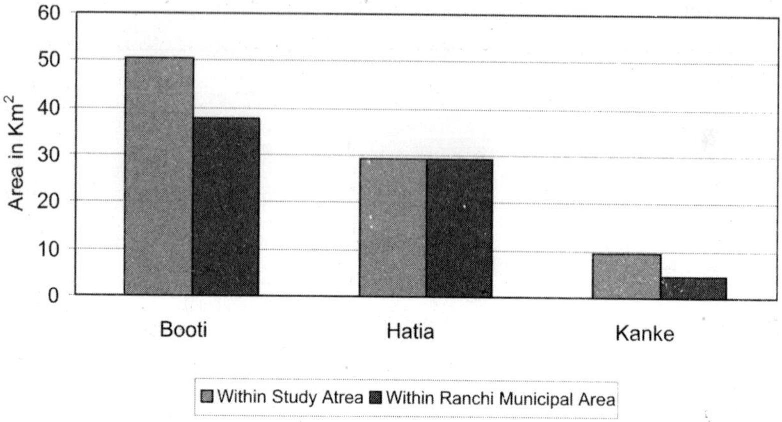


Total water supplied by three pump houses in Ranchi Municipal Corporation is shown in Table 2. Figure 3 shows the water supply facilities available within the Ward (RMC). Figure 4 and Map 5 show the water distribution area of the different reservoirs from different pump houses (PH).

Table 3 is shown below in which the areas of water supply in Ranchi city and Ranchi Municipal Corporation (RMC) are given.

**Table 3**

*Contribution of Water Supply by Different Pump Houses*

| *S.N.* | *Pump House* | *Water supply (mgd)* | *Reservoir* | *Total area (sq.km)* | *Total supply within RMC (sq.km)* |
|---|---|---|---|---|---|
| 1. | Hatiya | 8.0 | Hatiya | 29.41 | 29.41 |
| 2. | Kanke | 4.32 | Kanke | 9.60 | 4.50 |
| 3. | Booti | 19.5 | Getalsud | 50.38 | 37.89 |
| | Total | | | 89.49 | 71.80 |

*Note:* These data correspond to the year 2002 and any additions or expansion in water supply in and around Ranchi made after the year 2002 have not been considered in this work.

## Water Demand Forecasting for Ranchi Municipal Area

The water demand for future (up to 2051) is shown in Figure 2. Water Demand Forecasting for Ranchi Municipal Ward for a period of 50 years.

**Table 4**

*Requirement of Water Demand by Ranchi Municipal Ward up to Year 2051*

| *S. No.* | *Year* | *Population* | *Supplied water* | *Actual water demand in (MGD)* | *Shortage of water* |
|---|---|---|---|---|---|
| 1 | 2001 | 8, 63180 | 31.86 (MGD) | 47.46 | 15.6 MGD |
| 2 | 2011 | 1243238 | — | 68.3698 | — |
| 3 | 2021 | 1790635 | — | 98.4731 | — |
| 4 | 2031 | 2579051 | — | 141.83078 | — |
| 5 | 2041 | 3714607 | — | 204.27887 | — |
| 6 | 2051 | 5350148 | — | 294.2228 | — |

The above figure emphasizes the pressing problems that are faced by the expansion of any metropolis. The planners should take a cue from this and appropriately plan the expansion of any city.

## Conclusion

Most of the water requirements are fulfilled from the surface water sources. The river Subarnarekha and three reservoirs namely Gatalsud, Hatiya, and Kanke are being largely used for public water supply in Ranchi city. The water supply of Hatiya, Kanke, Booti (Gatalsud) indicated that the area of 29.41 km$^2$, 4.50 km$^2$, and 37.89 km$^2$, (Map 5) is covered by three pump houses respectively. The actual water demand is 47.46 mgd within RMC area but pipeline water supply is 31.86 mgd. Hence, the shortage of water was found to be 15.60 mgd.

Presently, water is not being supplied in all the Wards under Ranchi Municipal Corporation. Also, the water demand of Ranchi city is not being fulfilled completely.

The water demand for next fifty years had been projected and it is concluded that to meet this large water demand the concerned agencies will have to do proper planning and will have to take necessary actions.

In view of the groundwater scarcity and depleting water table the dependability on surface water will increase in coming years. The environmental changes like, climate change, deforestation, soil erosion, sedimentation in rivers, etc., add to the problem of water scarcity.

The Kanke reservoir capacity has decreased due to accelerated siltation from sewage and other effluents draining from the surrounding urban load which has tremendously grown and it has to supply more water in RMC area. To restore its optimum capacity, deposited silt from the reservoir has to be removed. Secondly, the reservoirs capacity is required to be increased for fulfilment of futuristic water demand.

In order to fulfil the growing demand of water in and around Ranchi city, more emphasis should be paid on the conjunctive use of surface water and groundwater. Detailed investigation and modelling should be done to help the decision makers in selecting different alternatives and also to evaluate the impact of implementation of any such decisions.

## References

Census of India 1991, Series –5, Bihar, Part A, Directorate of Census Operations, Bihar. www.censusofindia.nic.in.

District Census Hand Book, Primary Census Abstract, District Ranchi, Series –4, Part XIII –A and B.

Garg, S.K., (1982), *Water Supply Engineering*, Khanna Pub., Delhi.

Khan, S.H.A. and Sinha, S.K. (2002) "Hydrology and Groundwater Potential of Jharkhand". Proceeding of Water Resources Planning in Jharkhand, State IE (I), Jharkand State Centre, Department of Water Resources, Government of Jharkhand.

McGhee, Terence J. (1991), *Water Supply and Sewerage*, McGraw Hill International Editions (6th ed.), Civil Engineering series, Princeton, NJ 08520.

Prasad, Lakshmi Shankar (2002) "Status of Water Resources in Jharkhand". Proceeding of the Water Resources Planning in Jharkhand State, IE (I), Jharkhand State Centre, Water Resources Department, Government of Jharkhand.

Rao, D.P. (1991) IRS – 1A Applications for District Level Planning, *Current Science*, Indian Academy of Science, Vol. 61, No. 3 and 4, pp. 260-65.

Sinha, S.C. (2002) "A case study of implementation of Subarnarekha Multi purpose projects (SMP)". Proceeding of the Water Resources Planning in Jharkhand State IE (I), Jharkhand State Centre, Water Resources Department Government of Jharkhand.

# 38

# Geomorphological Mapping in Parts of Dikrong Basin, Papumpare District, Arunachal Pradesh, Using IRS IC/ID PAN Data

Swapna Acharjee

In the recent times high resolution images such as IKONOS and IRS IC/ID PAN have become a powerful source for mapping of natural resources and creating user specific database in GIS environment for micro-level planning and sustainable development. These data provide high quality topographic information which help in locating various units and their spatial distribution. In the present study IRS IC/ID PAN image on 1:12, 500 scale of 5.8m resolution was used to map various geomorphic units along Dikrong river and its tributaries surrounding the capital complex of Arunachal Pradesh.

## Methodology

The IRS PAN image was geo-referenced using the ground control points in ERDAS 8.6. The major drainage networks were digitized from the image. Visual analytical techniques were used to interpret the geomorphic features. Besides satellite data, information from Survey of India toposheets, ground truth and collateral data published in journals and reports were also collected. After digitization, the vector layer topologies was created using clean and build command in the vector module of ERDAS. For preparation of DEM, the contours from the Survey of India toposheets were digitized with 20m contour interval for entire study area.

## Geological Set-up

Geologically, the area under investigation falls under the Siwalik Group of Miocene to Pliocene age. The Siwalik hills also locally known as Dafla Hills are composed of rocks belonging to the Subansiri (=Middle Siwalik) and Kimin (=Upper Siwalik) formations exposed in the form of an anticline along EW axis with Subansiri formation in its core and flanked by rocks of the Kimin formation. The general strike of the formation varies from NW-SE to E-W through NE-SW with low to steep northerly dips in the northern part of the area and southerly dips in the southern part of the study area. The Subansiri formation is repeated in the north overlying the Kimin formation along a faulted contact (Singh and Jain, 1988).

The Subansiri formation is characterized by salt and pepper appearance, soft current bedded sandstone. The sandstone is bluish grey in colour, medium to coarse grained with nodular silty shale, clay, at places pebbly with calcareous concretions and coaly matter.

The Kimin formation includes soft friable grey coloured sandstone with intercalations of pebble/gravel, shale, clay and gritty sandstone.

## Geomorphological Units

The major landform includes Structural hills (Siwalik upland), river terraces and Cuesta slope along the major river Dikrong and Pachin, Chimpu and Dokhoso nalah (Figure 1).

**Figure 1**

*Geomorphological Map of the Study Area*

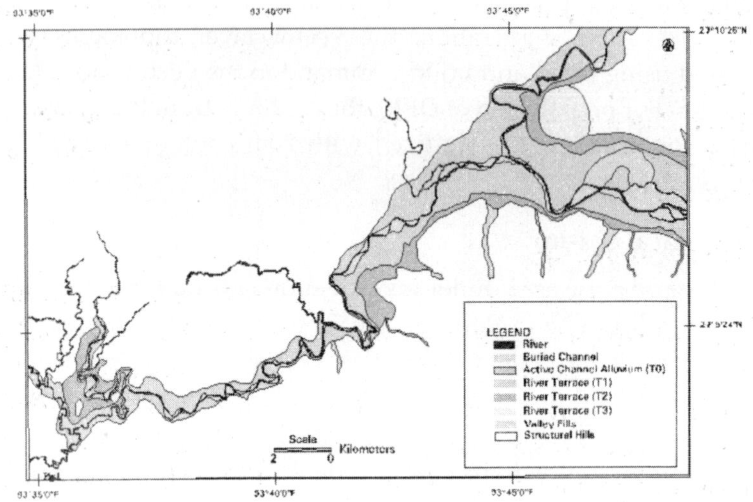


The hills are at places bounded by steep escarpments. At some places fresh landslide is seen to occur along the escarpment. In the Karsingsa area, the river Dikrong is downcutting the

**Figure 2**

*Contour Map of the Study*

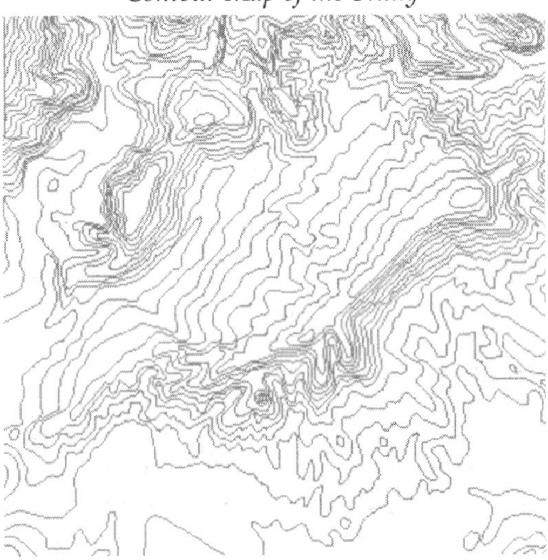

**Figure 3**
*Relief Map of the Study Area*

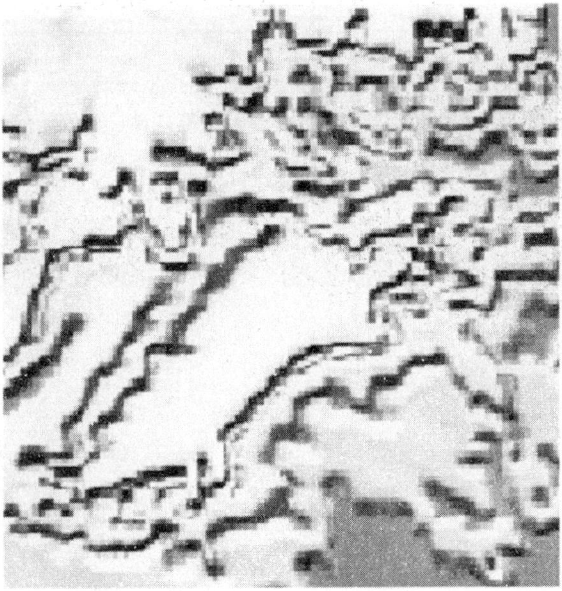

Siwalik basement along the south bank, underneath National Highway (52A) triggering frequent landslides. In the contour map and relief map it is observed that a major portion of Itanagar township is situated on a Cuesta slope (Figure 2 and 3). In a comparative study of drainage system from Survey of India toposheet and fresh collects of IRS data, it was observed that the Dikrong river nearing Banderdewa exit has moved considerably southwards within a span of 20-25 years.

The terraces occur as uplifted surfaces which terminates against the Siwalik rocks. The terraces which represent the former valley floors are found to occur along all the major rivers and nalahs. Along the small streams they occur as valley fills. Near Thero nalah the terraces overlie the basement; Siwaliks have also been incised by the nalah.

The terraces are grouped into T1, T2 and T3 terrace surface. The T1 unit/Flood plain is composed of boulder beds, unoxidized

**Figure 4**

*River Tenace (TO) Exposed along the River Bank*

sand and minor clay which at times is affected by flash floods while the T2 unit is composed of boulder bed oxidized sand and carbonaceous clay. The T3 surface is composed of boulder beds oxidized sand and sandy clay.

The active channel alluvium (channel bars and point bars) along the river is grouped under T0 terrace. This unit is composed of large boulders, cobbles, pebbles and coarse sand (Figure 4).

## Conclusion

In the present study the 5.8m resolution panchromatic data is found to be very useful in geomorphic mapping on 1:12, 500 scale. Among all the geomorphic units, structural hills is the major unit and occupy a major portion of the study area. The river terraces (RT1, RT2, RT3) are being studied in detail to understand the processes involved in their spatial distribution at the present day topographic level. The active channel alluvium (RT0), buried channel and valley fills are also very important units and are being studied in the perspective of groundwater exploitation programme.

*Acknowledgement*

The author is grateful to Department of Science and Technology, Government of India for financial assistance (Project No.SR/FTP/ESA-18/2002) and Director, State Remote Sensing Application Centre, Itanagar for providing necessary facilities for the study. Mr. Saumitra Deb, Project Scientist is thanked for his help in many ways.

## References

Singh, B.K. and Jain, V.K. (1988): Preliminary investigation for Placer gold in Siwalik rocks around Poma, Sonajuli and Balijan, Lower Subansiri District, *GSI Records*, Vol. 124, Part-4.

Sharma, K.K. (2002): Geoenvironmental Appraisal of Itanagar Capital Complex, Papumpare District, Arunachal Pradesh, *GSI Records*, Vol. 133, Part-4.

# Contributors

**Swapna Acharjee**, Department of Science & Technology, Itanagar.

**C.M.S. Adhikari**, NRDMS Centre, Department of Geography, Kumaun University Campus, Almora, Uttaranchal.

**Shalini Ajayan**, National Institute of Technology, Warangal.

**P.S. Arun**, Centre for Studies in Resource Engineering, Indian Institute of Technology-Bombay, Mumbai.

**Kirti Avishek**, Environmental Science & Engineering Group, Birla Institute of Technolgy, Mesra, Ranchi.

**S. Bandyopadhyay**, Indian Space Research Organisation (ISRO), Antariksh Bhavan, Bangalore.

**Vijay Kumar Baraik**, Jharkhand Space Application Centre, Ranchi.

**A. Bora**, En-Geo Consultancy & Research Centre, Guwahati, Assam.

**G.S. Chattopadhyay**, PAN Network Pvt. Ltd., Kolkata.

**S. Chattopadhyay**, Faculty of Forestry, Birsa Agricultural University, Kanke, Ranchi.

**Sujit Choudhury**, PAN Network Pvt. Ltd., Kolkata.

**S. Deb**, State Remote Sensing Application Centre, Itanagar, Arunachal Pradesh.

**A.M. Deshmukh**, Maharashtra Engineering Research Institute, Nashik.

**K.S. Gajbhiye**, National Bureau of Soil Survey & Land Use Planning, Nagpur.

**R.D. Garg**, PRS Division, Indian Institute of Remote Sensing, Dehradun.

**Subrata Ghosh**, Department of Pharmaceutical Sciences, Birla Institute of Technology, Mesra, Ranchi.

**Rahul Goel**, Geomatics Engineering Section, Department of Civil Engineering, Indian Institute of Technology, Roorkee.

**Hari Shankar Gupta**, Indian Institute of Forest Management, Bhopal.

**V.S. Hegde**, Indian Space Research Organisation (ISRO), Antariksh Bhavan, Bangalore.

**Hema Malini B.**, Department of Geography, Andhra University, Visakhapatnam.

**Kamal Jain**, Geomatics Engineering Section, Department of Civil Engineering, Indian Institute of Technology, Roorkee.

**R. Jana**, Centre for Studies in Resource Engineering, Indian Institute of Technology-Bombay, Mumbai.

**V. Jayaraman**, Indian Space Research Organisation (ISRO), Antariksh Bhavan, Bangalore.

**S.K. Kalvit**, Maharashtra Engineering Research Institute, Nashik.

**Ashok Kumar**, Remote Sensing Application Centre, IGSC-Planetarium Complex, Patna.

**Minakshi Kumar**, PRS Division, Indian Institute of Remote Sensing, Dehradun.

**Kiran Kumar P.**, National Institute of Technology, Warangal.

**Praveen Kumar**, Department of Remote Sensing, BIT, Mesra, Ranchi.

**A.A. Laskar**, Department of Geological Sciences, Gauhati University, Guwahati, Assam.

**A.K. Maji**, National Bureau of Soil Survey & Land Use Planning, Nagpur.

**Rupali S. Mathur**, Department of Remote Sensing, Birla Institute of Technology, Mesra, Ranchi.

**P.K. Mishra**, Agricultural Finance Corporation Ltd., Ranchi.

**Md. Surabuddin Mondal**, Department of Remote Sensing, Birla Institute of Technology, Mesra, Ranchi.

**S. Naithani**, NRDMS Centre, Department of Geography, Kumaun University Campus, Almora, Uttaranchal.

**Rajiv Narayan**, Department of Geography, R.L.S.Y. College, Ranchi.

**M.S. Nathawat**, Department of Remote Sensing, Birla Institute of Technology, Mesra, Ranchi.

**Hina Pande**, Indian Institute of Remote Sensing, Dehradun.

**A.C. Pandey**, Department of Remote Sensing, Birla Institute of Technology, Mesra, Ranchi.

**O.N. Pandey**, Faculty of Forestry, Birsa Agricultural University, Kanke, Ranchi.

**A.D. Patgiri**, Department of Geological Sciences, Gauhati University, Guwahati, Assam.

**S. Phukan**, Department of Geological Sciences, Gauhati University, Guwahati, Assam.

**P. Phukon**, Department of Geological Sciences, Gauhati University, Guwahati, Assam.

**Deva Pratap**, National Institute of Technology, Warangal.

**M.P. Punia**, Remote Sensing Division, B.M. Birla Science Centre, Jaipur.

**K.M.M. Rao**, National Remote Sensing Agency (NRSA), Hyderabad.

**P.S. Rathore**, Manager (Projects), ADCC Infocad Pvt. Ltd., Atrey Layout, Nagpur.

**V.S. Rathore**, Department of Remote Sensing, Birla Institute of Technology, Mesra, Ranchi.

**G. Rawat**, NRDMS Centre, Department of Geography, Kumaun University Campus, Almora, Uttaranchal.

**J.S. Rawat**, NRDMS Centre, Department of Geography, Kumaun University Campus, Almora, Uttaranchal.

**V.S. Rawat**, NRDMS Centre, Department of Geography, Kumaun University Campus, Almora, Uttaranchal.

**G.P. Obi Reddy**, National Bureau of Soil Survey & Land Use Planning, Nagpur.

**P. Sreedhar Reddy**, National Institute of Technology, Warangal.

**Siva Sankara Reddy**, National Institute of Technology, Warangal.

**R. Saha**, Agricultural Finance Corporation Ltd., Ranchi.

**U.C. Sahoo**, Lecturer, Department of Civil Engineering, Birla Institute of Technology, Mesra, Ranchi.

**Tanima Sarkar**, Department of Remote Sensing, Birla Institute of Technology, Mesra, Ranchi.

**Poonam Seth**, Indian Institute of Remote Sensing, Dehradun.

**Arabinda Sharma**, Centre for Regional Studies, T.M. Bhagalpur University, Bhagalpur.

**R.N. Sharma**, En-Geo Consultancy & Research Centre, Guwahati.

**Richa N.K. Sharma**, Department of Remote Sensing, Birla Institute of Technology, Mesra, Ranchi.

**Jitendra Shukla**, P.G. Department, Geography, Ranchi University, Ranchi.

**A.K. Singh**, Department of Remote Sensing, Birla Institute of Technology, Mesra, Ranchi.

**Kumar Ramjee Prasad Singh**, Remote Sensing Application Centre, Patna.

**Sushil Kumar Singh**, Department of Remote Sensing, Birla Institute of Technology, Mesra, Ranchi.

**A.K. Sinha**, Department of Civil Engineering, Birla Institute of Technology, Mesra, Ranchi.

**R.K. Sinha**, Centre for Regional Studies, T.M. Bhagalpur University, Bhagalpur.

**Deepak Srivastava**, Bihar Remote Sensing Application Centre, Taramandal, Patna.

**S.K. Srivastava**, Indian Space Research Organisation (ISRO), Antariksh Bhavan, Bangalore.

**Anuja Tigga**, Department of Geography, Andhra University, Visakhapatnam.

**S.R. Vaijapurkar**, Maharashtra Engineering Research Institute, Nashik.

**Kiran Kumar Vemula**, National Institute of Technology, Warangal.

**K. Vindhya**, National Institute of Technology, Warangal.

**Hira Lal Yadav**, Department of Geography, D.D.U. Gorakhpur University, Gorakhpur.

**Sushma Yadav**, Indian Institute of Remote Sensing, Dehradun.